Large Deviations

This is Volume 137 in
PURE AND APPLIED MATHEMATICS

A complete list of titles in this series appears at the end of this volume.

Large Deviations

Jean-Dominique Deuschel
Department of Mathematics
Cornell University
Ithaca, New York

Daniel W. Stroock
Department of Mathematics
Massachusetts Institute of Technology
Cambridge, Massachusetts

ACADEMIC PRESS, INC.
Harcourt Brace Jovanovich, Publishers
Boston San Diego New York
Berkeley London Sydney
Tokyo Toronto

ACADEMIC PRESS, INC.
1250 Sixth Avenue, San Diego, CA 92101

United Kingdom Edition published by
ACADEMIC PRESS INC. (LONDON) LTD.
24–28 Oval Road, London NW1 7DX

Library of Congress Cataloging-in-Publication Data

Deuschel, Jean-Dominique, Date–
Large deviations / Jean-Dominique Deuschel, Daniel W. Stroock.
p. cm.—(Pure and applied mathematics; v. 137)
Rev. ed. of: An introduction to the theory of large deviations /
D.W. Stroock. c1984.
Bibliography: p.
Includes index.
ISBN 0-12-213150-9
1. Large deviations. I. Stroock, Daniel W. Introduction to the
theory of large deviations. II. Title. III. Series: Pure and
applied mathematics (Academic Press); 137.
QA3.P8 vol. 137
[QA273.67]
510 s—dc19 89-397
[519.5′34] CIP

Printed in the United States of America
89 90 91 92 9 8 7 6 5 4 3 2 1

For

Monroe D. Donsker

who has always liked it best in function space

Preface

The title of this book to the contrary not withstanding, there is no more a "theory" of large deviations than there is a "theory" of partial differential equations; and what passes for the "theory" is, in reality, little more than a grab-bag of techniques which have been successfully applied to special situations and are therefore worth trying in sufficiently closely related settings. Thus, even though the title implies that a master key is contained herein, the reader will discover that reading this book prepares him to analyze large deviations in the same sense as the manual for his computer prepared him to write his first program; that is, hardly at all! In spite of the preceding admission, we have written this book in the belief that even (and, perhaps, particularly) when a field possesses no "CAUCHY integral formula," a useful purpose can be served by a book which surveys a few outstanding successes and attempts to codify some of the principles on which those successes are based. In the present case, the examples of success are plentiful but the underlying principles are few and somewhat illusive. We hope that the brief synopsis given below will help the reader spot and understand these few principles, at least in so far as we have recognized and understood them ourselves.

After attempting, in Section 1.1, a heuristic explanation of the ideas on which the theory of large deviations rests, the remainder of Chapter I is devoted to a detailed account of two basic examples. The first of these, which is the content of Section 1.2, is CRAMÈR's renowned theorem on the large deviations of the CESÀRO means of independent $\mathbb{R}$-valued random variables from the Law of Large Numbers. In order to emphasize, as soon as possible, that large deviations can be successfully analyzed even in an infinite dimensional context, for our second example we have chosen

SCHILDER's Theorem for re-scaled WIENER's measure. The derivation is carried out in Section 1.3, and applications to first STRASSEN's Law of the Iterated Logarithm and second to the estimates of VENTCEL and FREIDLIN are given in Section 1.4. In connection with the VENTCEL-FREIDLIN estimates, we have assumed that the reader is familiar with the elements of ITÔ's theory of stochastic differential equations; however, because the rest of the book relies on neither the contents of Section 1.4 nor a knowledge of ITÔ's calculus, readers who are not acquainted with the quirks of stochastic integration need not (on that account) be too concerned about what lies ahead.

Armed with the examples from Chapter I, we turn in Chapter II to the formulation of two of the guiding principles on which the rest of the book is more or less based. The first of these is contained in Lemma 2.1.4 which provides a reasonably general statement of the "covariant" nature of large deviations results under mappings which are sufficiently continuous. (The treatment given in Section 1.4 of the VENTCEL-FREIDLIN estimates should be ample evidence of the potential power of this principle.) In order to formulate the second general principle set forth in this chapter, we start in Section 2.1 with VARADHAN's version of the LAPLACE asymptotic formula (cf. Theorem 2.1.10) and combine this in Section 2.2 with a little elementary convex analysis to arrive at the conclusion (drawn in Theorem 2.2.21) that when large deviations are governed by a convex rate function then that rate function must be the LEGENDRE transform of the logarithmic moment generating function. Since, as we saw in Chapter I, the rate functions produced in both CRAMÈR's and SCHILDER's Theorems are in fact LEGENDRE transforms of the corresponding logarithmic moment generating functions, this observation leads one to guess that there may be circumstances in which the easiest approach to large deviation results will consist of two steps: one being an abstract existential proof that the large deviations are governed by a convex rate function and the second being the "computation" of a LEGENDRE transform. (Such a procedure is reminiscent of the time-honored technique to describe the solution to a partial differential equation by first invoking some abstract existence principle and only then trying to actually say something concrete about its properties.)

The contents of Chapters III and IV may be viewed as a sequence of examples to which the principles developed in Chapter II can be applied. In Chapter III, all the examples concern partial sums of independent random variables. After introducing, in Section 3.1, a general argument (cf. Theorem 3.1.6 and its Corollary 3.1.7) for carrying out an abstract existential proof that large deviation results for such sums are governed by convex rate functions, we return in the rest of the chapter to CRAMÈR's Theorem; this time in its full glory as a statement about random variables taking values either in a space of probability measures or in a BANACH space. Thus, Section 3.2 contains a proof of SANOV's Theorem (cf. Theorem 3.2.17) for empirical distributions; and Section 3.3 is devoted to the BANACH space version of CRAMÈR's Theorem. (In connection with the derivation of these results, we introduce in Lemma 3.2.7 a somewhat technical mini-principle which turns out to play an important role throughout the rest of the book.) Finally, in Section 3.4, we show that SCHILDER's Theorem is a special case of the BANACH space statement of CRAMÈR's Theorem and, in fact, that a SCHILDER-like result can be proved for general GAUSSian measures.

As we said before, Chapter IV is again an application of the principles laid down in Chapter 2. In particular, we now take up the study of SANOV-type theorems for MARKOV processes which do not necessarily have independent increments. In order to make the development here mimic the one in Chapter III, we impose extremely strong hypotheses to guarantee that the processes with which we are dealing possess ergodic properties which are nearly as good as those possessed by processes with independent increments. As a result, basically the same ideas as those in Chapter III apply to nice additive functionals of such processes and allow us to prove (cf. Theorems 4.1.14 and 4.2.16) that these functionals have large deviations which are governed by a convex rate function. In particular, after identifying the rate functions involved, we use these considerations to obtain a variant of the original DONSKER-VARADHAN theory for the large deviations of the normalized occupation time distribution (i.e. the empirical distribution of the position) of a MARKOV process (cf. Theorems 4.1.43 and 4.2.43). Because it is technically the simpler, we do MARKOV chains (i.e., MARKOV processes with a discrete time-parameter) in Section 4.1 and move to the continuous-time setting in Section 4.2; and in Section 4.4 we show how, under the hypotheses used in Sections 4.1 and 4.2, one can realize the large deviation theory for the empirical distribution of the whole process as the projective limit of the theory for the position. Section 4.3, which is somewhat a digression from the main theme and should probably be skipped on first reading, contains DONSKER and VARADHAN's analysis of the WIENER sausage problem.

To some extent, Chapter V represents to retreat from the pattern set in Chapters III and IV and a return to the more "hands-on" approach of Chapter I. Thus, just as in Chapter I, the approach in Chapter V is to first get an upper bound, basically as an application CHEBYSHEV's inequality; then a lower bound via ergodic considerations; and finally a reconciliation the two. A rather general treatment of the upper bound is given in Section 5.1, where, in Theorem 5.1.6 and Corollary 5.1.11, we sharpen results obtained earlier in Theorem 2.2.4. In preparation for the derivation of the lower bound, we digress in Section 5.2 and give a brief resumé of a few more or less familiar results from ergodic theory. As a first application of these considerations, we present in Section 5.3 a very general large deviation result for the empirical distribution of the position of a symmetric MARKOV process (cf. Theorem 5.3.10). Our second application is the content of Section 5.4, where we prove CHIYONOBU and KUSUOKA's recent theorem about the process level large deviations of a (not necessarily MARKOV) hypermixing process (cf. Theorem 5.4.27); and, in Section 5.5, we discuss the hypermixing property for processes which are ϵ-MARKOV.

The motivation behind Chapter V has been our desire to get away from the extremely strong ergodic assumptions on which the techniques in Chapters III and IV depend and to replace them with assumptions which have a better chance of holding in either non-compact or infinite dimensional situations. In order to test and compare the scope of the various techniques which are contained in Chapters IV and V, we describe in Chapter VI some analytic results with which one can see, at least in the context of diffusion processes, the relative position of these results as measured on the scale of elliptic coercivity.

The contents of Chapters I through IV constitute a reasonably thorough introduction to the basic ideas of the theory and more or less record lectures given by the second author during the fall of 1987. Thus, we consider these four chapters as a suitable package on which to base a semester length course for advanced graduate students with a strong background in analysis and some knowledge of probability theory. In this connection, we point out that each section ends with a large selection of exercises. Although some of these exercises are quite routine and do not require any particular ingenuity on the part of the student, others are more demanding. Indeed, we have not hesitated to include in the exercises a good deal of important material. In particular, it is only in the exercises that one can find most of the applications.

Finally, a word about the history of this book may be in order. In 1983, the second author gave a course, at the University of Colorado, in which he taught himself and one or two others something about the modern theory of large deviations. Having expended considerable effort on the task, he decided to set down everything which he then knew about the subject in a little book [**101**]. That was five years ago. In the intervening years, both the subject as well as his understanding of it have grown; and, with the aid and comfort provided by a fellow sufferer, he took on the more ambitious project of basing a full blown exposition on the course which he gave in fall of 1987 at M.I.T. Thus, the present book is a great deal longer: both because it contains more material and because the exposition is more detailed. Unfortunately, in the process of removing some of the more glaring imperfections and omissions in [**101**], we are confident that we have introduced a sufficient number of new flaws to keep our readers somewhat annoyed and, occasionally, thoroughly confounded. However, the responsibility for these flaws is entirely ours and not that of the ever patient students in 18.158, who struggled with the class notes out of which this final version evolved. In particular, we take this opportunity to thank STEVE FROMM for goading us into addressing several of the more perplexing inanities in those class notes. Also, we are indebted to MICHAEL SHARPE who saved us many harrowing hours manipulating TEX into doing our bidding (cf. the similarity between the format, if not the content, of the present volume and volume # 133 in the same series); and, last but not least, it is a pleasure for us to thank our typist for Eir beautiful work.

Cambridge, MA
December 31, 1988

Contents

I Some Examples

1.1 The General Idea

Let E be a **Polish space** (i.e., a complete, separable metric space) and suppose that $\{\mu_\epsilon : \epsilon > 0\}$ is a family of probability measures on E with the property that $\mu_\epsilon \Longrightarrow \delta_p$ as $\epsilon \to 0$ for some $p \in E$ (i.e., μ_ϵ tends weakly to the point mass δ_p). Then, for each open set $U \ni p$, we have that $\mu_\epsilon(U^c) \longrightarrow 0$; and so we can reasonably say that, as $\epsilon \to 0$, the measures μ_ϵ "see p as being typical." Equivalently, one can say that events $\Gamma \subseteq E$ lying outside of a neighborhood of p describe increasingly "deviant" behavior. What is often an important and interesting problem is the determination of just how "deviant" a particular event is. That is, given an event Γ for which $p \notin \overline{\Gamma}$, one wants to know the rate at which $\mu_\epsilon(\Gamma)$ is tending to 0. In general, a detailed answer to this question is seldom available. However, if one restricts ones attention to events which are "very deviant" in the sense that $\mu_\epsilon(\Gamma)$ goes to zero exponentially fast and if one only asks about the exponential rate, then one has a much better chance of finding a solution and one is studying the *large deviations* of the family $\{\mu_\epsilon : \epsilon > 0\}$. In order to understand why the analysis of large deviations ought to be relatively easy and what one should expect such an analysis to yield, consider the case in which all of the measures μ_ϵ are absolutely continuous with respect to some fixed reference measure m. Since $\mu_\epsilon \Longrightarrow \delta_p$, it is reasonable to suppose that

$$\frac{d\mu_\epsilon}{dm} = g_\epsilon \exp\bigl[-I/\epsilon\bigr]$$

where $\epsilon \log g_\epsilon \longrightarrow 0$ uniformly fast as $\epsilon \to 0$ and I is a non-negative function which vanishes only at the point p. One then has, for any Γ with $m(\Gamma) < \infty$,

$$\begin{aligned}\epsilon\log\big(\mu_\epsilon(\Gamma)\big) &= \log\left(\int_\Gamma g_\epsilon\cdot\exp\big[-I/\epsilon\big]\,dm\right)^\epsilon\\ &= \log\left(\int_\Gamma \exp\big[-I/\epsilon\big]\,dm\right)^\epsilon + o(1);\end{aligned}$$

and so (since $m(\Gamma)<\infty$)

$$\left(\int_\Gamma \exp\big[-I/\epsilon\big]\,dm\right)^\epsilon \longrightarrow -\text{ess.}\sup\Big\{\exp\big[-I(q)\big] : q\in\Gamma\Big\}$$

as $\epsilon\to 0$. (The "essential" here refers to the measure m.)

Hence, in the situation described above, we have, at least when $m(\Gamma)<\infty$:

$$\lim_{\epsilon\to 0}\epsilon\log\mu_\epsilon(\Gamma) = -\text{ess.}\inf\big\{I(q) : q\in\Gamma\big\}. \tag{1.1.1}$$

In particular, the factor g_ϵ plays no role in the analysis of large deviations; and it is this fact which accounts for the relative simplicity of this sort of analysis. Moreover, it is often easy to extend (1.1.1) to cover all Γ's. For instance, such an extension can certainly be made if one knows that for each $L>0$ there is a Γ_L such that

$$m(\Gamma_L)<\infty \quad\text{and}\quad \overline{\lim_{\epsilon\to 0}}\,\epsilon\log\big(\mu_\epsilon(\Gamma_L^c)\big)\le -L. \tag{1.1.2}$$

In particular, we see that if $E=\mathbb{R}^d$, $\lambda_{\mathbb{R}^d}$ is LEBESGUE's measure on $\mathbb{R}^d$, and

$$\gamma_\epsilon(dq) = (2\pi\epsilon)^{-d/2}\exp\left[-\frac{|q|^2}{2\epsilon}\right]\lambda_{\mathbb{R}^d}(dq), \tag{1.1.3}$$

then

$$\lim_{\epsilon\to 0}\epsilon\log\big(\gamma_\epsilon(\Gamma)\big) = -\text{ess.}\inf\{|q|^2/2 : q\in\Gamma\} \tag{1.1.4}$$

for all measurable Γ in $\mathbb{R}^d$.

Although the preceding gives some insight into the phenomena of large deviations, it relies entirely on the existence of the reference measure m and therefore does not apply to many situations of interest (e.g., it will nearly never apply when E is an infinite dimensional space). When there is no reference measure, it is clear that (1.1.1) has got to be replaced by an expression in which m does not appear. Taking a hint from the theory of weak convergence, one is tempted to guess that a reasonable replacement for (1.1.1) in more general situations is the statement that there exists a function $I : E \longrightarrow [0,\infty]$ with the property that

$$(1.1.5) \qquad -\inf_{\Gamma^\circ} I \le \varliminf_{\epsilon\to 0} \epsilon \log\bigl(\mu_\epsilon(\Gamma)\bigr) \le \varlimsup_{\epsilon\to 0} \epsilon \log\bigl(\mu_\epsilon(\Gamma)\bigr) \le -\inf_{\overline{\Gamma}} I.$$

For instance, it is easy to pass from (1.1.4) to (1.1.5) with $\mu_\epsilon = \gamma_\epsilon$ and $I(q) = |q|^2/2$.

With the preceding in mind, we will adopt the attitude that the study of large deviations for $\{\mu_\epsilon : \epsilon > 0\}$ centers around the identification of an appropriate I for which (1.1.5) holds. Before attempting to lay out a general strategy, we will begin by presenting two classical cases in which such a program can be successfully carried to completion.

1.1.6 Exercise.

Let $E = [0,\infty)$ and define

$$\mu_\epsilon(dq) = \frac{1}{\epsilon} \exp\bigl[-q/\epsilon\bigr]\, \lambda_{[0,\infty)}(dq)$$

for $\epsilon \in (0,\infty)$. Show that (1.1.5) holds with $I(q) = q$, $q \in [0,\infty)$.

1.2 The Classical Cramèr Theorem

Let μ be a probability measure on $\mathbb{R}$ and, for $n \ge 1$, let μ^n on $\mathbb{R}^n$ denote the n-fold tensor product of μ with itself. Next, let μ_n on $\mathbb{R}$ denote the distribution of $\mathbf{x} \in \mathbb{R}^n \longmapsto \frac{1}{n}\sum_1^n x_i$ under μ^n. Assuming that $\int_{\mathbb{R}} |x|\,\mu(dx) < \infty$, the weak law of large numbers says that $\mu_n \Longrightarrow \delta_p$, where $p = \int x\,\mu(dx)$. Thus, $\{\mu_n : n \ge 1\}$ is a candidate for a theory of large deviations (take $\mu_\epsilon = \mu_n$ for $n - 1 < 1/\epsilon \le n$ in order to make the notation here conform with that in Section 1.1). Moreover, in the case when $\mu(dx) = \gamma_1(dx)$ (cf. (1.1.3) and take the d there to be 1), we have that $\mu_n = \gamma_{1/n}$. Hence, at least for this special case, we know the theory of large deviations. Namely, we know that we can take $I(x) = |x|^2/2$. The purpose of the present section is to find the large deviation theory for other choices of μ.

We begin our program by introducing the **logarithmic moment generating function**

$$(1.2.1) \qquad \Lambda_\mu(\lambda) \equiv \log\left(\int_{\mathbb{R}} \exp\bigl[\lambda q\bigr]\,\mu(dq)\right), \qquad \lambda \in \mathbb{R}.$$

Note that $\lambda \in \mathbb{R} \longmapsto \Lambda_\mu(\lambda) \in [0,\infty]$ is a lower semi-continuous convex function. Indeed, by truncation, it is easy to write Λ_μ as the non-decreasing limit of smooth functions, and the convexity of Λ_μ follows from HÖLDER's inequality. Next, let Λ_μ^* be the **Legendre transform** of Λ_μ:

$$\Lambda_\mu^*(x) \equiv \sup\{\lambda x - \Lambda_\mu(\lambda) : \lambda \in \mathbb{R}\}, \quad x \in \mathbb{R}. \tag{1.2.2}$$

Note that, by its definition as the point-wise supremum of linear functions, Λ_μ^* is necessarily lower semi-continuous and convex. In order to develop some feeling for the relationship between Λ_μ, Λ_μ^*, and μ, we present the following elementary lemma.

1.2.3 Lemma. *Let μ be a probability measure on $\mathbb{R}$. Then $\Lambda_\mu^* \geq 0$. Moreover:*

(i) *If $\int_{\mathbb{R}} |x|\,\mu(dx) < \infty$ and $p = \int_{\mathbb{R}} x\,\mu(dx)$, then $\Lambda_\mu^*(p) = 0$, Λ_μ^* is non-decreasing on $[p,\infty)$ and non-increasing on $(-\infty,p]$. In addition, for $q \geq p$, $\Lambda_\mu^*(q) = \sup\{\lambda q - \Lambda_\mu(\lambda) : \lambda \geq 0\}$ and $\mu([q,\infty)) \leq \exp[-\Lambda_\mu^*(q)]$; and, for $q \leq p$, $\Lambda_\mu^*(q) = \sup\{\lambda q - \Lambda_\mu(\lambda) : \lambda \leq 0\}$ and $\mu((-\infty,q]) \leq \exp[-\Lambda_\mu^*(q)]$.*

(ii) *If $\Lambda_\mu(\lambda) < \infty$ for all λ 's in a neighborhood of 0, then $\Lambda_\mu^*(x) \longrightarrow \infty$ as $|x| \to \infty$.*

(iii) *If $\Lambda_\mu(\lambda) < \infty$ for all $\lambda \in \mathbb{R}$, then $\Lambda_\mu \in C^\infty(\mathbb{R})$ and $\Lambda_\mu^*(x)/|x| \longrightarrow \infty$ as $|x| \to \infty$.*

PROOF: We begin by noting that, since $\lambda x - \Lambda_\mu(\lambda) = 0$ for $\lambda = 0$ and every $x \in \mathbb{R}$, $\Lambda_\mu^*(x) \geq 0$.

Now suppose that $\int_{\mathbb{R}} |x|\,\mu(dx) < \infty$ and set $p = \int_{\mathbb{R}} x\,\mu(dx)$. To see that $\Lambda_\mu^*(p) = 0$, we use JENSEN's inequality to obtain

$$\Lambda_\mu(\lambda) \geq \lambda p \quad \text{for all } \lambda \in \mathbb{R}. \tag{1.2.4}$$

In particular, this shows that $\lambda p - \Lambda_\mu(\lambda) \leq 0$ for all $\lambda \in \mathbb{R}$ and so $\Lambda_\mu^*(p) \leq 0$. Since Λ_μ^* is non-negative and convex, this proves that $\Lambda_\mu^*(p) = 0$, Λ_μ^* is non-decreasing on $[p,\infty)$, and Λ_μ^* is non-increasing on $(-\infty,p]$. To complete the proof of **i**), we first note that, as a consequence of (1.2.4), if $q \geq p$ then $\Lambda_\mu^*(q) = \sup\{\lambda q - \Lambda_\mu(\lambda) : \lambda \geq 0\}$ and if $q \leq p$ then $\Lambda_\mu^*(q) = \sup\{\lambda q - \Lambda_\mu(\lambda) : \lambda \leq 0\}$. Hence, if $q \geq p$, then, since (by CHEBYCHEV's inequality)

$$\mu([q,\infty)) \leq \exp\bigl[-\bigl(\lambda q - \Lambda_\mu(\lambda)\bigr)\bigr], \quad \lambda \geq 0,$$

we see that

$$\mu([q,\infty)) \leq \exp\bigl[-\Lambda_\mu^*(q)\bigr].$$

Similarly, if $q \leq p$, then

$$\mu((-\infty,q]) \leq \exp\bigl[-\Lambda_\mu^*(q)\bigr].$$

We next turn to the proof of **(ii)** and **(iii)**. To this end, note that if $\lambda > 0$ ($\lambda < 0$) and $\Lambda_\mu(\lambda) < \infty$, then $\overline{\lim}_{x\to\infty} \Lambda^*_\mu(x)/x \geq \lambda$ ($\underline{\lim}_{x\to-\infty} \Lambda^*_\mu(x)/x \leq -\lambda$). Hence, the only assertion left to be proved is that $\Lambda_\mu \in C^\infty(\mathbb{R})$ if $\Lambda_\mu(\lambda) < \infty$ for all λ. But, by TAYLOR's Theorem and the LEBESGUE Dominated Convergence Theorem, it is easy to check that $\lambda \in (-\delta, \delta) \longmapsto \Lambda_\mu(\lambda)$ is, in fact, real-analytic as long as $\Lambda_\mu(\pm\delta) < \infty$. ∎

As a consequence of part **(i)** of Lemma 1.2.3 we have the following.

1.2.5 Lemma. *If $\int_{\mathbb{R}} |x|\, \mu(dx) < \infty$ then for every closed set $F \subseteq \mathbb{R}$*

$$\overline{\lim_{n\to\infty}} \frac{1}{n} \log\bigl(\mu_n(F)\bigr) \leq -\inf_F \Lambda^*_\mu.$$

PROOF: Let $p = \int_{\mathbb{R}} x\, \mu(dx)$ and note that $\int_{\mathbb{R}} |x|\, \mu_n(dx) \leq \int_{\mathbb{R}} |x|\, \mu(dx) < \infty$ and $\int_{\mathbb{R}} x\, \mu_n(dx) = p$ for all $n \geq 1$. Next, observe that if $\Lambda_n = \Lambda_{\mu_n}$, then $\Lambda_n(\lambda) = n\Lambda_\mu(\lambda/n)$, and therefore that $\Lambda^*_n = n\Lambda^*_\mu$. Now suppose that $q \geq p$ ($q \leq p$). Then, by **(i)** applied to μ_n we see that $\mu_n([q,\infty)) \leq \exp\bigl[-n\Lambda^*_\mu(q)\bigr]$ ($\mu_n((-\infty,q]) \leq \exp\bigl[-n\Lambda^*_\mu(q)\bigr]$). Since Λ^*_μ is non-decreasing (non-increasing) on $[p,\infty)$ (on $(-\infty,p]$), this proves the result when either $F \subseteq [p,\infty)$ or $F \subseteq (-\infty,p]$. On the other hand, if both $F \cap [p,\infty) \neq \emptyset$ and $F \cap (-\infty,p] \neq \emptyset$, let $q_+ = \inf\{x \geq p : x \in F\}$ and $q_- = \sup\{x \leq p : x \in F\}$. Then

$$\mu_n(F) \leq \exp\bigl[-n\Lambda^*_\mu(q_-)\bigr] + \exp\bigl[-n\Lambda^*_\mu(q_+)\bigr] \leq 2\exp\bigl[-n \inf_F \Lambda^*_\mu\bigr],$$

and so the result holds in this case also. ∎

1.2.6 Theorem. (CRAMÈR) *Assume that $\Lambda_\mu(\lambda) < \infty$ for every $\lambda \in \mathbb{R}$. Then for every measurable $\Gamma \subseteq \mathbb{R}$ one has that*

$$-\inf_{\Gamma^\circ} \Lambda^*_\mu \leq \underline{\lim_{n\to\infty}} \frac{1}{n} \log\bigl(\mu_n(\Gamma)\bigr) \leq \overline{\lim_{n\to\infty}} \frac{1}{n} \log\bigl(\mu_n(\Gamma)\bigr) \leq -\inf_{\overline{\Gamma}} \Lambda^*_\mu.$$

(We adopt here, and throughout, the convention that the infimum over the null set is $+\infty$.)

PROOF: In view of Lemma 1.2.5, we need only show that if $q \in \mathbb{R}$ and $\delta > 0$,

$$\underline{\lim_{n\to\infty}} \frac{1}{n} \log\bigl[\mu_n((q-\delta, q+\delta))\bigr] \geq -\Lambda^*_\mu(q). \tag{1.2.7}$$

In proving (1.2.7), we first suppose that there is a $\lambda \in \mathbb{R}$ for which $\Lambda^*_\mu(q) = \lambda q - \Lambda_\mu(\lambda)$. Consider the probability measure

$$\tilde{\mu}(dx) = \frac{\exp\bigl[\lambda x\bigr]}{\exp\bigl[\Lambda_\mu(\lambda)\bigr]} \mu(dx),$$

and define the measures $\tilde{\mu}_n$ accordingly. Note that $\int_{\mathbb{R}} |x| \, \tilde{\mu}(dx) < \infty$ and that

$$\int_{\mathbb{R}} x \, \tilde{\mu}(dx) = \frac{1}{\exp\left[\Lambda_\mu(\lambda)\right]} \int_{\mathbb{R}} x \exp\left[\lambda x\right] \mu(dx) = \frac{d}{dt} \Lambda_\mu(t)\Big|_{t=\lambda}.$$

At the same time, note that $\frac{d}{dt}\left(tq - \Lambda_\mu(t)\right)\Big|_{t=\lambda} = 0$ since $t \in \mathbb{R} \longmapsto tq - \Lambda_\mu(t)$ achieves its maximum value at λ. Combining these, we conclude that $q = \int_{\mathbb{R}} x \, \tilde{\mu}(dx)$ and therefore (by the Weak Law of Large Numbers) that $\tilde{\mu}_n((q-\delta, q+\delta)) \longrightarrow 1$ as $n \to \infty$. Assuming that $\lambda \geq 0$, note that

$$\begin{aligned}
\mu_n((q-\delta, q+\delta)) &= \mu^n(\Delta_n) \\
&\geq \exp\left[-n\lambda(q+\delta)\right] \int_{\Delta_n} \exp\left[\lambda \sum_1^n y_k\right] \mu^n(dy) \\
&= \exp\left[-n\left(\lambda(q+\delta) - \Lambda_\mu(\lambda)\right)\right] \tilde{\mu}_n\left((q-\delta, q+\delta)\right),
\end{aligned}$$

where

$$\Delta_n = \left\{ \mathbf{y} \in \mathbb{R}^n : \left| \frac{1}{n} \sum_1^n y_k - q \right| < \delta \right\}.$$

From this and the preceding comments, we conclude that

$$\varliminf_{n \to \infty} \frac{1}{n} \log\left[\mu_n((q-\delta, q+\delta))\right] \geq -\Lambda_\mu^*(q) - \lambda\delta$$

for every $\delta > 0$. Since the left hand side of the above is clearly non-decreasing as a function of $\delta > 0$, we have now proved (1.2.7) for the case when there is a $\lambda \geq 0$ for which $\Lambda_\mu^*(q) = \lambda q - \Lambda_\mu(\lambda)$. Clearly, the same argument (with $q - \delta$ replacing $q + \delta$) works when $\Lambda_\mu^*(q) = \lambda q - \Lambda_\mu(\lambda)$ for some $\lambda \leq 0$.

We must now handle the case in which $\Lambda_\mu^*(q) > \lambda q - \Lambda_\mu(\lambda)$ for all $\lambda \in \mathbb{R}$. If $q \geq \int_{\mathbb{R}} x \mu(dx)$, then (cf. (**i**) of Lemma 1.2.3) there exists a sequence $\lambda_\ell \nearrow \infty$ such that $\lambda_\ell q - \Lambda_\mu(\lambda_\ell) \nearrow \Lambda_\mu^*(q)$. Since it is clear that

$$\int_{(-\infty,q)} \exp\left[\lambda_\ell \cdot (x-q)\right] \mu(dx) \longrightarrow 0,$$

we have that

$$\int_{[q,\infty)} \exp\left[\lambda_\ell \cdot (x-q)\right] \mu(dx) \longrightarrow \exp\left[-\Lambda_\mu^*(q)\right].$$

But this is possible only if $\mu((q,\infty)) = 0$ and $\mu(\{q\}) = \exp\left[-\Lambda_\mu^*(q)\right]$. Hence, $\mu^n(\{(q,\ldots,q)\}) = \exp\left[-n\Lambda_\mu^*(q)\right]$, and so $\mu_n(\{q\}) \geq \exp\left[-n\Lambda_\mu^*(q)\right]$. Clearly, this implies (1.2.7) holds for every $\delta > 0$. An analogous argument can be used in the case when $q \leq \int_{\mathbb{R}} x \, \mu(dx)$. ∎

1.2.8 Remark.

The reader should take note of the structure of the preceding line of reasoning. Namely, the upper bound comes from optimizing over a family of CHEBYCHEV inequalities; while the lower bound comes from introducing a RADON–NIKODYM factor in order to make what was originally "deviant" behavior look like typical behavior. This pattern of proof is one of the two most powerful tools in the theory of large deviations. In particular, it will be used in the next section as well as Sections 5.3 and 5.4.

1.2.9 Exercise.

Assuming that $\int_{\mathbb{R}} |x|\,\mu(dx) < \infty$, show that

$$\int_{\mathbb{R}} \exp\left[\alpha\Lambda_\mu^*(x)\right]\,\mu(dx) \le \frac{2}{1-\alpha}, \quad \alpha \in (0,1). \tag{1.2.10}$$

Hint: Set $p = \int_{\mathbb{R}} x\,\mu(dx)$ and show that if $\Lambda_\mu^*(q) < \infty$, then

$$\int_{[p,q]} \exp\left[\alpha\Lambda_\mu^*(x)\right]\,\mu(dx) \le \frac{1}{1-\alpha} \quad \text{or} \quad \int_{[q,p]} \exp\left[\alpha\Lambda_\mu^*(x)\right]\,\mu(dx) \le \frac{1}{1-\alpha}$$

according to whether $q \ge p$ or $q \le p$.

1.2.11 Exercise.

(i) Show that for every $p \in \mathbb{R}$: $\Lambda_{\mu_p}^*(x) = \Lambda_\mu^*(x-p)$, $x \in \mathbb{R}$, where $\mu_p \equiv \delta_p * \mu$ and we use $\nu * \mu$ to denote the convolution of ν with μ.

(ii) If $\mu = \alpha\delta_a + (1-\alpha)\delta_b$, where $a < b$ and $\alpha \in (0,1)$, show that

$$\Lambda_\mu^*(x) = \begin{cases} \infty & \text{for } x \notin [a,b] \\ \frac{x-a}{b-a}\log\frac{x-a}{(1-\alpha)(b-a)} + \frac{b-x}{b-a}\log\frac{b-x}{\alpha(b-a)} & \text{for } x \in [a,b], \end{cases}$$

where $0\log 0 \equiv 0$.

(iii) If $\mu(dx) = \chi_{[0,\infty)}(x)e^{-x}\,dx$, show that

$$\Lambda_\mu^*(x) = \begin{cases} \infty & \text{for } x \le 0 \\ x - 1 - \log x & \text{for } x > 0. \end{cases}$$

(iv) If $\mu(dx) = (2\pi\sigma^2)^{-1/2}\exp\left[-(x-a)^2/2\sigma^2\right]dx$, where $a \in \mathbb{R}$ and $\sigma > 0$, show that

$$\Lambda_\mu^*(x) = \frac{(x-a)^2}{2\sigma^2}.$$

1.3 Schilder's Theorem

In this section we give an example of a large deviation result for a certain family of measures on an infinite dimensional space.

Let $d \in \mathbb{Z}^+$ be given and set

$$(1.3.1) \qquad \Theta = \left\{ \theta \in C\big([0,\infty);\mathbb{R}^d\big) : \theta(0) = 0 \quad \text{and} \quad \lim_{t\to\infty} \frac{|\theta(t)|}{t} = 0 \right\}.$$

For $\theta \in \Theta$ define

$$\|\theta\|_\Theta = \sup_{t\ge 0} \frac{|\theta(t)|}{1+t},$$

and observe that $\big(\Theta, \|\cdot\|_\Theta\big)$ is a separable real BANACH space. In order to represent the dual Θ^* of Θ, note that Θ is naturally isometric to the space of continuous paths on $[0,\infty)$ which vanish at 0 and at ∞ (namely, map θ to the path $t \longmapsto (1+t)^{-1}\theta(t)$); and use this isometry to identify Θ^* with the space of $\mathbb{R}^d$-valued, BOREL measures λ on $[0,\infty)$ with the properties that $\lambda(\{0\}) = 0$ and $\int_{[0,\infty)}(1+t)\,|\lambda|(dt) < \infty$, where $|\lambda|$ denotes the variation measure associated with λ. With this identification, the duality relation ${}_{\Theta^*}\langle\lambda,\theta\rangle_\Theta$ is given by $\int_{[0,\infty)} \theta(t)\cdot\lambda(dt)$ (the "$\cdot$" here standing for the ordinary inner product in $\mathbb{R}^d$) and $\|\lambda\|_{\Theta^*} = \int_{[0,\infty)}(1+t)\,|\lambda|(dt)$.

Let $\mathcal{B} = \mathcal{B}_\Theta$ denote the BOREL field over Θ; and, for $t \ge 0$, let $\mathcal{B}_t$ denote the smallest σ-algebra over Θ with respect to which all of the maps $\theta \longmapsto \theta(s)$, $s \in [0,t]$, are measurable. As is easy to check, $\mathcal{B} = \sigma\big(\bigcup_{t\ge 0}\mathcal{B}_t\big)$. The following remarkable existence theorem is due to N. WIENER [**112**]. We have added a few small embellishments to WIENER's original statement.

1.3.2 Theorem. (WIENER) *There is a unique probability measure $\mathcal{W}$ on $(\Theta, \mathcal{B})$ with the property that*

$$(1.3.3) \qquad \int_\Theta \exp\Big[\sqrt{-1}\;{}_{\Theta^*}\langle\lambda,\theta\rangle_\Theta\Big]\,\mathcal{W}(d\theta) = \exp\big[-\Lambda_{\mathcal{W}}(\lambda)\big], \quad \lambda \in \Theta^*,$$

where

$$(1.3.4) \qquad \Lambda_{\mathcal{W}}(\lambda) \equiv \frac{1}{2}\int_{[0,\infty)^2} s\wedge t\;\lambda(ds)\cdot\lambda(dt).$$

Moreover, if P is a probability measure on $(\Theta,\mathcal{B})$, then $P = \mathcal{W}$ if and only if any one of the following holds:

(**i**) *For all $0 \le s < t$, the random variable $\theta \longmapsto \theta(t) - \theta(s)$ under P is independent of $\mathcal{B}_s$ and is GAUSSian with mean 0 and covariance $(t-s)I_{\mathbb{R}^d}$.*

(ii) *For all* $0 \le s < t$ *and* $\Gamma \in \mathcal{B}_{\mathbb{R}^d}$,

$$P\big(\{\theta : \theta(t) \in \Gamma\}\big|\mathcal{B}_s\big)(\psi) = \gamma_{t-s}(\psi(s) + \Gamma) \tag{1.3.5}$$

for P-almost all $\psi \in \Theta$. (*The measure* γ_{t-s} *in* (1.3.5) *is the one described in* (1.1.3).)

(iii) *For every* $n \in \mathbb{Z}^+$, $0 \le t_1 < \cdots < t_n$, *and* $\xi_1, \ldots, \xi_n \in \mathbb{R}^d$,

$$\int_\Theta \exp\left[\sqrt{-1}\sum_{m=1}^n \xi_m \cdot \theta(t_m)\right] \mathcal{W}(d\theta) = \exp\left[-\frac{1}{2}\sum_{m,m'=1}^n t_m \wedge t_{m'} \xi_m \cdot \xi_{m'}\right].$$

Finally, $\mathcal{W}$ *has the properties that*

(iv) *For each* $\alpha \in (0,\infty)$, $\mathcal{W}$ *is invariant under the* **scaling transformation** $\theta \longmapsto \alpha^{1/2}\theta(\cdot/\alpha)$ *and the* **time shift transformation** $\theta \longmapsto \theta(\cdot + \alpha) - \theta(\alpha)$.

(v) $\mathcal{W}$ *is invariant under the* **time inversion transformation** $\theta \longmapsto \cdot\,\theta(1/\cdot)$ ($\equiv 0$ *at* $t = 0$).

(vi) *For each* $\alpha \in (0, 1/2)$ *and* $T > 0$,

$$\mathcal{W}\left(\left\{\theta : \sup_{0 \le s < t \le T} \frac{|\theta(t) - \theta(s)|}{|t-s|^\alpha} < \infty\right\}\right) = 1.$$

The reader is assumed to be familiar with some form of WIENER's basic existence theorem and with the basic properties of **Wiener's measure** $\mathcal{W}$. In particular, he is advised to reconcile the statement which he knows with the one given above.

We can now describe the family of measures which we want to study in this section. Namely, for each $\epsilon > 0$, let $\mathcal{W}_\epsilon$ denote the distribution of $\theta \longmapsto \epsilon^{1/2}\theta$ under $\mathcal{W}$. Clearly $\mathcal{W}_\epsilon \Longrightarrow \delta_0$, where δ_0 is the point mass at the path which never leaves 0. Hence, we are again dealing with a family for which it is reasonable to ask about large deviations. Before getting into the details, it may be helpful to make a couple of remarks. In the first place, it should be noted that, at least formally, we are dealing here with a situation like the one discussed in Section 1.1. Indeed, an often useful heuristic representation of WIENER's measure is the formula

$$\mathcal{W}(d\theta) = c\exp\left[-\frac{1}{2}\int_0^\infty |\dot\theta(t)|^2\,dt\right]\,d\theta. \tag{1.3.6}$$

The expression in (1.3.6) is somewhat fanciful. Indeed, none of quantities on the right hand side makes sense by itself. In particular, "$d\theta$" stands for the (non-existent) translation invariant measure on Θ, "$\dot{\theta}$" denotes the derivative (which $\mathcal{W}$-almost surely fails to exist) of θ, and the constant "c" is infinite. Thus, (1.3.6) is at best just a schematic representation of what one gets by formally passing to the limit in the expression for the $\mathcal{W}$-measure of a subset of Θ whose description involves a continuum of times. Leaving such technicalities aside, one has to admit that to whatever degree one accepts (1.3.6), one has to grant the expression

$$\mathcal{W}_\epsilon(d\theta) = c_\epsilon \exp\left[-\frac{1}{2\epsilon}\int_0^\infty |\dot{\theta}(t)|^2\,dt\right]\,d\theta$$

an equal degree of acceptance; and, on the basis of this expression combined with the discussion in Section 1.1, one is led to predict that the function I governing the large deviations of $\{\mathcal{W}_\epsilon : \epsilon > 0\}$ ought to be $\frac{1}{2}\int_0^\infty |\dot{\theta}(t)|^2\,dt$.

A second remark, and the one on which our analysis will be based, is that the family $\{\mathcal{W}_\epsilon : \epsilon > 0\}$ is related to the sort of family which was handled in Section 1.2; and, as we will see later (cf. Sections 3.3 and 3.4), the result which we are about to obtain can be considered as a consequence of CRAMÈR's Theorem for measures on Θ. To understand the relationship with the situation dealt with in CRAMÈR's Theorem, note that the measure $\mathcal{W}_{1/n}$ here is precisely the distribution under $\mathcal{W}^n$ of the random variable

$$(\theta_1, \ldots, \theta_n) \in \Theta^n \longmapsto \frac{1}{n}\sum_1^n \theta_k \in \Theta.$$

Hence, on the basis of CRAMÈR's Theorem, we should predict that the I governing the large deviations of $\{\mathcal{W}_\epsilon : \epsilon > 0\}$ is the LEGENDRE transform of the logarithmic moment generation function for $\mathcal{W}$. In this connection, one should observe that the quantity $\Lambda_{\mathcal{W}}$ introduced in WIENER's Theorem above *is* the logarithmic moment generating function

$$\lambda \in \Theta^* \longmapsto \log\left(\int_\Theta \exp\left[{}_{\Theta^*}\langle\lambda, \theta\rangle_\Theta\right]\,\mathcal{W}(d\theta)\right)$$

for $\mathcal{W}$. Thus, what we are now predicting is that the function

$$\text{(1.3.7)} \qquad \Lambda^*_{\mathcal{W}}(\psi) \equiv \sup\Big\{{}_{\Theta^*}\langle\lambda, \psi\rangle_\Theta - \Lambda_{\mathcal{W}}(\lambda) : \lambda \in \Theta^*\Big\}, \quad \psi \in \Theta,$$

is the function which governs the large deviations of $\{\mathcal{W}_\epsilon : \epsilon > 0\}$.

We begin our rigorous analysis with a lemma which shows, among other things, that the two predictions made above are at least consistent.

1.3.8 Lemma. *Given $\lambda \in \Theta^*$ define $\psi_\lambda \in \Theta$ by*

$$\psi_\lambda(t) = \int_0^t \lambda((s,\infty))\,ds, \quad t \geq 0. \tag{1.3.9}$$

Then, for all $\lambda, \eta \in \Theta^*$,

$$\begin{aligned}\int_{[0,\infty)^2} s \wedge t\, \lambda(ds) \cdot \eta(dt) &= \int_{[0,\infty)} \lambda((s,\infty)) \cdot \eta((s,\infty))\,ds \\ &= {}_{\Theta^*}\langle \eta, \psi_\lambda \rangle_\Theta.\end{aligned} \tag{1.3.10}$$

In particular,

$$\Lambda_{\mathcal{W}}(\lambda) = \frac{1}{2}\int_{[0,\infty)} |\lambda((s,\infty))|^2\,ds = \frac{1}{2}\,{}_{\Theta^*}\langle \lambda, \psi_\lambda \rangle_\Theta. \tag{1.3.11}$$

Next, define $H^1 = H^1([0,\infty);\mathbb{R}^d)$ to be the space of $\psi \in \Theta$ with the property that $\psi(t) = \int_0^t \dot{\psi}(s)\,ds$, $t \geq 0$, for some $\dot{\psi} \in L^2([0,\infty);\mathbb{R}^d)$; and set $\|\psi\|_{H^1} = \|\dot{\psi}\|_{L^2([0,\infty);\mathbb{R}^d)}$ for $\psi \in H^1$. Then

$$\Lambda^*_{\mathcal{W}}(\psi) = \begin{cases} \infty & \text{if } \psi \notin H^1 \\ \frac{1}{2}\|\psi\|^2_{H^1} & \text{if } \psi \in H^1. \end{cases} \tag{1.3.12}$$

In particular,

$$\Lambda^*_{\mathcal{W}}(\psi_\lambda) = \Lambda_{\mathcal{W}}(\lambda), \quad \lambda \in \Theta^*, \tag{1.3.13}$$

and for each $L \geq 0$, $\{\psi \in \Theta : \Lambda^*_{\mathcal{W}}(\psi) \leq L\} \subset\subset \Theta$ (i.e., is a compact subset).

PROOF: We first observe that the second equality in (1.3.10) is an elementary integration by parts. Second, we note that it suffices to prove the first equality in (1.3.10) when $\lambda = \eta$ (since the general case then follows from this by polarization) and that, by an elementary approximation argument, we need only handle $\lambda \in \Theta^*$ which are non-atomic and compactly supported. But in this case we have:

$$\begin{aligned}&\int_{[0,\infty)^2} s \wedge t\, \lambda(ds) \cdot \lambda(dt) = 2\int_{[0,\infty)} \lambda(dt) \cdot \int_{[0,t)} s\,\lambda(ds) \\ &= -\int_{[0,\infty)} t\,d|\lambda((t,\infty))|^2 = \int_{[0,\infty)} |\lambda((t,\infty))|^2\,dt.\end{aligned}$$

Turning to the proof of (1.3.12), first suppose that $\psi \in H^1$. Then

$$\int_{[0,\infty)} \psi(t)\,\lambda(dt) = \int_{[0,\infty)} \dot{\psi}(t)\lambda((t,\infty))\,dt,$$

and therefore, by (1.3.10), $\Lambda^*_{\mathcal{W}}(\psi)$ is equal to

$$\sup\left\{\int_{[0,\infty)} \dot{\psi}(t)\cdot\lambda((t,\infty))\,dt - \frac{1}{2}\int_{[0,\infty)} |\lambda((t,\infty))|^2\,dt : \lambda \in \Theta^*\right\}$$

$$= \sup\left\{\left(\dot{\psi},\phi\right)_{L^2([0,\infty);\mathbb{R}^d)} - \frac{1}{2}\|\phi\|^2_{L^2([0,\infty);\mathbb{R}^d)} : \phi \in L^2([0,\infty);\mathbb{R}^d)\right\}$$

$$= \frac{1}{2}\|\dot{\psi}\|^2_{L^2([0,\infty);\mathbb{R}^d)}.$$

Hence, the proof of (1.3.12) will be complete once we show that $\psi \in H^1$ whenever $\Lambda^*_{\mathcal{W}}(\psi) < \infty$. But if $\phi \in C^\infty_c([0,\infty);\mathbb{R}^d)$ (i.e., it is a smooth path with compact support) and we define $\lambda \in \Theta^*$ by $\phi(t) = \lambda((t,\infty))$, $t \geq 0$, then

$$-\int_{[0,\infty)} \psi(t)\cdot\dot{\phi}(t)\,dt - \frac{1}{2}\|\phi\|^2_{L^2([0,\infty);\mathbb{R}^d)} = {}_{\Theta^*}\langle\lambda,\psi\rangle_{\Theta} - \Lambda_{\mathcal{W}}(\lambda) \leq \Lambda^*_{\mathcal{W}}(\psi),$$

and so there exists a unique $\dot{\psi} \in L^2([0,\infty);\mathbb{R}^d)$ such that

$$-\int_{[0,\infty)} \psi(t)\cdot\dot{\phi}(t)\,dt = \int_{[0,\infty)} \dot{\psi}(t)\cdot\phi(t)\,dt, \quad \phi \in C^\infty_c([0,\infty);\mathbb{R}^d).$$

From here it is an easy step to the conclusion that $\psi(t) = \int_0^t \dot{\psi}(s)\,ds$, $t \geq 0$, and therefore that $\psi \in H^1$. Given (1.3.12), (1.3.13) is clearly a consequence of (1.3.11).

To complete the proof, note first that, directly from its definition in (1.3.7), $\Lambda^*_{\mathcal{W}}$ is lower semi-continuous. Thus, the fact that $\{\psi : \Lambda^*_{\mathcal{W}}(\psi) \leq L\}$ is compact follows immediately from (1.3.12) and the easily verified observation that bounded subsets of H^1 are relatively compact in Θ. ∎

We will now prove a slightly deficient form of the right hand side of (1.1.5) with $I = \Lambda^*_{\mathcal{W}}$. The reader should remark the similarity of this argument with the proof of Lemma 1.2.5.

1.3.14 Lemma. *Let $\psi \in \Theta$ be given. Then for each $\delta > 0$ there exists an $r > 0$ such that*

$$\text{(1.3.15)} \qquad \mathcal{W}_\epsilon\left(\overline{B}(\psi,r)\right) \leq \begin{cases} \exp\left[-\frac{1}{\epsilon\delta}\right] & \text{if } \Lambda^*_{\mathcal{W}}(\psi) = \infty \\ \exp\left[-\frac{\Lambda^*_{\mathcal{W}}(\psi)-\delta}{\epsilon}\right] & \text{if } \Lambda^*_{\mathcal{W}}(\psi) < \infty. \end{cases}$$

(Here, and throughout, $B(x,r)$ denotes the open ball of radius r around a point x in a metric space; and $\overline{B}(x,r)$ denotes the corresponding closed ball.) In particular, if $K \subset\subset \Theta$, then

$$\text{(1.3.16)} \qquad \overline{\lim_{\epsilon\to 0}}\, \epsilon \log\left(\mathcal{W}_\epsilon(K)\right) \leq -\inf_K \Lambda^*_{\mathcal{W}}.$$

PROOF: To prove (1.3.15), note that

$$
\begin{aligned}
\mathcal{W}_\epsilon\big(\overline{B}(\psi, r)\big) &= \mathcal{W}\big(\overline{B}(\psi/\epsilon^{1/2}, r/\epsilon^{1/2})\big) \\
&\leq \sup_{\phi \in \overline{B}(\psi, r)} \exp\left[-\frac{{}_{\Theta^*}\langle \lambda, \phi \rangle_\Theta}{\epsilon}\right] \\
&\quad \times \int_{\overline{B}(\psi/\epsilon^{1/2}, r/\epsilon^{1/2})} \exp\left[{}_{\Theta^*}\langle \lambda/\epsilon^{1/2}, \theta \rangle_\Theta\right] \mathcal{W}(d\theta) \\
&\leq \exp\left[-\frac{1}{\epsilon}\left({}_{\Theta^*}\langle \lambda, \psi \rangle_\Theta - \Lambda_{\mathcal{W}}(\lambda) - \|\lambda\|_{\Theta^*} r\right)\right]
\end{aligned}
$$

for all $\lambda \in \Theta^*$. If $\Lambda^*_{\mathcal{W}}(\psi) = \infty$, choose $\lambda \in \Theta^*$ so that ${}_{\Theta^*}\langle \lambda, \psi \rangle_\Theta - \Lambda_{\mathcal{W}}(\lambda) \geq 1 + 1/\delta$ and $r = 1/(1 + \|\lambda\|_{\Theta^*})$. If $\Lambda^*_{\mathcal{W}}(\psi) < \infty$, choose $\lambda \in \Theta^*$ so that ${}_{\Theta^*}\langle \lambda, \psi \rangle_\Theta - \Lambda_{\mathcal{W}}(\lambda) \geq \Lambda^*_{\mathcal{W}}(\psi) - \delta/2$ and $r = \frac{\delta}{2(1+\|\lambda\|_{\Theta^*})}$.

To prove (1.3.16), set $\ell = \inf_K \Lambda^*_{\mathcal{W}}$ and, for given $\delta > 0$, use (1.3.15) and the compactness of K to choose $\psi_1, \ldots, \psi_n \in K$ and $r_1, \ldots, r_n \in (0, \infty)$ so that $K \subseteq \bigcup_1^n B(\psi_k, r_k)$ and

$$
\mathcal{W}_\epsilon\big(B(\psi_k, r_k)\big) \leq \begin{cases} \exp\left[-\frac{1}{\epsilon\delta}\right] & \text{if } \Lambda^*_{\mathcal{W}}(\psi_k) = \infty \\ \exp\left[-\frac{\Lambda^*_{\mathcal{W}}(\psi_k) - \delta}{\epsilon}\right] & \text{if } \Lambda^*_{\mathcal{W}}(\psi_k) < \infty. \end{cases}
$$

Then

$$
\mathcal{W}_\epsilon(K) \leq n \exp\left[-\frac{1}{\epsilon}\left(\frac{1}{\delta} \wedge (\ell - \delta)\right)\right],
$$

and so

$$
\varlimsup_{\epsilon \to 0} \epsilon \log\left(\mathcal{W}_\epsilon(K)\right) \leq -\left[\frac{1}{\delta} \wedge (\ell - \delta)\right].
$$

Finally, let $\delta \searrow 0$. ∎

1.3.17 Remark.

Suppose that $\{\mu_\epsilon : \epsilon > 0\}$ is a family of probability measures on a BANACH space $(X, \|\cdot\|)$ and let Λ_{μ_ϵ} denote the logarithmic moment generating function for μ_ϵ (i.e.,

$$
\Lambda_{\mu_\epsilon}(\lambda) = \log\left(\int_X \exp\left[{}_{X^*}\langle \lambda, x \rangle_X\right] \mu_\epsilon(dx)\right)
$$

for $\lambda \in X^*$). Further, assume that

$$
\Lambda(\lambda) \equiv \lim_{\epsilon \to 0} \epsilon \Lambda_{\mu_\epsilon}(\lambda/\epsilon) \tag{1.3.18}
$$

exists for every $\lambda \in X^*$. Then the argument used to prove Lemma 1.3.14 leads to the conclusion that for any $K \subset\subset X$

$$\overline{\lim_{\epsilon \to 0}} \, \epsilon \log\bigl(\mu_\epsilon(K)\bigr) \le -\inf_K \Lambda^*,$$

where

$$\Lambda^*(x) \equiv \sup\left\{{}_{X^*}\langle \lambda, x\rangle_X - \Lambda(\lambda) : \lambda \in X^*\right\}$$

is the LEGENDRE transform of Λ. In the particular case treated in Lemma 1.3.14, we had that $\epsilon\Lambda_{\mathcal{W}_\epsilon}(\lambda/\epsilon) = \Lambda_{\mathcal{W}}(\lambda)$, and so (1.3.18) was trivial. See Theorem 2.2.4 for more details on this subject.

Although the result obtained in Lemma 1.3.14 is restricted to compact subsets and is therefore less than we really want, we will turn to the left hand side of (1.1.5) before addressing the problem of extending (1.3.16) to all closed sets. Just as in the proof of Theorem 1.2.6, the key to proving the left hand side of (1.1.5) is the use of an efficient method for moving the "center" (or mean) of the measures $\mathcal{W}_\epsilon$. In the present setting, this key is contained in the following important **quasi-invariance** property of WIENER's measure.

1.3.19 Lemma. (CAMERON & MARTIN) *Given $\lambda \in \Theta^*$, let $\mathcal{W}^\lambda$ denote the distribution of $\theta \longmapsto \theta + \psi_\lambda$ under $\mathcal{W}$, where ψ_λ is the element of Θ described in* (1.3.9). *Then $\mathcal{W}^\lambda << \mathcal{W}$ and*

$$\frac{d\mathcal{W}^\lambda}{d\mathcal{W}}(\theta) = R_\lambda(\theta) \equiv \exp\left[{}_{\Theta^*}\langle \lambda, \theta\rangle_\Theta - \Lambda_{\mathcal{W}}(\lambda)\right], \quad \theta \in \Theta. \tag{1.3.20}$$

PROOF: Define the measure P on Θ by $P(d\theta) = \bigl(1/R_\lambda(\theta)\bigr)\,\mathcal{W}^\lambda(d\theta)$. Then the required result is equivalent to the statement that $P = \mathcal{W}$. But, for any $\eta \in \Theta^*$,

$$\begin{aligned}
&\int_\Theta \exp\left[{}_{\Theta^*}\langle \eta, \theta\rangle_\Theta\right] P(d\theta) \\
&\quad = \int_\Theta \frac{1}{R_\lambda(\theta + \psi_\lambda)} \exp\left[{}_{\Theta^*}\langle \eta, \theta + \psi_\lambda\rangle_\Theta\right] \mathcal{W}(d\theta) \\
&\quad = \exp\left[{}_{\Theta^*}\langle \eta - \lambda, \psi_\lambda\rangle_\Theta + \Lambda_{\mathcal{W}}(\lambda)\right] \int_\Theta \exp\left[{}_{\Theta^*}\langle \eta - \lambda, \theta\rangle_\Theta\right] \mathcal{W}(d\theta) \\
&\quad = \exp\left[{}_{\Theta^*}\langle \eta, \psi_\lambda\rangle_\Theta - \Lambda_{\mathcal{W}}(\lambda) + \Lambda_{\mathcal{W}}(\eta - \lambda)\right] = \exp\left[\Lambda_{\mathcal{W}}(\eta)\right],
\end{aligned}$$

where we have made repeated use of (1.3.10) and (1.3.11). From the preceding, we see that the function

$$\zeta \in \mathbb{R} \longmapsto \int_\Theta \exp\left[\zeta \; {}_{\Theta^*}\langle \eta, \theta\rangle_\Theta\right] P(d\theta)$$

extends to an entire function on the whole complex plane, and, in particular, that

$$\int_{\Theta} \exp\left[\sqrt{-1}\ {}_{\Theta^*}\langle \eta, \theta \rangle_{\Theta}\right] P(d\theta) = \exp\left[-\Lambda_{\mathcal{W}}(\eta)\right].$$

That is, $P = \mathcal{W}$. ∎

1.3.21 Lemma. *For every open $G \subseteq \Theta$,*

$$\varliminf_{\epsilon \to 0} \epsilon \log\left(\mathcal{W}_\epsilon(G)\right) \geq -\inf_G \Lambda^*_{\mathcal{W}}.$$

PROOF: What we must show is that

$$\varliminf_{\epsilon \to 0} \epsilon \log\left(\mathcal{W}_\epsilon(G)\right) \geq -\Lambda^*_{\mathcal{W}}(\psi) \tag{1.3.22}$$

for every $\psi \in G$. By (1.3.12), we need only check (1.3.22) for $\psi \in G \cap H^1$. But if $\psi \in G \cap H^1$, then we can find $\{\psi_n\}_1^\infty \subseteq C_c^\infty([0,\infty); \mathbb{R}^d)$ such that $\|\psi_n - \psi\|_{H^1} \longrightarrow 0$. In particular, $\psi_n \in G$ for all sufficiently large n's and $\Lambda^*_{\mathcal{W}}(\psi_n) \longrightarrow \Lambda^*_{\mathcal{W}}(\psi)$. Hence, we need only prove (1.3.22) for $\psi \in C_c^\infty([0,\infty); \mathbb{R}^d)$. Given such a ψ, define $\lambda \in \Theta^*$ by $\lambda((t,\infty)) = \dot{\psi}(t)$ and choose $r > 0$ so that $B(\psi, r) \subseteq G$. Then $\psi = \psi_\lambda$ and so, for $0 < \delta < r$,

$$\begin{aligned}
\mathcal{W}_\epsilon(G) &\geq \mathcal{W}_\epsilon(B(\psi,\delta)) = \mathcal{W}^{-\lambda/\epsilon^{1/2}}\left(B(0, \delta/\epsilon^{1/2})\right) \\
&= \int_{B(0,\delta/\epsilon^{1/2})} \exp\left[-{}_{\Theta^*}\langle \lambda/\epsilon^{1/2}, \theta \rangle_{\Theta} - \frac{1}{\epsilon}\Lambda_{\mathcal{W}}(\lambda)\right] \mathcal{W}(d\theta) \\
&\geq \exp\left[-\frac{1}{\epsilon}\left(\Lambda_{\mathcal{W}}(\lambda) + \delta\|\lambda\|_{\Theta^*}\right)\right].
\end{aligned}$$

Since, by (1.3.13), $\Lambda_{\mathcal{W}}(\lambda) = \Lambda^*_{\mathcal{W}}(\psi)$, we see that (1.3.22) holds. ∎

We must now return to the problem of removing the compactness restriction from Lemma 1.3.14. Our idea will be to produce a family of compact sets K_L, $L > 0$ with the property that

$$\varlimsup_{\epsilon \to 0} \epsilon \log\left(\mathcal{W}_\epsilon\left(K_L^c\right)\right) \leq -L, \quad L > 0. \tag{1.3.23}$$

What (1.3.23) says is that, as $L \nearrow \infty$, the events K_L^c become so "deviant" that they cannot even be seen on the scale at which we are looking; and, therefore, they cannot contribute to our calculation (cf. the proof of Theorem 1.3.27 below).

There are several ways in which one can go about constructing the sets K_L. The method which we will adopt here will be to construct a function $\Phi : \Theta \longrightarrow [0, \infty]$ with the properties that

(1) Φ is sub-additive and $\Phi(\alpha\theta) = |\alpha|\Phi(\theta)$ for all $\alpha \in \mathbb{R}$ and $\theta \in \Theta$,
(2) $\{\theta : \Phi(\theta) \leq L\} \subset\subset \Theta$ for each $L > 0$,
(3) $\mathcal{W}\big(\{\theta : \Phi(\theta) < \infty\}\big) = 1$.

In order to construct such a Φ and to pass from the fact that it exists to (1.3.23), we will make use of the following beautiful and powerful estimate due to X. FERNIQUE [45].

1.3.24 Theorem. (FERNIQUE) *Let X be a real, separable* FRÉCHET *space and $\Phi : X \longrightarrow [0, \infty]$ a measurable sub-additive function with the property that $\Phi(\alpha x) = |\alpha|\Phi(x)$ for all $\alpha \in \mathbb{R}$ and $x \in X$. Next, let μ be a probability measure on $(X, \mathcal{B}_X)$ with the property that μ^2 on $(X^2, \mathcal{B}_{X^2})$ is invariant under the transformation*

$$(x_1, x_2) \in X^2 \longmapsto \left(\frac{x_1 - x_2}{2^{1/2}}, \frac{x_1 + x_2}{2^{1/2}}\right) \in X^2.$$

If $\mu\big(\{x : \Phi(x) < \infty\}\big) = 1$, then there exists an $\alpha > 0$ for which

$$\int \exp\big[\alpha\Phi(x)^2\big]\, \mu(dx) < \infty.$$

PROOF: Given $0 < s < t$, we have

$$\begin{aligned}
\mu\big(\{x : &\Phi(x) \leq s\}\big)\, \mu\big(\{x : \Phi(x) \geq t\}\big) \\
&= \mu^2\big(\{(x_1, x_2) \in X^2 : \Phi(x_1) \leq s \text{ and } \Phi(x_2) \geq t\}\big) \\
&= \mu^2\big(\{(x_1, x_2) : \Phi(x_1 - x_2) \leq 2^{1/2}s \\
&\qquad\qquad\qquad \text{and } \Phi(x_1 + x_2) \geq 2^{1/2}t\}\big) \\
&\leq \mu^2\big(\{(x_1, x_2) : |\Phi(x_1) - \Phi(x_2)| \leq 2^{1/2}s \\
&\qquad\qquad\qquad \text{and } \Phi(x_1) + \Phi(x_2) \geq 2^{1/2}t\}\big) \\
&\leq \mu^2\big(\{(x_1, x_2) : \Phi(x_1) \wedge \Phi(x_2) \geq 2^{-1/2}(t - s)\}\big) \\
&= \Big(\mu\big(\{x : \Phi(x) \geq 2^{-1/2}(t - s)\}\big)\Big)^2.
\end{aligned}$$

Hence, if $s > 0$ is chosen so that $\mu\big(\{x : \Phi(x) < s\}\big) > \frac{1}{2}$ and $\{t_n\}_0^\infty$ is defined by $t_0 = s$ and $t_n = s + 2^{1/2}t_{n-1}$, $n \in \mathbb{Z}^+$, then

$$\mu\big(\{x : \Phi(x) \geq t_n\}\big)\, \mu\big(\{x : \Phi(x) \leq s\}\big) \leq \big[\mu\big(\{x : \Phi(x) \geq t_{n-1}\}\big)\big]^2, \quad n \geq 1;$$

and therefore

$$\frac{\mu(\{x : \Phi(x) \geq t_n\})}{\mu(\{x : \Phi(x) \leq s\})} \leq \left(\frac{\mu(\{x : \Phi(x) \geq t_{n-1}\})}{\mu(\{x : \Phi(x) \leq s\})} \right)^2, \quad n \geq 1.$$

Working by induction, we conclude from this that

$$\frac{\mu(\{x : \Phi(x) \geq t_n\})}{\mu(\{x : \Phi(x) \leq s\})} \leq \left(\frac{\mu(\{x : \Phi(x) \geq s\})}{\mu(\{x : \Phi(x) \leq s\})} \right)^{2^n};$$

and so

$$\mu(\{x : \Phi(x)^2 \geq 2^n \beta\}) \leq \exp[-2^n \sigma], \quad n \geq 0,$$

where

$$\sigma \equiv -\log \left[\frac{\mu(\{x : \Phi(x) \geq s\})}{\mu(\{x : \Phi(x) \leq s\})} \right] > 0 \quad \text{and} \quad \beta \equiv \left[\frac{2s}{2^{1/2} - 1} \right)^2.$$

Thus if $\alpha < \sigma/(2\beta)$, then

$$\begin{aligned}\int_{\{x:\Phi(x)^2 \geq \beta\}} \exp[\alpha \Phi(x)^2] \, \mu(dx) &\leq \sum_{n=0}^{\infty} \exp[2^{n+1} \alpha \beta] \, \mu(\{x : \Phi(x)^2 \geq 2^n \beta\}) \\ &\leq \sum_{0}^{\infty} \exp[-2^n (\sigma - 2\alpha\beta)] < \infty. \blacksquare\end{aligned}$$

1.3.25 Lemma. *For $\theta \in \Theta$ set*

$$\Phi(\theta) = \sum_{n=1}^{\infty} \frac{1}{2^n} \sup_{0 \leq s < t \leq n} \frac{|\theta(t) - \theta(s)|}{|t - s|^{1/4}} + \sup_{t \geq 1} \frac{|\theta(t)|}{t^{3/4}}.$$

Then $\{\theta \in \Theta : \Phi(\theta) \leq L\} \subset\subset \Theta$, for each $L > 0$; and there exists an $\alpha > 0$ such that

$$\int_{\Theta} \exp[\alpha \Phi(\theta)^2] \, \mathcal{W}(d\theta) < \infty. \tag{1.3.26}$$

In particular, if $K_L = \{\theta : \Phi(\theta)^2 \leq L/\alpha\}$, then $K_L \subset\subset \Theta$ and (1.3.23) holds.

PROOF: The proof that, for every $R \in (0, \infty)$, $\{\theta : \Phi(\theta) \leq R\} \subset\subset \Theta$ is a standard application of the ASCOLI-ARZELÀ criterion combined with a diagonalization argument. The details are left to the reader.

To prove that (1.3.26) holds for some $\alpha > 0$, we first observe that $\mathcal{W}^2$ has the invariance property required in FERNIQUE's Theorem. (Indeed, any centered GAUSSian measure on a FRÉCHET space will have this property.) Thus, the existence of α will follow once we show that $\mathcal{W}(\{\theta : \Phi(\theta) < \infty\}) = 1$. To this end, note that, by parts (**iv**) and (**vi**) of Theorem 1.3.2 combined with FERNIQUE's Theorem,

$$\int_\Theta \left(\sup_{0 \le s < t \le n} \frac{|\theta(t) - \theta(s)|}{|t-s|^{1/4}} \right)^8 \mathcal{W}(d\theta) = A_d n^2, \quad n \ge 1,$$

for some $A_d < \infty$. At the same time, again as a consequence of FERNIQUE's Theorem and elementary properties of $\mathcal{W}$, we see that

$$\int_\Theta \left(\sup_{n \le t \le n+1} \frac{|\theta(t) - \theta(n)|}{n^{3/4}} \right)^8 \mathcal{W}(d\theta) = \frac{B_d}{n^6}, \quad n \ge 1,$$

for some $B_d < \infty$. Finally, since, by (**iv**) in Theorem 1.3.2,

$$\int_\Theta \left(\frac{|\theta(n)|}{n^{3/4}} \right)^8 \mathcal{W}(d\theta) = \frac{C_d}{n^2}$$

for some $C_d < \infty$, we can combine these into the estimate that

$$\int_\Theta \big(\Phi(\theta)\big)^8 \mathcal{W}(d\theta) < \infty,$$

which is more than enough for our purposes.

Knowing (1.3.26), we can proceed to prove (1.3.23) as follows:

$$\begin{aligned} \mathcal{W}_\epsilon\big(K_L^c\big) &= \mathcal{W}\Big(\big\{\theta : \alpha\Phi(\theta)^2 > L/\epsilon\big\}\Big) \\ &\le \exp\big[-L/\epsilon\big] \int_\Theta \exp\big[\alpha\Phi(\theta)^2\big]\, \mathcal{W}(d\theta). \end{aligned}$$

Together with (1.3.26), this surely leads to (1.3.23). ∎

1.3.27 Theorem. (SCHILDER) *For every* $\Gamma \in \mathcal{B}_\Theta$:

$$(1.3.28) \quad -\inf_{\Gamma^\circ} \Lambda^*_{\mathcal{W}} \le \varliminf_{\epsilon \to 0} \epsilon \log\big(\mathcal{W}_\epsilon(\Gamma)\big) \le \varlimsup_{\epsilon \to 0} \epsilon \log\big(\mathcal{W}_\epsilon(\Gamma)\big) \le -\inf_{\overline{\Gamma}} \Lambda^*_{\mathcal{W}}.$$

PROOF: In view of Lemma 1.3.21, all that we have to do is show that

$$\varlimsup_{\epsilon \to 0} \epsilon \log\big(\mathcal{W}_\epsilon(F)\big) \le -\inf_F \Lambda^*_{\mathcal{W}}$$

for each closed sets F. To this end, let $\ell = \inf_F \Lambda^*_{\mathcal{W}}$, and, for $L > 0$, set $F_L = F \cap K_L$, where K_L is the compact subset produced in the preceding lemma. Then $\mathcal{W}_\epsilon(F) \le \mathcal{W}_\epsilon\big(F_L\big) + \mathcal{W}_\epsilon\big(K_L^c\big)$; and so, by Lemma 1.3.14 and (1.3.23),

$$\varlimsup_{\epsilon \to 0} \epsilon \log\big(\mathcal{W}_\epsilon(F)\big) \le -(\ell \wedge L).$$

After letting $L \nearrow \infty$, we arrive at the required result. ∎

1.3.29 Exercise.

Given $\psi \in \Theta$ and $n \geq 1$, define

$$V_n(\psi) = n \sum_{k=1}^{n^2} \left| \psi\left(\frac{k}{n}\right) - \psi\left(\frac{k-1}{n}\right) \right|^2. \tag{1.3.30}$$

Show that

$$V(\psi) \equiv \sup_n V_n(\psi) = \lim_{n\to\infty} V_n(\psi), \tag{1.3.31}$$

$V(\psi) < \infty$ if and only if $\psi \in H^1$, and $V(\psi) = \|\psi\|_{H^1}^2$ for $\psi \in H^1$.

1.3.32 Exercise.

The Lemma 1.3.19 is not a complete statement of CAMERON and MARTIN's result [**15**]. Indeed, suppose that $\psi \in H^1$ and choose

$$\{\psi_n\}_1^\infty \subseteq C_c^\infty\left([0,\infty);\mathbb{R}^d\right)$$

so that $\|\psi_n - \psi\|_{H^1} \longrightarrow 0$. Set

$$\Phi_n(\theta) = {}_{\Theta^*}\langle \lambda_n, \theta \rangle_\Theta, \quad \theta \in \Theta,$$

where λ_n is the element of Θ^* defined by $\lambda_n((t,\infty)) = \dot{\psi}_n(t)$, $t \geq 0$. Show that $\Phi_n \longrightarrow \Phi$ in $L^2(\mathcal{W})$, where Φ under $\mathcal{W}$ is GAUSSian with mean 0 and variance $\frac{1}{2}\|\psi\|_{H^1}^2$. Next, show that $\exp\left[\Phi_n - \Lambda_{\mathcal{W}}(\lambda_n)\right] \longrightarrow \exp\left[\Phi - \frac{1}{2}\|\psi\|_{H^1}^2\right]$ in $L^1(\mathcal{W})$. Finally, conclude from this that if $\mathcal{W}^\psi$ denotes the distribution of $\theta \longmapsto \theta + \psi$ under $\mathcal{W}$, then $\mathcal{W}^\psi << \mathcal{W}$ and

$$\frac{d\mathcal{W}^\psi}{d\mathcal{W}} = \exp\left[\Phi - \frac{1}{2}\|\psi\|_{H^1}\right] \quad \text{(a.s., } \mathcal{W}). \tag{1.3.33}$$

The expression (1.3.33) is often called the **Cameron–Martin formula**.

1.3.34 Exercise.

The purpose of this exercise is to show that the result obtained in Exercise 1.3.32 is optimal. The proof outlined below relies on a knowledge of the DOOB's Martingale Convergence Theorem. In particular, one needs to know that if P and Q are probability measures on a measurable space $(E,\mathcal{F})$, $\{\mathcal{F}_n\}_0^\infty$ is a non-decreasing sequence of sub-σ-algebras of $\mathcal{F}$ with the property that $\mathcal{F} = \sigma\left(\bigcup_0^\infty \mathcal{F}_n\right)$, and the restriction Q_n of Q to $\mathcal{F}_n$ is absolutely continuous with respect to the restriction P_n of P to $\mathcal{F}_n$, then $R_n \equiv \frac{dQ_n}{dP_n} \longrightarrow R$ (a.s., P) where R is the RADON–NIKODYM derivative of the absolutely continuous part of Q with respect to P.

(**i**) Given $\psi \in \Theta$ and $n \geq 0$, let ψ_n be the element of Θ such that $\psi_n\left(\frac{k}{2^n}\right) = \psi\left(\frac{k}{2^n}\right)$ for $0 \leq k \leq 4^n$, $\psi_n(t) = \psi(2^n)$ for $t \geq 2^n$, and ψ_n is linear on each of the intervals $\left[\frac{k-1}{2^n}, \frac{k}{2^n}\right]$, $1 \leq k \leq 4^n$. Show that $\psi_n \in H^1$ and that $\|\psi_n\|_{H^1}^2 = V_{2^n}(\psi)$ (cf. (1.3.30)).

(**ii**) Given $n \geq 0$, define $\mathcal{F}_n = \sigma\left(\theta(k/2^n) : 0 \leq k \leq 4^n\right)$ (i.e., the smallest σ-algebra over Θ with respect to which all the functions $\theta \longmapsto \theta(k/2^n)$, $0 \leq k \leq 4^n$ are measurable). Note that $\mathcal{F}_n \subseteq \mathcal{F}_{n+1}$ and show that $\mathcal{B}_\Theta = \sigma\left(\bigcup_0^\infty \mathcal{F}_n\right)$.

(**iii**) Given $\psi \in \Theta$, let $\mathcal{W}^\psi$ denote the distribution of $\theta \longmapsto \theta + \psi$ under $\mathcal{W}$. Referring to (**i**) and (**ii**) above, let $\mathcal{W}_n^\psi$ and $\mathcal{W}_n$ be the restriction of $\mathcal{W}^\psi$ and $\mathcal{W}$, respectively, to $\mathcal{F}_n$; and define $R_n = \exp\left[\Phi_n - \frac{1}{2}\|\psi_n\|_{H^1}\right]$, where Φ_n is the function corresponding to ψ_n as in Exercise 1.3.32. Show that $\mathcal{W}_n^\psi << \mathcal{W}_n$ and that $\frac{d\mathcal{W}_n^\psi}{d\mathcal{W}_n} = R_n$ (a.s., $\mathcal{W}_n$).

(**iv**) Referring to (**iii**), show that

$$\int_\Theta R_n^{1/2}\, d\mathcal{W} = \exp\left[-\frac{1}{4}V_{2^n}(\psi)\right].$$

Now suppose that $\psi \in \Theta \setminus H^1$ and define $\mathcal{W}^\psi$ accordingly. Using the preceding, show that $\mathcal{W}^\psi \perp \mathcal{W}$ (i.e., is singular with respect to $\mathcal{W}$.) This is the sense in which Exercise 1.3.32 gives the optimal result.

1.3.35 Exercise.

Given $T > 0$, define $I_T : \Theta \longrightarrow [0, \infty]$ by

$$I_T(\psi) = \begin{cases} \infty & \text{if } \psi \notin H^1 \\ \frac{1}{2}\int_{[0,T]} |\dot{\psi}(t)|^2\, dt & \text{if } \psi \in H^1. \end{cases} \tag{1.3.36}$$

Show that if $\Gamma \in \mathcal{B}_T$, then

$$-\inf_{\Gamma^\circ} I_T \leq \underline{\lim}_{\epsilon \to 0}\, \epsilon \log\left(\mathcal{W}_\epsilon(\Gamma)\right) \leq \overline{\lim}_{\epsilon \to 0}\, \epsilon \log\left(\mathcal{W}_\epsilon(\Gamma)\right) \leq -\inf_{\overline{\Gamma}} I_T. \tag{1.3.37}$$

1.4 Two Applications of Schilder's Theorem

We continue in this section with the notation introduced in Section 1.3.

Perhaps the single most striking application of SCHILDER's Theorem is to the derivation of V. STRASSEN's renowned **Law of the Iterated Logarithm** [100]. Thus, our first goal in this section will be to show how SCHILDER's Theorem provides the key estimates in the proof of the following statement.

1.4.1 Theorem. (STRASSEN) *For $n \geq 3$, define*

$$\xi_n(t,\theta) = \frac{\theta(nt)}{\beta(n)}, \quad (t,\theta) \in [0,\infty) \times \Theta,$$

where $\beta(n) \equiv \left(2n\log(\log n)\right)^{1/2}$; and set $K = \{\psi \in H^1 : \|\psi\|_{H^1} \leq 1\}$. Then for $\mathcal{W}$-almost every $\theta \in \Theta$ the sequence $\{\xi_n(\theta)\}_{n=3}^{\infty}$ has the properties:

(**i**) *$\{\xi_n(\theta)\}_3^{\infty}$ is relatively compact in Θ and every limit point is an element of K.*

(**ii**) *For every $\psi \in K$ there is a subsequence of $\{\xi_n(\theta)\}_3^{\infty}$ which converges in Θ to ψ.*

In particular, for every $\Phi \in C(\Theta;\mathbb{R})$,

$$\mathcal{W}\left(\{\theta : \varlimsup_{n\to\infty} \Phi(\xi_n(\theta)) = \sup_K \Phi\}\right) = 1. \tag{1.4.2}$$

In the proof of (**i**) we will use the following elementary observation.

1.4.3 Lemma. *Let $S \subseteq (1,2)$ and assume that $1 \in \overline{S}$. If $\theta \in \Theta$ has the property that*

$$\varlimsup_{m\to\infty} \|\xi_{[s^m]}(\theta) - K\|_\Theta = 0 \tag{1.4.4}$$

for every $s \in S$ ($[r]$ denotes the greatest integer less than or equal $r \in \mathbb{R}$), then $\varlimsup_{n\to\infty} \|\xi_n(\theta) - K\|_\Theta = 0$.

PROOF: Assume that (1.4.4) holds for every $s \in S$ and set

$$\psi_n = \xi_n(\theta) \quad \text{and} \quad \psi_{s,m} = \psi_{[s^m]}$$

for $n \geq 3$, $s \in S$, and m for which $s^m \geq 3$. Given $\delta > 0$, choose $s \in S$ so that

$$\left(1 - \frac{1}{s}\right)^{1/2} < \delta \quad \text{and} \quad (s-1) < \frac{\delta}{1+\delta}.$$

Next, choose $M \in \mathbb{Z}^+$ so that

$$s^M \geq 3, \; \frac{\log(\log sr)}{\log(\log r)} < s \text{ for } r \geq s^M, \text{ and } \|\psi_{s,m} - K\|_\Theta < \delta \text{ for } m \geq M.$$

Then for $m \geq M+1$ and $s^{m-1} \leq n \leq s^m$ we have that

$$\|\psi_n - K\|_\Theta \leq \delta + \|\psi_n - \psi_{s,m}\|_\Theta$$

and that

$$\|\psi_n - \psi_{s,m}\|_\Theta \leq \frac{\beta([s^m])}{\beta(n)} \left\| \psi_{s,m}\left(\frac{n}{[s^m]}\cdot\right) - \psi_{s,m} \right\|_\Theta + \left(\frac{\beta([s^m])}{\beta(n)} - 1\right) \|\psi_{s,m}\|_\Theta.$$

Note that

$$1 \leq \frac{\beta([s^m])}{\beta(n)} \leq \left(s \frac{\log(\log sn)}{\log(\log n)}\right)^{1/2} < s < 2.$$

Also, since

$$\|\psi\|_\Theta \leq 1 \quad \text{and} \quad \left\|\psi\left(\frac{n}{[s^m]}\cdot\right) - \psi\right\|_\Theta \leq \left(1 - \frac{n}{s^m}\right)^{1/2} < \delta$$

for all $\psi \in K$, we see that

$$\|\psi_{s,m}\|_\Theta \leq 1 + \delta \quad \text{and} \quad \left\|\psi_{s,m}\left(\frac{n}{[s^m]}\cdot\right) - \psi_{s,m}\right\|_\Theta < 4\delta.$$

Combining these, we conclude that $\|\psi_n - K\|_\Theta < 10\delta$ as long as $n \geq s^{M+1}$. ∎

PROOF OF THEOREM 1.4.1: We begin by proving (**i**). Because $K \subset\subset \Theta$, it suffices for us to show that $\xi_n(\theta) \longrightarrow K$ for $\mathcal{W}$-almost every $\theta \in \Theta$. Moreover, because of Lemma 1.4.3, we will know this as soon as we show that

$$\mathcal{W}\left(\left\{\theta : \overline{\lim_{m\to\infty}} \left\|\xi_{[s^m]}(\theta) - K\right\|_\Theta = 0\right\}\right) = 1 \tag{1.4.5}$$

for every $s > 1$. Indeed, if (1.4.5) holds for every $s > 1$ and we take $S = \{1 + 1/n : n \geq 1\}$, then we can choose one $\mathcal{W}$-null set A so that

$$\overline{\lim_{m\to\infty}} \left\|\xi_{[s^m]}(\theta) - K\right\|_\Theta = 0$$

for every $\theta \notin A$ and $s \in S$. To prove (1.4.5), let $\delta > 0$ be given and set $K^{(\delta)} = \{\psi \in \Theta : \|\psi - K\|_\Theta < \delta\}$. Then $\gamma \equiv \inf\{\|\psi\|_{H^1}^2 : \psi \notin K^{(\delta)}\} > 1$; and therefore, by SCHILDER's Theorem and $\mathcal{W}$-scaling invariance (cf. part (**iv**) of Theorem 1.3.2),

$$\mathcal{W}\left(\{\theta : \xi_{[s^m]}(\theta) \notin K^{(\delta)}\}\right) = \mathcal{W}_{\epsilon_s(m)}\left(\left(K^{(\delta)}\right)^{\mathrm{c}}\right) \leq \exp\left[-\frac{\gamma}{2\epsilon_s(m)}\right]$$

for all sufficiently large m, where

$$\epsilon_s(m) \equiv \frac{1}{2\log\left(\log[s^m]\right)}.$$

Since γ and s are strictly larger than 1, it follows immediately from the preceding that

$$\sum_{s^m \geq 3} \mathcal{W}\left(\{\theta : \xi_{[s^m]}(\theta) \notin K^{(\delta)}\}\right) < \infty;$$

and therefore, by the BOREL–CANTELLI Lemma,

$$\mathcal{W}\left(\left\{\theta : \xi_{[s^m]}(\theta) \in K^{(\delta)} \text{ for infinitely many } m \in \mathbb{Z}^+\right\}\right) = 1.$$

Since this is true for every $\delta > 0$, (1.4.5) and, therefore, (**i**) have now been proved.

Set $\Theta' = \{\theta : \overline{\lim}_{n\to\infty} \|\xi_n(\theta) - K\|_\Theta = 0\}$. We now know that $\mathcal{W}(\Theta') = 1$. Note that for any $\theta \in \Theta'$ and $\psi \in K$,

$$\lim_{k\to\infty} \sup_{n\geq 3} \sup_{t\vee\frac{1}{t}\geq k} \frac{1}{1+t}\left(\left|\xi_n(t,\theta)\right| + |\psi(t)|\right) = 0. \tag{1.4.6}$$

Thus, in proving (**ii**), we need only show that for each $k \geq 3$ and $\psi \in K$

$$\underline{\lim}_{n\to\infty} \sup_{\frac{1}{k}\leq t\leq k} \frac{1}{1+t}\left|\xi_n(t,\theta) - \psi(t)\right| = 0$$

for $\mathcal{W}$-almost every $\theta \in \Theta$. Moreover, by another application of (1.4.6), this will be shown if we can prove that

$$\mathcal{W}\left(\left\{\theta : \underline{\lim}_{m\to\infty} \left\|\eta_{k,m}(\theta) - \psi_k\right\|_\Theta = 0\right\}\right) = 1, \tag{1.4.7}$$

where

$$\eta_{k,m}(t,\theta) = \begin{cases} 0 & \text{if } t \in \left[0, \frac{1}{k}\right] \\ \left(\theta(k^m t) - \theta(k^{m-1})\right)/\beta(k^m) & \text{if } t \in \left(\frac{1}{k}, k\right) \\ \left(\theta(k^{m+1}) - \theta(k^{m-1})\right)/\beta(k^m) & \text{if } t \in [k, \infty) \end{cases}$$

and

$$\psi_k(t) = \begin{cases} 0 & \text{if } t \in \left[0, \frac{1}{k}\right] \\ \psi(t) - \psi\left(\frac{1}{k}\right) & \text{if } t \in \left(\frac{1}{k}, k\right) \\ \psi(k) - \psi\left(\frac{1}{k}\right) & \text{if } t \in [k, \infty). \end{cases}$$

The advantage gained by dealing with the functions $\eta_{k,m}$ instead of the original ξ_n 's is that, for fixed k, the $\eta_{k,m}$'s are mutually independent under $\mathcal{W}$. Hence, by the BOREL–CANTELLI Lemma, we will have proved (1.4.7) once we show that

$$\text{(1.4.8)} \qquad \sum_{m=1}^{\infty} \mathcal{W}\Big(\{\theta : \left\|\eta_{k,m}(\theta) - \psi_k\right\|_{\Theta} < \delta\}\Big) = \infty$$

for each $\delta > 0$.

To prove (1.4.8), note that, by **(iv)** in Theorem 1.3.2,

$$\mathcal{W}\left(\{\theta : \left\|\eta_{k,m}(\theta) - \psi_k\right\|_{\Theta} < \delta\}\right) \geq \mathcal{W}_{\epsilon_k(m)}\big(B(\tilde{\psi}_k, \delta)\big),$$

where $\tilde{\psi}_k(t) \equiv \psi_k\left(t + \frac{1}{k}\right)$, $t \geq 0$. Since $\left\|\tilde{\psi}_k\right\|_{H^1} \leq \|\psi\|_{H^1} \leq 1$ and therefore

$$\inf_{\phi \in B(\tilde{\psi}_k, \delta)} \|\phi\|_{H^1} < 1$$

for every $\delta > 0$, we can use SCHILDER's Theorem to find a $\gamma < 1$ such that

$$\mathcal{W}_{\epsilon_k(m)}\big(B(\tilde{\psi}_k, \delta)\big) \geq \exp\left[-\frac{\gamma}{2\epsilon_s(m)}\right]$$

for all sufficiently large m 's. When combined with the preceding, this clearly proves (1.4.8) and therefore **(ii)**.

Given **(i)** and **(ii)**, the proof of (1.4.2) is easy and is left to the reader. ∎

Our second application of SCHILDER's theorem will be to VENTCEL and FREIDLIN's estimate on the large deviations of randomly perturbed dynamical systems [**109**]. Our approach is based on the ideas of R. AZENCOTT [**1**].

Given $T > 0$, let $\left(\Omega_T, \|\cdot\|_{\Omega_T}\right)$ be the BANACH space $C([0,T]; \mathbb{R}^d)$ with the uniform norm. The theory of VENTCEL and FREIDLIN deals with families of measures $\{P_\epsilon : \epsilon > 0\}$ on $\left(\Omega_T, \mathcal{B}_{\Omega_T}\right)$ of which the following is a typical example. For a given bounded, uniformly LIPSCHITZ continuous function $b : \mathbb{R}^d \longrightarrow \mathbb{R}^d$, define $\theta \in \Theta \longmapsto X(\theta) \in \Omega_T$ by

$$X(t, \theta) = \theta(t) + \int_0^t b\big(X(s, \theta)\big)\, ds, \quad t \in [0, T];$$

and for $\epsilon > 0$ let $P_\epsilon = \mathcal{W}_\epsilon \circ X^{-1}$ be the distribution of $\theta \longmapsto X(\theta)$ under $\mathcal{W}_\epsilon$. Since an equivalent description of P_ϵ is as the distribution of $\theta \longmapsto X(\epsilon^{1/2}\theta)$ under $\mathcal{W}$, it is clear that $P_\epsilon \Longrightarrow \delta_{X_0}$, where $X_0 \in \Omega_T$ is the integral curve of b which starts at 0. Moreover, it is easy to see that $\theta \in \Theta \longmapsto X(\theta) \in \Omega_T$ is a continuous mapping. Thus, if $G \subseteq \Omega_T$ is open, then SCHILDER's Theorem (cf. Exercise 1.3.35) says that

$$\begin{aligned}
\underline{\lim}_{\epsilon\to 0} \epsilon \log\left[P_\epsilon(G)\right] &= \underline{\lim}_{\epsilon\to 0} \epsilon \log\left[\mathcal{W}_\epsilon\big(X^{-1}(G)\big)\right] \\
&\geq -\inf\big\{I_T(\psi) : X(\psi) \in G\big\} \\
&= -\inf\big\{I_T \circ X^{-1}(\phi) : \phi \in G \text{ and } \phi(0) = 0\big\} \\
&= -\inf\left\{\frac{1}{2}\int_0^T \big|\dot{X}(t) - b\big(X(t)\big)\big|^2\, dt : X \in G \text{ and } X(0) = 0\right\} \\
&= -\inf_G I_T^b
\end{aligned}$$

where

$$I_T^b(X) = \begin{cases} \infty & \text{if } X \notin H_T^1 \\ \frac{1}{2}\int_0^T \big|\dot{X}(t) - b\big(X(t)\big)\big|^2\, dt & \text{if } X \in H_T^1 \end{cases}$$

and

$$H_T^1 = H^1(([0,T];\mathbb{R}^d) = \big\{\psi|_{[0,T]} : \psi \in H^1\big\}. \tag{1.4.9}$$

Similarly, if $F \subseteq \Omega_T$ is closed, then

$$\overline{\lim_{\epsilon\to 0}} \log\left[P_\epsilon(F)\right] \leq -\inf_F I_T^b.$$

In other words, SCHILDER's Theorem leads directly to a large deviation result for $\{P_\epsilon : \epsilon > 0\}$.

The preceding example of VENTCEL and FREIDLIN's theory is as simple as it is because the map $\theta \longmapsto X(\theta)$ is especially pleasant; in particular, it is continuous and its inverse is easy to compute. In general, the maps involved are not only more complicated but are not even continuous. To be precise, let $a : \mathbb{R}^d \longrightarrow \mathbb{R}^d \otimes \mathbb{R}^d$ be symmetric matrix valued, $b : \mathbb{R}^d \longrightarrow \mathbb{R}^d$, and assume that there exists an $M \in [1,\infty)$ such that

$$\frac{1}{M} I_{\mathbb{R}^d} \leq a(x) \leq M I_{\mathbb{R}^d} \quad \text{and} \quad |b(x)| \leq M, \quad x \in \mathbb{R}^d, \tag{1.4.10}$$

and

$$\|a(y) - a(x)\|_{\text{H.S.}} \vee |b(y) - b(x)| \leq M|y - x|, \quad x, y \in \mathbb{R}^d. \tag{1.4.11}$$

(In (1.4.11) and elsewhere, $\|\cdot\|_{\text{H.S.}}$ stands for the HILBERT-SCHMIDT norm.) Next, for $x \in \mathbb{R}^d$ and $\epsilon > 0$, let $X_\epsilon^x : [0,T] \times \Theta \longrightarrow \mathbb{R}^d$ be the $\mathcal{W}$-almost surely unique $\{\mathcal{B}_t : t \in [0,T]\}$-progressively measurable solution to the ITÔ stochastic integral equation

$$(1.4.12) \qquad \begin{aligned} X_\epsilon^x(t,\theta) = x + \epsilon^{1/2} \int_0^t \sigma\big(X_\epsilon^x(s,\theta)\big)\, d\theta(s) \\ + \int_0^t b\big(X_\epsilon^x(s,\theta)\big)\, ds, \qquad t \in [0,T], \end{aligned}$$

where $\sigma \equiv a^{1/2}$; and define

$$P_\epsilon^x = \mathcal{W} \circ \left(X_\epsilon^x\right)^{-1}$$

on $\left(\Omega_T, \mathcal{B}_{\Omega_T}\right)$ (since $X_\epsilon^x(\cdot,\theta) \in \Omega_T$ for $\mathcal{W}$-almost every $\theta \in \Theta$, there is no problem with considering P_ϵ^x on Ω_T). Once again, $P_\epsilon^x \Longrightarrow \delta_{X_0^x}$, where $X_0^x \in \Omega_T$ is the integral curve

$$X_0^x(t) = x + \int_0^t b\big(X_0^x(s)\big)\, ds, \quad t \in [0,T].$$

Moreover, if one pretends that (1.4.12) means that

$$\dot{X}_\epsilon^x(t,\theta) = \epsilon^{1/2}\sigma\big(X_\epsilon^x(s,\theta)\big)\dot{\theta}(s) + b\big(X_\epsilon^x(s,\theta)\big), \quad t \in [0,T],$$

(this is not even formally correct, since we are dealing with ITÔ and not STRATONOVICH integral; however this error becomes negligible as $\epsilon \to 0$) and one ignores all continuity questions, then one can repeat the argument given in the preceding paragraph and thereby arrive at the conjecture that the large deviations of $\{P_\epsilon^x : \epsilon > 0\}$ are governed by the function

$$(1.4.13) \qquad I_{T,x}^{a,b}(X) = \begin{cases} \infty \\ \frac{1}{2}\int_0^T \left| a^{-1/2}\big(X(t)\big)\left(\dot{X}(t) - b\big(X(t)\big)\right)\right|^2 dt, \end{cases}$$

according to whether $X - x \notin H_T^1$ or $X - x \in H_T^1$. Considering all the objections which one can raise to the above naïve line of reasoning, it is somewhat remarkable that the conjecture to which it leads is, nonetheless, absolutely correct.

In order to get around the most serious flaw in our heuristic argument (namely, our treatment of the maps $\theta \in \Theta \longmapsto X_\epsilon^x(\theta) \in \Omega_T$ as if they were continuous), we introduce EULER approximations. Namely, set

$$\mathbf{T}_n(t) = \frac{[nt]}{n}, \quad n \in \mathbb{Z}^+ \text{ and } t \in [0,\infty)$$

(recall that $[r]$ is the integer part of $r \in \mathbb{R}$), and consider the maps $\theta \in \Theta \longmapsto X^x_{n,\epsilon}(\theta) \in \Omega_T$ given by

$$(1.4.14)\quad \begin{aligned} X^x_{n,\epsilon}(t,\theta) = x + \epsilon^{1/2} \int_0^t \sigma\Big(X^x_{n,\epsilon}(\mathbf{T}_n(t),\theta)\Big)\, d\theta(s) \\ + \int_0^t b\left(X^x_{n,\epsilon}(s,\theta)\right)\, ds, \end{aligned}$$

for $t \in [0,T]$. Clearly the maps $\theta \in \Theta \longmapsto X^x_{n,\epsilon}(\theta) \in \Omega_T$ are continuous. Moreover, $X^x_{n,\epsilon}(\theta) = X^x_{n,1}(\epsilon^{1/2}\theta)$; and so, just as in the original case considered, we can apply SCHILDER's Theorem to deduce that

$$(1.4.15)\quad \begin{aligned} -\inf_{\Gamma^\circ} I^{a,b}_{n,T,x} &\le \varliminf_{\epsilon\to 0} \epsilon \log\left[\mathcal{W}\big(\{\theta : X^x_{n,\epsilon}(\theta) \in \Gamma\}\big)\right] \\ &\le \varlimsup_{\epsilon\to 0} \epsilon \log\left[\mathcal{W}\big(\{\theta : X^x_{n,\epsilon}(\theta) \in \Gamma\}\big)\right] \le -\inf_{\overline{\Gamma}} I^{a,b}_{n,T,x}, \end{aligned}$$

where

$$(1.4.16)\quad I^{a,b}_{n,T,x}(X) = \begin{cases} \infty \\ \frac{1}{2}\int_0^T \left|a^{-1/2}\big(X(\mathbf{T}_n(t))\big)\Big(\dot{X}(t) - b\big(X(t)\big)\Big)\right|^2 dt, \end{cases}$$

according to whether $X - x \notin H^1_T$ or $X - x \in H^1_T$. Since it is clear that $X^x_{n,\epsilon} \longrightarrow X^x_\epsilon$ in $\mathcal{W}$-measure and that $I^{a,b}_{n,T,x} \longrightarrow I^{a,b}_{T,x}$, as $n \to \infty$, all that stands between us and the conjectured result are estimates which allow us to exchange the order in which n-limits and ϵ-limits are taken.

The following lemma takes care of the required facts about the convergence of $\{I^{a,b}_{n,T,x}\}_1^\infty$ to $I^{a,b}_{T,x}$.

1.4.17 Lemma. *For each $x \in \mathbb{R}^d$, $\{X : I^{a,b}_{T,x}(X) \le L\} \subset\subset \Omega_T$ for all $L \ge 0$ and $\inf_\Gamma I^{a,b}_{n,T,x} \longrightarrow \inf_\Gamma I^{a,b}_{T,x}$ as $n \to \infty$ for every $\Gamma \subseteq \Omega_T$.*

PROOF: Assume that $x = 0$ and set $I = I^{a,b}_{T,0}$ and $I_n = I^{a,b}_{n,T,0}$. Because $\{X : I(X) \le L\}$ is a bounded subset of H^1_T, we will know that it is compact in Ω_T as soon as we show that it is closed there. To this end, suppose that $\{X_n\}_1^\infty$ is a sequence of elements in Ω_T with the properties that $X_n \longrightarrow X$ in Ω_T and $\sup_n I(X_n) \le L$. Then $X \in H^1_T$ and $X_n \longrightarrow X$ weakly in H^1_T. Since this means that

$$\sigma^{-1}(X_n(\cdot))\Big(\dot{X}_n(\cdot) - b\big(X_n(\cdot)\big)\Big) \longrightarrow \sigma^{-1}(X(\cdot))\Big(\dot{X}(\cdot) - b\big(X(\cdot)\big)\Big)$$

weakly in $L^2([0,T];\mathbb{R}^d)$, it follows that $I(X) \le \varliminf_{n\to\infty} I(X_n) \le L$. Thus, $\{X : I(X) \le L\} \subset\subset \Omega_T$. To prove the convergence assertion, first note that $\inf_\Gamma I = \infty$ if and only if $\inf_\Gamma I_n = \infty$ for every $n \geq 1$. Next, note that if B is a bounded subset of H^1_T, then

$$\lim_{n\to 0} \sup_{X\in B} \left|I_n(X) - I(X)\right| = 0. \tag{1.4.18}$$

In particular, this proves that $\inf_\Gamma I \geq \overline{\lim}_{n\to\infty} \inf_\Gamma I_n$. Finally, if $\ell = \inf_\Gamma I < \infty$, then we can choose a bounded subset B of H^1_T so that $I(X) \wedge \inf_{n\geq 1} I_n(X) \geq \ell + 1$ for $X \notin B$. Hence, because $\inf_\Gamma I_n \leq \ell + 1$ for all sufficiently large n's, we can use (1.4.18) to conclude that

$$\inf_\Gamma I = \inf_{\Gamma\cap B} I = \lim_{n\to\infty} \inf_{\Gamma\cap B} I_n = \lim_{n\to\infty} \inf_\Gamma I_n. \quad \blacksquare$$

As a preliminary to our estimate on the rate of convergence of the $X^x_{n,\epsilon}$'s to X^x_ϵ, we present the following standard estimate for stochastic integrals.

1.4.19 Lemma. *Let $\alpha : [0,\infty)\times\Theta \longrightarrow \mathbb{R}^N \otimes \mathbb{R}^d$ and $\beta : [0,\infty)\times\Theta \longrightarrow \mathbb{R}^N$ be bounded $\{\mathcal{B}_t\}$-progressively measurable functions and set*

$$Y(t,\theta) = \int_0^t \alpha(s,\theta)\, d\theta(s) + \int_0^t \beta(s,\theta)\, ds, \quad t \geq 0.$$

Define $A = \sup_{(t,\theta)} \sup_{\xi\in\mathbf{S}^{N-1}} \left|\alpha^{\mathrm{T}}(t,\theta)\xi\right|^2_{\mathbb{R}^d}$ and $B = \sup_{(t,\theta)} \left|\beta(t,\theta)\right|_{\mathbb{R}^N}$. Then, for every $s \geq 0$, $T > 0$ and $r > BT$,

$$\begin{aligned} &\mathcal{W}\left(\left\{\theta : \sup_{s\leq t\leq s+T} \left|Y(t,\theta) - Y(s,\theta)\right| \geq r\right\}\right) \\ &\leq \quad 2N \exp\left[-\frac{(r - BT)^2}{2ANT}\right]. \end{aligned} \tag{1.4.20}$$

PROOF: Set $\overline{Y}(t,\theta) = Y(t,\theta) - Y(s,\theta) - \int_s^t \beta(\tau,\theta)\, d\tau$ for $t \geq s$. For $\rho > 0$ and $\xi \in S^{N-1}$, define

$$E_{\rho,\xi}(t,\theta) = \exp\left[\rho\xi\cdot\overline{Y}(t,\theta) - \frac{\rho^2}{2}\int_s^t \left|\alpha^{\mathrm{T}}(\tau,\theta)\xi\right|^2_{\mathbb{R}^d} d\tau\right], \quad t \geq s$$

and

$$\zeta_{\xi,r}(\theta) = \inf\left\{t \geq s : \xi\cdot\overline{Y}(t,\theta) \geq r\right\}.$$

By ITÔ's formula and DOOB's Stopping Time Theorem,

$$\left(E_{\rho,\xi}\left((t\vee s)\wedge\zeta_{\xi,r},\theta\right), \mathcal{B}_t, \mathcal{W}\right)$$

is a bounded martingale. Therefore

$$\begin{aligned}
&\mathcal{W}\big(\{\theta : \zeta_{\xi,r}(\theta) \le s+T\}\big) \\
&\qquad \le \exp[-\rho r] \int_\Theta \exp\big[\xi \cdot \overline{Y}\big((s+T)\wedge \zeta_{\xi,r}(\theta),\theta\big)\big]\, \mathcal{W}(d\theta) \\
&\qquad \le \exp\left[-\rho r + \frac{AT\rho^2}{2}\right] \int_\Theta E_{\rho,\xi}\big((s+T)\wedge \zeta_{\xi,r}(\theta),\theta\big)\, \mathcal{W}(d\theta) \\
&\qquad = \exp\left[-\rho r + \frac{AT\rho^2}{2}\right].
\end{aligned}$$

Hence, after minimizing with respect to $\rho > 0$, we see that

$$\mathcal{W}\left(\left\{\theta : \sup_{s\le t\le s+T} \xi \cdot \overline{Y}(t,\theta) \ge r\right\}\right) \le \exp\left[-\frac{r^2}{2AT}\right];$$

and from this it is an easy step to

$$\mathcal{W}\left(\left\{\theta : \sup_{s\le t\le s+T} |\overline{Y}(t,\theta)| \ge r\right\}\right) \le 2N \exp\left[-\frac{r^2}{2ANT}\right]$$

and thence to (1.4.20). ∎

We next show that $X^x_{n,\epsilon}$ approximates X^x_ϵ sufficiently fast as $n \to \infty$.

1.4.21 Lemma. *For each $x \in \mathbb{R}^d$ and all $\delta > 0$,*

$$\text{(1.4.22)} \quad \varlimsup_{\epsilon\to 0} \epsilon \log\left[\mathcal{W}\left(\left\{\theta : \sup_{t\in[0,T]} \left|X^x_{n,\epsilon}(t,\theta) - X^x_\epsilon(t,\theta)\right| > \delta\right\}\right)\right] \longrightarrow -\infty$$

as $n \longrightarrow \infty$.

PROOF: We assume that $x = 0$ and $T = 1$. (The reduction to this case is left to the reader.) Set $X_\epsilon = X^0_\epsilon$, $X_{n,\epsilon} = X^0_{n,\epsilon}$, and $Y_{n,\epsilon} = X_{n,\epsilon} - X_\epsilon$; and, for $\rho > 0$, define

$$\tau^\rho_{n,\epsilon}(\theta) = \Big(\inf\{t \ge 0 : |X_{n,\epsilon}(t,\theta) - X_{n,\epsilon}(\mathbf{T}_n(t),\theta)| \ge \rho\}\Big) \wedge 1,$$

$$\zeta^\rho_{n,\epsilon}(\theta) = \Big(\inf\Big\{t \ge 0 : \big|Y_{n,\epsilon}(t,\theta)\big| \ge \delta\Big\}\Big) \wedge \tau^\rho_{n,\epsilon}(\theta),$$

and

$$u^\rho_{n,\epsilon}(t) = \int_\Theta \Big(\rho^2 + \big|Y_{n,\epsilon}\big(t\wedge \zeta^\rho_{n,\epsilon}(\theta),\theta\big)\big|\Big)^{1/\epsilon} \mathcal{W}(d\theta).$$

Clearly,

$$\mathcal{W}\left(\left\{\theta : \sup_{t\in[0,T]} \left|Y_{n,\epsilon}(t,\theta)\right| > \delta\right\}\right)$$
$$\leq \mathcal{W}\left(\left\{\theta : \tau^{\rho}_{n,\epsilon}(\theta) < 1\right\}\right) + \mathcal{W}\left(\left\{\theta : \zeta^{\rho}_{n,\epsilon}(\theta) < 1\right\}\right);$$

and so it suffices for us to prove that

$$\lim_{n\to\infty} \overline{\lim_{\epsilon\to 0}}\, \epsilon \log\left[\mathcal{W}\left(\left\{\theta : \tau^{\rho}_{n,\epsilon}(\theta) < 1\right\}\right)\right] = -\infty \tag{1.4.23}$$

for each $\rho > 0$ and that

$$\lim_{\rho\to 0} \sup_{n\geq 1} \overline{\lim_{\epsilon\to 0}}\, \epsilon \log\left[\mathcal{W}\left(\left\{\theta : \zeta^{\rho}_{n,\epsilon}(\theta) < 1\right\}\right)\right] = -\infty. \tag{1.4.24}$$

To prove (1.4.23) we use Lemma 1.4.19 to obtain

$$\mathcal{W}\left(\left\{\theta : \tau^{\rho}_{n,\epsilon}(\theta) < 1\right\}\right)$$
$$\leq \sum_{k=0}^{n-1} \mathcal{W}\left(\left\{\theta : \sup_{k\leq nt\leq k+1} \left|X_{n,\epsilon}(t,\theta) - X_{n,\epsilon}\left(\frac{k}{n},\theta\right)\right| \geq \rho\right\}\right)$$
$$\leq 2dn \exp\left[-\frac{n\left(\rho - \frac{M}{n}\right)^2}{2Md\epsilon}\right], \quad \text{for } n > \frac{M}{\rho};$$

from which (1.4.23) is immediate.

The proof of (1.4.24) goes as follows. Set $f_{\epsilon,\rho}(y) = \left(\rho^2 + |y|^2\right)^{1/\epsilon}$. Note that there is a $K < \infty$, which is independent of n, ϵ, and ρ, such that

$$\left\|\sigma\left(X_{n,\epsilon}(\mathbf{T}_n(t),\theta)\right) - \sigma\left(X_\epsilon(t,\theta)\right)\right\|_{\text{H.S.}} \vee \left|b\left(X_{n,\epsilon}(t,\theta)\right) - b\left(X_\epsilon(t,\theta)\right)\right|$$
$$\leq K\left(\rho^2 + \left|Y_{n,\epsilon}(t,\theta)\right|^2\right)^{1/2}, \quad t \in \left[0, \tau^{\rho}_{n,\epsilon}(\theta)\right],$$
$$\left|\nabla f_{\epsilon,\rho}(y)\right| \leq \frac{K}{\epsilon}\left(\rho^2 + |y|^2\right)^{-1/2} f_{\epsilon,\rho}(y), \quad y \in \mathbb{R}^d,$$
$$\left\|\nabla^2 f_{\epsilon,\rho}(y)\right\|_{\text{H.S.}} \leq \frac{K}{\epsilon^2}\left(\rho^2 + |y|^2\right)^{-1} f_{\epsilon,\rho}(y), \quad y \in \mathbb{R}^d.$$

($\nabla^2 f$ denotes the HESSian matrix of f.) Hence, an application of ITÔ's formula to $f_{\epsilon,\rho}\left(Y_{n,\epsilon}(t,\theta)\right)$ together with DOOB's Stopping Time Theorem shows that there is a $C < \infty$, which is independent of n, ϵ, and ρ, such that

$$u^{\rho}_{n,\epsilon}(t) \leq \rho^{2/\epsilon} + \frac{C}{\epsilon}\int_0^t u^{\rho}_{n,\epsilon}(s)\, ds, \quad t \in [0,1];$$

and therefore,

$$u_{n,\epsilon}^{\rho}(1) \le \exp\left[\frac{1}{\epsilon}(C + \log \rho^2)\right].$$

Since

$$\mathcal{W}\left(\{\theta : \zeta_{n,\epsilon}^{\rho}(\theta) < 1\}\right) \le \left(\rho^2 + \delta^2\right)^{-1/\epsilon} u_{n,\epsilon}^{\rho}(1),$$

this completes the proof of (1.4.24). ∎

We have, at last, made all the preparations necessary to prove the VENTCEL and FREIDLIN's estimate.

1.4.25 Theorem. (VENTCEL & FREIDLIN) *Let X_ϵ^x be the solution to (1.4.12) and define $\{P_\epsilon^x : \epsilon > 0\}$ on $(\Omega_T, \mathcal{B}_{\Omega_T})$ accordingly. Then, for all $\Gamma \in \mathcal{B}_{\Omega_T}$,*

$$(1.4.26) \quad -\inf_{\Gamma^\circ} I_{T,x}^{a,b} \le \varliminf_{\epsilon \to 0} \epsilon \log\left(P_\epsilon^x(\Gamma)\right) \le \varlimsup_{\epsilon \to 0} \epsilon \log\left(P_\epsilon^x(\Gamma)\right) \le -\inf_{\overline{\Gamma}} I_{T,x}^{a,b},$$

where $I_{T,x}^{a,b}$ is defined in (1.4.13). (*See* Exercise *2.1.25 below for a slightly more general statement.*)

PROOF: Let F be a closed subset of Ω_T. Then, for any $\delta > 0$ and $n \ge 1$,

$$P_\epsilon^x(F) \le \mathcal{W}\left(\left\{\theta : X_{n,\epsilon}^x(\theta) \in \overline{F^{(\delta)}}\right\}\right) + \mathcal{W}\left(\left\{\theta : \sup_{t \in [0,T]} \left|X_{n,\epsilon}^x(t,\theta) - X_\epsilon^x(t,\theta)\right| > \delta\right\}\right).$$

($F^{(\delta)}$ denotes the open δ-neighborhood around F.) Thus, by (1.4.15), Lemma 1.4.17, and Lemma 1.4.21, we see that

$$(1.4.27) \qquad \varlimsup_{\epsilon \to 0} \epsilon \log\left(P_\epsilon^x(F)\right) \le -\inf_{\overline{F^{(\delta)}}} I_{T,x}^{a,b}$$

for every $\delta > 0$. Finally, set $\ell_\delta = \inf_{\overline{F^{(\delta)}}} I_{T,x}^{a,b}$ for $\delta \ge 0$. It is clear that $\ell_\delta \nearrow \ell \le \ell_0$ as $\delta \searrow 0$. Suppose that $\ell < \ell_0$. We could then find $\{X_n\}_1^\infty$ and $\ell < L < \ell_0$ so that $X_n \in \overline{F^{(1/n)}}$ and $I_{T,x}^{a,b}(X_n) \le L$. Further, by Lemma 1.4.17, we could assume that $X_n \longrightarrow X$. But clearly this would mean that $X \in F$ and, again by Lemma 1.4.17, that $\inf_F I_{T,x}^{a,b} \le I_{T,x}^{a,b}(X) < \ell_0$. Hence we can let $\delta \nearrow 0$ in (1.4.27) and thereby get the right hand side of (1.4.26).

Next, let G be an open set in Ω_T. Then, for each $X \in G$ and $n \ge 1$, we see that

$$\mathcal{W}\left(\{\theta : X_{n,\epsilon}^x(\theta) \in B(X,\delta)\}\right) \le P_\epsilon^x(G) + \mathcal{W}\left(\left\{\theta : \sup_{t \in [0,T]} \left|X_{n,\epsilon}^x(t,\theta) - X^x(t,\theta)\right| \ge \delta\right\}\right)$$

as long as $B(X, 2\delta) \subseteq G$. Using (1.4.15), Lemma 1.4.17, and Lemma 1.4.21, we conclude from this that

$$\varliminf_{\epsilon \to 0} \epsilon \log\left(P_\epsilon^x(G)\right) \geq -I_{T,x}^{a,b}(X). \quad \blacksquare$$

1.4.28 Exercise.

STRASSEN's Theorem is the function space version of the **Classical Law of the Iterated Logarithm**. That is, given real-valued, identically distributed, independent random variables $X_1, \dots, X_n, \dots$ with mean 0 and variance 1, set $S_n = \sum_1^n X_m$, $n \geq 1$. Then the Classical Law of the Iterated Logarithm is the statement that

$$\varlimsup_{n \to \infty} \frac{S_n}{\beta(n)} = 1 \quad \text{almost surely.} \tag{1.4.29}$$

When the X_n's are standard GAUSSian random variables, (1.4.29) is an immediate consequence of (1.4.2) with $\Phi(\psi) = \psi(1)$ since, in this case, $\{S_n\}_1^\infty$ has the same distribution as the distribution of $\theta \longmapsto \{\theta(n)\}_1^\infty$ under $\mathcal{W}$. It turns out that the general classical result can also be seen as a consequence of STRASSEN's Theorem. The proof entails the use of the SKOROKHOD Representation Theorem [**97**]. We outline below how this argument proceeds in the special case when the X_n's are standard BERNOULLI random variables (i.e., $P(X_n = 1) = P(X_n = -1) = \frac{1}{2}$). Throughout the rest of this exercise, the X_n's are BERNOULLI and $d = 1$.

(**i**) Define $\tau_0(\theta) = 0$ and

$$\tau_{n+1}(\theta) = \inf\left\{t - \Sigma_n(\theta) : t \geq \Sigma_n(\theta) \text{ and } |\theta(t) - \theta(\Sigma_n(\theta))| \geq 1\right\}, \quad n \geq 0$$

where $\Sigma_n(\theta) = \sum_{m=0}^n \tau_m(\theta)$. Show that the τ_n's under $\mathcal{W}$ are identically distributed, independent, and have mean 1. Next, set $\Sigma_n = \sum_1^n \tau_m$ and define $Y_n(\theta) = \theta(\Sigma_n) - \theta(\Sigma_{n-1})$ for $n \geq 1$, and show that the Y_n's (under $\mathcal{W}$) are independent standard BERNOULLI random variables. (Both of these assertions turn on the fact that if τ is a $\{\mathcal{B}_t\}$-stopping time with $\mathcal{W}(\tau < \infty) = 1$, then $\theta \in \Theta \longmapsto \theta(\cdot \vee \tau(\theta)) - \theta(\tau(\theta)) \in \Theta$ under $\mathcal{W}$ is independent of $\mathcal{B}_\tau$ and has distribution $\mathcal{W}$.) Conclude that $\left\{S_n/\beta(n)\right\}_3^\infty$ has the same distribution as the distribution of

$$\theta \longmapsto \left\{\xi_n\left(\Sigma_n(\theta)/n, \theta\right)\right\}_3^\infty$$

under $\mathcal{W}$. In particular, (1.4.29) for BERNOULLI random variables is equivalent to

$$\varlimsup_{n \to \infty} \xi_n\left(\frac{\Sigma_n(\theta)}{n}, \theta\right) = 1 \tag{1.4.30}$$

for $\mathcal{W}$-almost every θ.

(**ii**) Use the Strong Law of Large Numbers to show that $\Sigma_n(\theta)/n \longrightarrow 1$ $\mathcal{W}$-almost surely; and from this, together with Theorem 1.4.1, conclude that (1.4.30) holds for $\mathcal{W}$-almost every θ.

The construction of the τ_n 's for more general random variables is more difficult. (The content of SKOROKHOD's Theorem is that such τ_n 's always exist.) However, once their existence has been established, the rest of the argument is the same as the one just given for the BERNOULLI case.

1.4.31 Exercise.

There is a more direct approach which can be taken to prove the left hand side of (1.4.26). Namely, given $\psi \in H^1_T$, let $\theta \longmapsto X^{x,\psi}_\epsilon(\theta)$ be the $\mathcal{W}$-almost surely unique $\{\mathcal{B}_t : t \in [0,T]\}$-progressively measurable solution to

$$X^{x,\psi}_\epsilon(t,\theta) = x + \epsilon^{1/2} \int_0^t \sigma\big(X^{x,\psi}_\epsilon(s,\theta)\big)\, d\theta(s) + \int_0^t \Big[b\big(X^{x,\psi}_\epsilon(s,\theta)\big) + \sigma\big(X^{x,\psi}_\epsilon(s,\theta)\big)\dot{\psi}(s)\Big]\, ds$$

for $t \in [0,T]$.

(**i**) Show that the distribution of $\theta \longmapsto X^{x,\psi}_\epsilon(\theta)$ under $\mathcal{W}$ is the same as that of $\theta \longmapsto X^x_\epsilon(\theta)$ under $\mathcal{W}^{\psi/\epsilon^{1/2}}$. (See Exercise 1.3.32 for the notation, and think of ψ as being the element of H^1 with $\psi(t) = \psi(T)$ for $t \geq T$.)

(**ii**) Define $Y^x(\psi) \in \Omega_T$ by

$$Y^x(t,\psi) = x + \int_0^t \Big[b\big(Y^x(s,\psi)\big) + \sigma\big(Y^x(s,\psi)\big)\dot{\psi}(s)\Big]\, ds, \quad t \in [0,T].$$

Using (**i**) above, Exercise 1.3.32, and HÖLDER's inequality, show that for every $q \in [1,\infty)$ and $r > 0$,

$$\mathcal{W}\Big(\Big\{\theta : X^{x,\psi}_\epsilon(\theta) \in B\big(Y^x(\psi),r\big)\Big\}\Big) \leq \exp\left[\frac{q-1}{\epsilon} I_T(\psi)\right] P^x_\epsilon\Big(B(Y^x(\psi),r)\Big)^{1-\frac{1}{q}}.$$

Conclude from this that

$$\varliminf_{\epsilon \to 0} \epsilon \log\Big[P^x_\epsilon\Big(B\big(Y^x(\psi),r\big)\Big)\Big] \geq -I_T(\psi)$$

for all $r > 0$.

(iii) From **(ii)**, show that for every open G in Ω_T,

$$\varliminf_{\epsilon \to 0} \epsilon \log\left(P_\epsilon^x(G)\right) \geq -\inf\left\{ I_T(\psi) : \psi \in H_T^1 \text{ and } Y^x(\psi) \in G \right\}; \tag{1.4.32}$$

and show that this is equivalent to the left hand side of (1.4.26).

It should be noted that the preceding derivation does not use in any way the strict positivity of $a(x)$ until the very end. Thus, (1.4.32) holds even if a is allowed to degenerate. However, when a can degenerate, it is not so easy to give as nice an expression as that in (1.4.14) for the quantity on the right hand side of (1.4.32). (cf. Exercise 2.1.25 below.)

1.4.33 Exercise.

Replace (1.4.10) and (1.4.11), respectively, by the assumptions that

$$0 < a(x) \leq M\left(1+|x|^2\right) I_{\mathbb{R}^d} \text{ and } |b(x)| \leq M\left(1+|x|^2\right)^{1/2}, \quad x \in \mathbb{R}^d \tag{1.4.34}$$

for some $M \in (0,\infty)$ and that, for each $r \in (0,\infty)$,

$$\|a(y) - a(x)\|_{\text{H.S.}} \vee |b(y) - b(x)| \leq M_r |y - x|, \quad x, y \in \overline{B}(0,r). \tag{1.4.35}$$

for some $M_r \in (0,\infty)$. Show that for each $x \in \mathbb{R}^d$, $\epsilon > 0$, and $T > 0$, there is a $\mathcal{W}$-almost surely unique $\{\mathcal{B}_t : t \in [0,T]\}$-progressively measurable solution $\theta \longmapsto X_\epsilon^x(\theta)$ to (1.4.12) and that both

$$\lim_{R\to\infty} \varlimsup_{\epsilon\to 0} \epsilon \log\left[\sup_{|x|\leq r} \mathcal{W}\left(\left\{\theta : \sup_{t\in[0,T]} \left|X_\epsilon^x(t,\theta)\right| \geq R\right\}\right)\right]$$

and

$$\lim_{R\to\infty} \varlimsup_{\epsilon\to 0} \epsilon \log\left[\sup_{|x|\vee|y|\leq r} \mathcal{W}\left(\left\{\theta : \sup_{t\in[0,T]} \left|X_\epsilon^x(t,\theta) - X_\epsilon^y(t,\theta)\right| \geq R\right\}\right)\right]$$

are $-\infty$ for every $r > 0$. Conclude from these not only that Theorem 1.4.25 continues to hold when (1.4.10) and (1.4.11) are replaced by (1.4.34) and (1.4.35) and also that (1.4.26) can be improved to the statement that

$$-\inf_{\Gamma^\circ} I_{T,x}^{a,b} \leq \varliminf_{\epsilon\to 0} \epsilon \log\left(P_\epsilon^{x_\epsilon}(\Gamma)\right) \leq \varlimsup_{\epsilon\to 0} \epsilon \log\left(P_\epsilon^{x_\epsilon}(\Gamma)\right) \leq -\inf_{\overline{\Gamma}} I_{T,x}^{a,b} \tag{1.4.36}$$

whenever $x_\epsilon \longrightarrow x$. Also, observe that it is still true that $\{X : I_{T,x}^{a,b}(X) \leq L\} \subset\subset \Omega_T$ for every $L \geq 0$.

II Some Generalities

2.1 The Large Deviation Principle

Having seen several examples for which it is possible to carry out a successful analysis of the large deviations, we will now attempt to formulate into general principles some of the ideas and techniques which proved useful in those examples. Because we never use completeness in this section, we will take E throughout this section to be a separable metric space.

A function $I : E \longrightarrow [0, \infty]$ is said to be a **rate function** if it is lower semi-continuous. Given a family $\{\mu_\epsilon : \epsilon > 0\} \subseteq \mathbf{M}_1(E)$ (we often use $\mathbf{M}_1(E)$ to denote the space of probability measures on $(E, \mathcal{B}_E)$), we will say that $\{\mu_\epsilon : \epsilon > 0\}$ **satisfies the full large deviation principle with rate function** I or, equivalently, that **the rate function** I **governs the large deviations of** $\{\mu_\epsilon : \epsilon > 0\}$ if (1.1.5) holds for every $\Gamma \in \mathcal{B}_E$. It is clear that if I is a rate function which governs the large deviations of some family $\{\mu_\epsilon : \epsilon > 0\}$ then it must be true that $\inf_E I = 0$.

The following result is elementary but reassuring.

2.1.1 Lemma. *For any given* $\{\mu_\epsilon : \epsilon > 0\} \subseteq \mathbf{M}_1(E)$ *there is at most one rate function governing the large deviations of* $\{\mu_\epsilon : \epsilon > 0\}$.

PROOF: Suppose there were two, and name them I_1 and I_2. Because of lower semi-continuity, we know that $I_j(p) = \lim_{r \searrow 0} \inf_{B(p,r)} I_j$ for every $p \in E$. Thus it suffices for us to show that, for each $p \in E$, $\inf_{B(p,r)} I_1 = \inf_{B(p,r)} I_2$ for each r in a dense subset of $(0, \infty)$. To this end, observe that

$$\inf_{B(p,r)} I_j = -\lim_{\epsilon \to 0} \epsilon \log \left[\mu_\epsilon \big(B(p, r) \big) \right]$$

for any $r > 0$ with the property that $\inf_{B(p,r)} I_j = \inf_{\overline{B}(p,r)} I_j$. In particular, this will be the case if $r > 0$ is a continuity point for the non-increasing function $r \in (0,\infty) \longmapsto \inf_{B(p,r)} I_j$; and therefore we see that $\inf_{B(p,r)} I_1 = \inf_{B(p,r)} I_2$ for all but a countable number of $r > 0$. ∎

In all our examples, the governing rate function was not only lower semi-continuous but also had the property that the level sets $\{q \in E : I(q) \leq L\}$ were compact for all $L \geq 0$. Because such rate functions play a prominent role and since the additional property is extremely useful, we will say that $I : E \longrightarrow [0,\infty]$ is a **good rate function** if $\{q \in E : I(q) \leq L\} \subset\subset E$ for all $L \geq 0$. Some elementary properties of good rate functions are listed in the next result.

2.1.2 Lemma. *Let I be a good rate function. Then, for each closed F in E,*

$$\inf_F I = \lim_{\delta \searrow 0} \inf_{F^{(\delta)}} I. \tag{2.1.3}$$

(Recall that $\Gamma^{(\delta)} = \{q \in E : \operatorname{dist}(q,\Gamma) < \delta\}$ for any subset Γ.) In addition, if $\Phi : E \longrightarrow [-\infty,\infty)$ is an upper semi-continuous function, then for any closed $F \subseteq E$ on which Φ is bounded above there is a $q \in F$ such that $\Phi(q) - I(q) = \sup_F(\Phi - I)$.

PROOF: The derivation of (2.1.3) in this general setting differs in no way from the one given for the special case handled at the end of the first paragraph in the proof of Theorem 1.4.25; thus, we will not repeat the argument here.

To prove the second assertion, first note that there is nothing to do if $\sup_F(\Phi - I) = -\infty$. Thus, we assume that $\ell \equiv \sup_F(\Phi - I) > -\infty$, in which case we know that $\ell \in (-\infty,\infty)$. Choose $\{q_n\}_1^\infty \in F$ so that $\Phi(q_n) - I(q_n) \geq \ell - \frac{1}{n}$. Because $\{q_n\}_1^\infty \subseteq \{q : I(q) \leq M - \ell + 1\}$, where $M \equiv \sup_F \Phi$, there is a convergent subsequence of $\{q_n\}_1^\infty$ which converges to some q; and, because $\Phi - I$ is upper semi-continuous, not only is $q \in F$ but also $\Phi(q) - I(q) \geq \ell$. ∎

Another advantage that good rate functions have is that the full large deviation principle is a **covariant** notion when the rate function is good. (In this connection, we use here and elsewhere the notation $\mu \circ f^{-1}$ to denote the covariant image of a measure μ under a measurable map f. Thus, $\mu \circ f^{-1}(\Gamma) \equiv \mu\big(f^{-1}(\Gamma)\big)$ for measurable subsets Γ of the image space.) That is, such principles can be "pushed forward" under mappings which are "nearly continuous." We have already seen an example of this when we discussed in Section 1.4 the passage from SCHILDER's Theorem to the estimate of VENTCEL and FREIDLIN. The next lemma provides a general statement of this technique. (See also Exercise 2.1.20 below.)

2.1.4 Lemma. *Let I be a good rate function on E, f a measurable map from E into a second separable metric space (E', ρ'), and assume that there exists a sequence $\{f_n\}_1^\infty \subseteq C(E; E')$ such that*

$$\lim_{n\to\infty} \sup\Big\{\rho'\big(f_n(q), f(q)\big) : q \in E \text{ with } I(q) \le L\Big\} = 0 \text{ for each } L \in (0,\infty).$$

Then the map $I' : E' \longrightarrow [0,\infty]$ given by

$$I'(q') = \inf\{I(q) : q \in E \text{ and } q' = f(q)\}, \quad q' \in E',$$

is a good rate function on E'. Moreover, if, in addition, $\{\mu_\epsilon : \epsilon > 0\} \subseteq \mathbf{M}_1(E)$ has the property that

$$\lim_{n\to\infty} \overline{\lim_{\epsilon\to 0}}\, \epsilon \log\Big[\mu_\epsilon\Big(\Big\{q \in E : \rho'\big(f_n(q), f(q)\big) \ge \delta\Big\}\Big)\Big] = -\infty$$

for each $\delta \in (0,\infty)$, then I' governs the large deviations of $\{\mu_\epsilon \circ f^{-1} : \epsilon > 0\}$ whenever I governs the large deviations of $\{\mu_\epsilon : \epsilon > 0\}$.

In particular, if $f \in C(E; E')$ and I is a good rate function on E which governs the large deviations of $\{\mu_\epsilon : \epsilon > 0\}$, then I' is a good rate function on E' which governs the large deviations of $\{\mu_\epsilon \circ f^{-1} : \epsilon > 0\}$.

PROOF: One should observe that the case when f is continuous everywhere on E is trivial and therefore really should not be thought of as a consequence of the general result.

First, observe that f is continuous on $K_L \equiv \{q \in E : I(q) \le L\}$ for each $L \in [0,\infty)$. Second, suppose that $q' \in E'$ with $I'(q') < \infty$. Then, for some $L \in [0,\infty)$,

$$I'(q') = \inf\{I(q) : q \in K_L \text{ and } q' = f(q)\};$$

and therefore, by Lemma 2.1.2, there is a $q \in f^{-1}(q')$ for which $I(q) = I'(q')$.

With these preliminaries, we can easily prove that I' is a good rate function. Indeed, if $L \in [0,\infty)$ and

$$\{q'_n\}_1^\infty \subseteq K'_L \equiv \{q' \in E' : I'(q') \le L\},$$

then there is a $\{q_n\}_1^\infty \subseteq K_L$ such that $q'_n = f(q_n)$ and $I'(q'_n) = I(q_n)$ for each $n \in \mathbb{Z}^+$. Thus, since $K_L \subset\subset E$, we can choose a subsequence $\{q_{n_m}\}_{m=1}^\infty$ so that $q_{n_m} \longrightarrow q \in K_L$. Because $f|_{K_L}$ is continuous, this means that

$$q' \equiv f(q) = \lim_{m\to\infty} q'_{n_m} \quad \text{and} \quad I'(q') \le I(q) \le L.$$

That is, $K'_L \subset\subset E'$.

In preparation for the second part of the proof, we next show that, for each closed F' in E',

$$\inf_{F'} I' = \lim_{\delta \searrow 0} \varliminf_{n \to \infty} \inf\Big\{ I(q) : \rho'\big(f_n(q), F'\big) \le \delta \Big\}.$$

To this end, first suppose that $p' \in F'$ with $I'(p') < \infty$ and $\delta \in (0,\infty)$ are given. Choose $p \in f^{-1}(p')$ so that $I(p) = I'(p')$. Noting that $f_n(p) \longrightarrow p'$ as $n \longrightarrow \infty$, we see that there is an $N \in \mathbb{Z}^+$ such that

$$I'(p') = I(p) \ge \inf\Big\{ I(q) : \rho'\big(f_n(q), f(q)\big) \le \delta \Big\}$$

for all $n \ge N$; and therefore, we now know that

$$\inf_{F'} I' \ge \lim_{\delta \searrow 0} \varliminf_{n \to \infty} \inf\Big\{ I(q) : \rho'\big(f_n(q), F'\big) \le \delta \Big\}.$$

To prove the opposite inequality, assume that

$$\ell = \lim_{\delta \searrow 0} \varliminf_{n \to \infty} \inf\Big\{ I(q) : \rho'\big(f_n(q), F'\big) \le \delta \Big\} < \infty.$$

We can then choose $\{q_m\}_1^\infty \subseteq K_{\ell+1}$ and $n_m \longrightarrow \infty$ so that

$$\rho'\Big(f_{n_m}(q_m), F'\Big) \le \frac{1}{m} \quad \text{and} \quad I(q_m) \le \ell + \frac{1}{m}$$

for each $m \in \mathbb{Z}^+$. Furthermore, because $K_{\ell+1} \subset\subset E$ and I is lower semi-continuous, we may and will assume that $q_m \longrightarrow q \in K_\ell$. Hence, since $f|K_{\ell+1}$ is continuous and therefore $q' \equiv f(q) \in F'$, we have that

$$\inf_{F'} I' \le I(q) \le \ell.$$

To complete the proof, assume that I governs the large deviations of $\{\mu_\epsilon : \epsilon > 0\}$ and that

$$\lim_{n \to \infty} \varlimsup_{\epsilon \to 0} \epsilon \log\Big[\mu_\epsilon\big(\Gamma(n;\delta)\big)\Big] = -\infty, \quad \delta \in (0,\infty),$$

where

$$\Gamma(n;\delta) \equiv \Big\{ q : \rho'\big(f_n(q), f(q)\big) \ge \delta \Big\}.$$

Given an open set G' in E' and $p' \in G'$ with $I'(p') < \infty$, choose $p \in f^{-1}(p')$ so that $I'(p') = I(p)$ and $\delta \in (0,\infty)$ so that

$$2\delta < \rho'\Big(p', (G')^{\mathrm{c}}\Big).$$

Then, since each f_n is continuous and $f_n(p) \longrightarrow f(p)$, there is an $N \in \mathbb{Z}^+$ and a sequence $\{r_n\}_{n=N}^{\infty} \subseteq (0,\infty)$ such that

$$B(p, r_n) \subseteq f_n^{-1}(B'(p', \delta)), \quad n \geq N,$$

where $B'(p', \delta)$ is the ρ'-ball in E' of radius δ around p'. Hence, for $n \geq N$,

$$B(p, r_n) \subseteq f^{-1}(G') \cup \Gamma(n; \delta);$$

and therefore, by choosing $n \geq N$ so that

$$\overline{\lim_{\epsilon \longrightarrow 0}} \, \epsilon \log \left[\mu_\epsilon(\Gamma(n; \delta))\right] \leq -I'(p') - 1,$$

we see, from the large deviation principle for $\{\mu_\epsilon : \epsilon > 0\}$, that

$$\begin{aligned}
-I'(p') = -I(p) &\leq \underline{\lim_{\epsilon \longrightarrow 0}} \, \epsilon \log \left[\mu_\epsilon(B(p, r_n))\right] \\
&\leq \underline{\lim_{\epsilon \longrightarrow 0}} \, \epsilon \log \left[\mu_\epsilon(f^{-1}(G')) + \mu_\epsilon(\Gamma(n; \delta))\right] \\
&\leq \left(\underline{\lim_{\epsilon \longrightarrow \infty}} \, \epsilon \log \left[\mu_\epsilon \circ f^{-1}(G')\right]\right) \vee \left(-I'(p') - 1\right);
\end{aligned}$$

from which we conclude that

$$\underline{\lim_{\epsilon \longrightarrow 0}} \, \epsilon \log \left[\mu_\epsilon \circ f^{-1}(G')\right] \geq -\inf_{G'} I'.$$

Finally, for closed F' in E', set

$$F(n; \delta) = \{q : \rho'(f_n(q), F') \leq \delta\}$$

and note that

$$f^{-1}(F') \subseteq f_n^{-1}(F(n; \delta)) \cup \Gamma(n; \delta)$$

for $n \in \mathbb{Z}^+$ and $\delta \in (0, \infty)$. Hence, for every $n \in \mathbb{Z}^+$ and $\delta > 0$,

$$\overline{\lim_{\epsilon \longrightarrow 0}} \, \epsilon \log \left[\mu_\epsilon \circ f^{-1}(F')\right] \leq \left(\overline{\lim_{\epsilon \longrightarrow 0}} \, \epsilon \log \left[\mu_\epsilon(F(n; \delta))\right]\right) \vee \left(-R(n; \delta)\right),$$

where

$$-R(n; \delta) \equiv \overline{\lim_{\epsilon \longrightarrow 0}} \, \epsilon \log \left[\mu_\epsilon(\Gamma(n; \delta))\right].$$

Since, by hypothesis, $R(n; \delta) \longrightarrow \infty$ as $n \longrightarrow \infty$ for each $\delta \in (0, \infty)$ and because, by the preceding paragraph,

$$\lim_{\delta \searrow 0} \underline{\lim_{n \longrightarrow \infty}} \inf \left\{I(q) : q \in F(n; \delta)\right\} = \inf_{F'} I',$$

the large deviation principle for $\{\mu_\epsilon : \epsilon > 0\}$ now leads to

$$\overline{\lim_{\epsilon\to 0}}\, \epsilon \log\left(\mu_\epsilon \circ f^{-1}(F')\right) \le -\inf_{F'} I'. \quad \blacksquare$$

Another situation which we encountered in Chapter I (cf. the proof of SCHILDER's Theorem) is that of a deficient large deviation principle; namely, one in which the right hand side of (1.1.5) has been proved only when the set Γ is relatively compact. As it was there, such a large deviation principle is usually a preliminary step on the way to proving a full large deviation principle. Nonetheless, it arises sufficiently often to warrant our giving it a name. Thus, if I is a rate function and $\{\mu_\epsilon : \epsilon > 0\} \subseteq \mathbf{M}_1(E)$ satisfies

$$\underline{\lim_{\epsilon\to 0}}\, \epsilon \log\left(\mu_\epsilon(G)\right) \ge -\inf_G I \qquad \text{for all open } G \text{ in } E$$

and

$$\overline{\lim_{\epsilon\to 0}}\, \epsilon \log\left(\mu_\epsilon(K)\right) \le -\inf_K I \qquad \text{for all } K \subset\subset E,$$

then we will say that $\{\mu_\epsilon : \epsilon > 0\}$ satisfies the **weak large deviation principle with rate function I.**

The passage from a weak to a full large deviation principle is often accomplished by an application of the following simple observation.

2.1.5 Lemma. *Let $\{\mu_\epsilon : \epsilon > 0\} \subseteq \mathbf{M}_1(E)$, and assume that, for each $L \ge 0$, there exists a $K_L \subset\subset E$ with the property that*

$$\overline{\lim_{\epsilon\to 0}}\, \epsilon \log\left(\mu_\epsilon(K_L^c)\right) \le -L. \tag{2.1.6}$$

If I is a rate function and $\{\mu_\epsilon : \epsilon > 0\}$ satisfies the weak large deviation principle with rate function I, then not only is I a good rate function, but it also governs the large deviations of $\{\mu_\epsilon : \epsilon > 0\}$.

PROOF: First note that

$$\inf_{K_L^c} I \ge -\underline{\lim_{\epsilon\to 0}}\, \epsilon \log\left(\mu_\epsilon(K_L^c)\right) \ge L;$$

and so $\{q : I(q) \le L\} \subseteq K_{L+1}$. Since I is lower semi-continuous, this proves that I is a good rate function.

Next, let F be a closed subset in E and set $F_L = F \cap K_L$ for $L \ge 0$. Then

$$\mu_\epsilon(F) \le \mu_\epsilon(F_L) + \mu_\epsilon(K_L^c),$$

and so

$$\overline{\lim_{\epsilon\to 0}}\,\epsilon\log\left(\mu_\epsilon(F)\right) \le -\left[\left(\inf_{F_L} I\right)\wedge L\right] \le -\left[\left(\inf_{F} I\right)\wedge L\right]$$

for every $L \ge 0$. Thus we get the required result upon letting $L \nearrow \infty$. ∎

We will say that a family $\{\mu_\epsilon : \epsilon > 0\} \subseteq \mathbf{M}_1(E)$ is **exponentially tight** if, for each $L > 0$, there is a $K_L \subset\subset E$ for which (2.1.6) holds.

We end this section with a result which, in its original version, was first proved by S.R.S. VARADHAN [**106**].

2.1.7 Lemma. *Let I be a rate function and suppose that $\{\mu_\epsilon : \epsilon > 0\}$ satisfies the weak large deviation principle with rate function I. If the function $\Phi : E \longrightarrow [-\infty, \infty]$ is lower semi-continuous, then*

$$\underline{\lim_{\epsilon\to 0}}\,\epsilon\log\left(\int \exp\left[\Phi/\epsilon\right] d\mu_\epsilon\right) \ge \sup\left\{\Phi(q) - I(q) : q \in E \text{ and } \Phi(q)\wedge I(q) < \infty\right\}.$$

(Throughout we adopt the convention that the supremum over the empty set is $-\infty$.)

PROOF: Let $q \in E$ satisfy $\Phi(q)\wedge I(q) < \infty$. Then, for each $r > 0$,

$$\underline{\lim_{\epsilon\to 0}}\,\epsilon\log\left(\int \exp\left[\Phi/\epsilon\right] d\mu_\epsilon\right) \ge \underline{\lim_{\epsilon\to 0}}\,\epsilon\log\left(\int_{B(q,r)} \exp\left[\Phi/\epsilon\right] d\mu_\epsilon\right) \ge \inf_{B(q,r)} \Phi - I(q).$$

Since $\Phi(q) = \lim_{r\to 0}\inf_{B(q,r)}\Phi$, we conclude that

$$\underline{\lim_{\epsilon\to 0}}\,\epsilon\log\left(\int \exp\left[\Phi/\epsilon\right] d\mu_\epsilon\right) \ge \Phi(q) - I(q). \quad ∎$$

2.1.8 Lemma. *Assume that I is a good rate function and that $\{\mu_\epsilon : \epsilon > 0\}$ satisfies the full large deviation principle with rate function I. If $\Phi : E \longrightarrow [-\infty, \infty)$ is an upper semi-continuous function which satisfies*

$$\text{(2.1.9)} \qquad \lim_{L\to\infty}\overline{\lim_{\epsilon\to 0}}\,\epsilon\log\left(\int_{\{q:\Phi(q)\ge L\}} \exp\left[\Phi(q)/\epsilon\right] \mu_\epsilon(dq)\right) = -\infty,$$

then

$$\overline{\lim_{\epsilon \to 0}} \epsilon \log \left(\int \exp[\Phi/\epsilon] \, d\mu_\epsilon \right) \le \sup_E (\Phi - I).$$

PROOF: We first work in the case when $\Phi \le M$ for some $M \in (0, \infty)$. Given $L > 0$, set $K_L = \{q : I(q) \le L\}$. Since Φ is upper semi-continuous and $K_L \subset\subset E$, we can choose, for given $\delta > 0$, a finite set $\{q_m\}_{m=1}^n \subseteq K_L$ and positive numbers $r_1, \ldots, r_n$ so that

$$K_L \subseteq \bigcup_{m=1}^n B(q_m, r_m), \quad \sup_{\overline{B}_m} \Phi \le \Phi(q_m) + \delta, \quad \text{and } \inf_{\overline{B}_m} I \ge I(q_m) - \delta$$

for $1 \le m \le n$, where $B_m \equiv B(q_m, r_m)$. Thus, if $G = \bigcup_{m=1}^n B_m$, then

$$\int \exp[\Phi/\epsilon] \, d\mu_\epsilon \le \exp\left[\frac{1}{\epsilon}\left(M + \epsilon \log\left(\mu_\epsilon(G^c)\right)\right)\right]$$
$$\times \sum_{m=1}^n \exp\left[\frac{1}{\epsilon}\left(\Phi(q_m) + \delta + \epsilon \log\left(\mu_\epsilon(\overline{B}_m)\right)\right)\right],$$

and so

$$\begin{aligned}
&\overline{\lim_{\epsilon \to 0}} \epsilon \log \left(\int \exp[\Phi/\epsilon] \, d\mu_\epsilon \right) \\
&\le \left(\max_{1 \le m \le n} \left(\Phi(q_m) - I(q_m) + 2\delta\right) \right) \vee (M - L) \\
&\le \left(\sup_E (\Phi - I) \right) \vee (M - L) + 2\delta.
\end{aligned}$$

Now let $\delta \searrow 0$ and $L \nearrow \infty$.

To treat the general case, set $\Phi_M = \Phi \wedge M$ for $M \in (0, \infty)$, and use the preceding to show that

$$\begin{aligned}
\overline{\lim_{\epsilon \to 0}} \epsilon \log \left(\int \exp[\Phi/\epsilon] \, d\mu_\epsilon \right) &\le \left(\sup_E (\Phi_M - I) \right) \vee A_M \\
&\le \left(\sup_E (\Phi - I) \right) \vee A_M,
\end{aligned}$$

where

$$A_M \equiv \overline{\lim_{\epsilon \to 0}} \epsilon \log \left(\int_{\Phi \ge M} \exp[\Phi/\epsilon] \, d\mu_\epsilon \right) \longrightarrow -\infty$$

as $M \longrightarrow \infty$. ∎

2.1.10 Theorem. (VARADHAN) *Let I be a good rate function and assume that $\{\mu_\epsilon : \epsilon > 0\} \subseteq \mathbf{M}_1(E)$ satisfies the full large deviation principle with rate function I. If $\Phi \in C(E;\mathbb{R})$ satisfies (2.1.9), then*

$$\lim_{\epsilon\to 0} \epsilon \log\left(\int \exp\big[\Phi/\epsilon\big]\, d\mu_\epsilon\right) = \sup_E(\Phi - I). \tag{2.1.11}$$

In particular, (2.1.11) *holds if $\Phi \in C(E;\mathbb{R})$ satisfies*

$$\sup_{0<\epsilon\le 1}\left(\int \exp\big[\alpha\Phi/\epsilon\big]\, d\mu_\epsilon\right)^{\epsilon} < \infty \tag{2.1.12}$$

for some $\alpha \in (1,\infty)$.

PROOF: In view of Lemma 2.1.7 and Lemma 2.1.8 , all that we have to do is check that (2.1.12) implies (2.1.9). But, by HÖLDER's inequality,

$$\begin{aligned}\int_{\Phi\ge L} \exp\big[\Phi/\epsilon\big]\, d\mu_\epsilon &\le \left(\int \exp\big[\alpha\Phi/\epsilon\big]\, d\mu_\epsilon\right)^{1/\alpha} \big(\mu_\epsilon(\Phi \ge L)\big)^{1-\frac{1}{\alpha}} \\ &\le \left(\int \exp\big[\alpha\Phi/\epsilon\big]\, d\mu_\epsilon\right)^{1/\alpha} \left(\exp\big[-\alpha L/\epsilon\big] \int \exp\big[\alpha\Phi/\epsilon\big]\, d\mu_\epsilon\right)^{1-\frac{1}{\alpha}} \\ &= \exp\big[(1-\alpha)L/\epsilon\big] \left(\int \exp\big[\alpha\Phi/\epsilon\big]\, d\mu_\epsilon\right),\end{aligned}$$

from which (2.1.9) follows immediately when (2.1.12) holds. ∎

2.1.13 Exercise.

(**i**) Define EULER's Gamma function by

$$\Gamma(\gamma) = \int_{(0,\infty)} t^{\gamma-1} e^{-t}\, dt, \quad \gamma \in (0,\infty).$$

Note that

$$\gamma^{-\gamma+1}\Gamma(\gamma) = \gamma \int_{(0,\infty)} t^{\gamma-1} e^{-\gamma t}\, dt;$$

and using Theorem 2.1.10 together with Exercise 1.1.6, conclude that

$$\lim_{\gamma\to\infty} \frac{1}{\gamma} \log\left(\gamma^{-\gamma+1}\Gamma(\gamma)\right) = -1.$$

This is, of course, a very weak version of STIRLING's formula and, as such, it serves as a good example of both the virtues and the deficiencies in the asymptotic theory with which we are dealing.

(ii) Let $\mathcal{W}$ be WIENER's measure on Θ with $d = 1$; and, for given $\beta \in \mathbb{R}$, define $X_\beta : [0,\infty) \times \Theta \longrightarrow \mathbb{R}$ by the equation

$$X_\beta(t,\theta) = \theta(t) + \beta \int_0^t X_\beta(s,\theta)\,ds, \quad t \in [0,\infty)$$

and $\sigma_\beta : \Theta \longrightarrow [0,\infty)$ by

$$\sigma_\beta(\theta) = \left(\int_0^1 \left(X_\beta(t,\theta)\right)^2 dt\right)^{1/2}.$$

If, for $\epsilon > 0$, $\mu_{\beta,\epsilon} \in \mathbf{M}_1\big([0,\infty)\big)$ is the distribution of $\theta \longmapsto \epsilon^{1/2}\sigma_\beta(\theta)$ under $\mathcal{W}$, show that $\{\mu_{\beta,\epsilon} : \epsilon > 0\}$ satisfies the full large deviation principle with the good rate function $I_\beta : [0,\infty) \longrightarrow [0,\infty]$ given by

$$I_\beta(\sigma) = \frac{\alpha_\beta \sigma^2}{2},$$

where

$$\alpha_\beta = \begin{cases} \beta^2 - \omega_\beta^2 & \text{for } \beta \geq 1 \\ \beta^2 + \omega_\beta^2 & \text{for } \beta < 1 \end{cases}$$

and

$$\omega_\beta = \begin{cases} \text{the } \omega \in (0,\pi) \text{ such that } \omega\cos\omega = \beta\sin\omega & \text{if } \beta \in (-\infty,1] \\ 0 & \text{if } \beta = 1 \\ \text{the } \omega \in (0,\infty) \text{ such that } \omega\cosh\omega = \beta\sinh\omega & \text{if } \beta \in (1,\infty). \end{cases}$$

Hint: Note that, by Lemma 2.1.4 combined with SCHILDER's Theorem, the desired large deviation principle holds with

$$\begin{aligned} I_\beta(\sigma) &\equiv \inf\left\{\frac{1}{2}\int_0^1 \left(\dot\phi(t) - \beta\phi(t)\right)^2 dt : \phi \in H^1 \text{ with } \int_0^1 \phi(t)^2\,dt = \sigma^2\right\} \\ &= I_\beta(1)\sigma^2; \end{aligned}$$

and use the calculus of variations to evaluate $I_\beta(1)$.

(iii) Next, define $Y_\beta : [0,\infty) \times \Theta^2 \longrightarrow [0,\infty)$ to be the solution of the ITÔ stochastic integral equation

$$Y_\beta(t,\theta,\theta') = 1 + \int_0^t X_\beta(s,\theta) Y_\beta(s,\theta,\theta')\,d\theta'(s), \quad t \in [0,\infty),$$

under $\mathcal{W}^2$, and note that

$$Y_\beta(t,\theta,\theta') = \exp\left[\int_0^t X_\beta(s,\theta)\,d\theta'(s) - \frac{1}{2}\int_0^t X_\beta^2(s,\theta)\,ds\right].$$

Letting $P_\beta \in \mathbf{M}_1\big([0,\infty)\big)$ denote the distribution of

$$(\theta,\theta') \in \Theta^2 \longmapsto Y_\beta(1,\theta,\theta')$$

under $\mathcal{W}^2$, check that $P_\beta(dy) = p_\beta(y)\,dy$ where $p_\beta(y) = \frac{1}{y}q_\beta\big(\log y\big)$ and, for $z \neq 0$,

$$q_\beta(z) = \epsilon \int_{(0,\infty)} \frac{1}{\big(2\pi\sigma^2\big)^{1/2}} \exp\left[-\frac{1}{\epsilon}\frac{\big(\mathrm{sgn}(z)+\sigma^2/2\big)^2}{2\sigma^2}\right] \mu_{\beta,\epsilon}(d\sigma),$$

where $\epsilon = 1/|z|$. Finally, use **ii)** above and VARADHAN's Theorem to show from this that

$$\begin{aligned}
-\lim_{y\to\infty}\frac{\log p_\beta(y)}{\log y} &= 1 - \lim_{\epsilon\to 0}\epsilon\log\left(\int_{(0,\infty)}\exp\left[-\frac{1}{\epsilon}\frac{\big(1+\sigma^2/2\big)^2}{2\sigma^2}\right]\mu_{\beta,\epsilon}(d\sigma)\right)\\
&= 1 + \inf_{\sigma\in(0,\infty)}\left\{\frac{\big(1+\sigma^2/2\big)^2}{2\sigma^2} + I_\beta(\sigma)\right\} = \frac{1}{2}\big[3 + \big(1+4\alpha_\beta\big)^{1/2}\big]
\end{aligned}$$

and that

$$\begin{aligned}
\lim_{y\to 0}\frac{\log p_\beta(y)}{\log y} &= 1 - \lim_{\epsilon\to 0}\epsilon\log\left(\int_{(0,\infty)}\exp\left[-\frac{1}{\epsilon}\frac{\big(1-\sigma^2/2\big)^2}{2\sigma^2}\right]\mu_{\beta,\epsilon}(d\sigma)\right)\\
&= 1 + \inf_{\sigma\in(0,\infty)}\left\{\frac{\big(-1+\sigma^2/2\big)^2}{2\sigma^2} + I_\beta(\sigma)\right\} = \frac{1}{2}\big[1 + \big(1+4\alpha_\beta\big)^{1/2}\big].
\end{aligned}$$

2.1.14 Exercise.

Let $\{\mu_\epsilon : \epsilon > 0\} \subseteq \mathbf{M}_1(E)$ and a rate function $I : E \longrightarrow [0,\infty]$ be given.

(i) If I is good and $\{\mu_\epsilon : \epsilon > 0\}$ satisfies the full large deviation principle with rate I, show that there is a $q \in E$ for which $I(q) = 0$.

(ii) Assuming that E is locally compact and that $\{\mu_\epsilon : \epsilon > 0\}$ satisfies the full large deviation principle with rate I, show that I is good if and only if $\{\mu_\epsilon : \epsilon > 0\}$ is exponentially tight.

(iii) If

$$\lim_{r\searrow 0}\varliminf_{\epsilon\to 0}\epsilon\log\left[\mu_\epsilon\big(B(q,r)\big)\right]\geq -I(q),\quad q\in E,$$

show that

$$\begin{aligned}\varliminf_{\epsilon\to 0}\epsilon\log&\left[\int\exp\left[\Phi/\epsilon\right]d\mu_\epsilon\right]\\ &\geq\sup\{\Phi(q)-I(q): q\in E\text{ and }\Phi(q)\wedge I(q)<\infty\}\end{aligned}$$

for every lower semi-continuous $\Phi: E\longrightarrow[-\infty,\infty]$.

(iv) If

$$\lim_{r\searrow 0}\varlimsup_{\epsilon\to 0}\epsilon\log\left[\mu_\epsilon\big(B(q,r)\big)\right]\leq -I(q),\quad q\in E,$$

show that

$$\lim_{\delta\searrow 0}\varlimsup_{\epsilon\longrightarrow 0}\log\left[\mu_\epsilon\big(K^{(\delta)}\big)\right]\leq -\inf_K I,\quad K\subset\subset E.$$

In particular, this means, of course, that

$$\varlimsup_{\epsilon\to 0}\epsilon\log\big(\mu_\epsilon(K)\big)\leq -\inf_K I,\quad K\subset\subset E.$$

Also, check that

$$\varlimsup_{\epsilon\to 0}\epsilon\log\left[\int\exp\left[\Phi/\epsilon\right]d\mu_\epsilon\right]\leq\sup\big(\Phi-I\big)$$

for every upper semi-continuous $\Phi: E\longrightarrow[-\infty,\infty)$ which satisfies not only (2.1.9) but also the condition that

$$\big\{q\in E: -\Phi(q)\leq L\big\}\subset\subset E,\quad L\in[0,\infty).$$

(v) Assume that

$$\begin{aligned}\lim_{r\searrow 0}\varliminf_{\epsilon\to 0}\epsilon\log\big(\mu_\epsilon\big(B(q,r)\big)\big)&=-I(q)\\ &=\lim_{r\searrow 0}\varlimsup_{\epsilon\to 0}\epsilon\log\big(\mu_\epsilon\big(B(q,r)\big)\big),\quad q\in E.\end{aligned}$$

Show that $\{\mu_\epsilon:\epsilon>0\}$ satisfies the weak large deviation principle with rate I and that

$$\lim_{\epsilon\to 0}\epsilon\log\left(\int\exp\left[\Phi/\epsilon\right]d\mu_\epsilon\right)=\sup\big(\Phi-I\big)$$

for $\Phi \in C(E;\mathbb{R})$ which satisfy (2.1.9) and the condition

$$\{q \in E : -\Phi(q) \le L\} \subset\subset E, \quad L \in [0,\infty).$$

2.1.15 Exercise.

For each i from an index set $\mathcal{I}$ let $\{\mu_{i,\epsilon} : \epsilon > 0\}$ be a family of probability measures on E. Assume that there exists a good rate function $I : E \longrightarrow [0,\infty]$ with the property that

$$\begin{aligned} -\inf_{\Gamma^\circ} I &\le \varliminf_{\epsilon\to 0} \epsilon \log\left(\inf_{i\in\mathcal{I}} \mu_{i,\epsilon}(\Gamma)\right) \\ &\le \varlimsup_{\epsilon\to 0} \epsilon \log\left(\sup_{i\in\mathcal{I}} \mu_{i,\epsilon}(\Gamma)\right) \le -\inf_{\overline{\Gamma}} I, \quad \Gamma \in \mathcal{B}_E. \end{aligned} \tag{2.1.16}$$

Show that for any $\Phi \in C(E;\mathbb{R})$ which satisfies

$$\lim_{L\to\infty} \varlimsup_{\epsilon\to 0} \epsilon \log\left(\sup_{i\in\mathcal{I}} \int_{\{q:\Phi(q)\ge L\}} \exp\big[\Phi(q)/\epsilon\big]\, \mu_{i,\epsilon}(dq)\right) = -\infty, \tag{2.1.17}$$

one has that

$$\begin{aligned} \lim_{\epsilon\to 0} \sup_{i\in\mathcal{I}} \Bigg| \epsilon \log\left(\int_E \exp\big[\Phi/\epsilon\big]\, \mu_{i,\epsilon}(dq)\right) \\ - \sup\{\Phi(q) - I(q) : q \in E\}\Bigg| = 0. \end{aligned} \tag{2.1.18}$$

In particular, show that (2.1.17), and therefore (2.1.18), holds if $\Phi \in C(E;\mathbb{R})$ satisfies

$$\sup_{0<\epsilon\le 1} \sup_{i\in\mathcal{I}} \left(\int_E \exp\big[\alpha\Phi/\epsilon\big]\, d\mu_{i,\epsilon}\right)^{\epsilon} < \infty \tag{2.1.19}$$

for some $\alpha \in (1,\infty)$.

2.1.20 Exercise.

This exercise contains several variations on the theme of Lemma 2.1.4. Throughout, $\{\mu_\epsilon : \epsilon > 0\} \subseteq \mathbf{M}_1(E)$, $I : E \longrightarrow [0,\infty]$ is a good rate function, and E' is a second separable metric space.

(i) Assume that $\{\mu_\epsilon : \epsilon > 0\}$ satisfies the full large deviation principle with rate I. Further, assume that there is a non-decreasing family $\{F_L : L \ge 0\}$

of closed sets in E with the properties that $\mu_\epsilon(F_\infty) = 1$, $\epsilon > 0$, where $F_\infty \equiv \bigcup_{L \geq 0} F_L$, and that

$$\overline{\lim_{\epsilon \to 0}}\, \epsilon \log\left(\mu_\epsilon(F_L^c)\right) \leq -L, \quad L \geq 0.$$

Finally, suppose that $f : F_\infty \longrightarrow E'$ is a function whose restriction to each F_L, $L \geq 0$, is continuous. If $\mu'_\epsilon \in \mathbf{M}_1(E')$ is defined by

$$\mu'_\epsilon(\Gamma') = \mu_\epsilon\left(\{q \in F_\infty : f(q) \in \Gamma'\}\right)$$

for $\epsilon > 0$ and $\Gamma' \in \mathcal{B}_{E'}$ and if

$$I'(q') = \inf\{I(q) : q \in F_\infty \text{ and } f(q) = q'\}, \quad q' \in E',$$

show that I' is a good rate function on E' and that it governs the large deviations of $\{\mu'_\epsilon : \epsilon > 0\}$.

(ii) Let $\{f_\epsilon : \epsilon \geq 0\}$ be a family of continuous maps from E into E', set

$$I'_0(q') = \inf\{I(q) : q \in E \text{ and } q' = f_0(q)\}, \quad q' \in E',$$

and assume that

$$\overline{\lim_{\epsilon \to 0}}\, \epsilon \log\left(\mu_\epsilon\left(\left\{q \in E : \rho'\left(f_\epsilon(q), f_0(q)\right) > \alpha\right\}\right)\right) = -\infty \quad \text{for } \alpha > 0,$$

where ρ' denotes the metric on E'. Assuming that $\{\mu_\epsilon : \epsilon > 0\}$ satisfies the full large deviation principle with rate I, show that $\{\mu_\epsilon \circ f_\epsilon^{-1} : \epsilon > 0\}$ satisfies the full large deviation principle with the good rate function I'_0.

(iii) Let $f : [0, \infty) \times E \longrightarrow E'$ be a measurable function for which there exists a sequence $\{f_n\}_1^\infty \subseteq C\left([0, \infty) \times E; E'\right)$ with the properties that $f_n(0, \cdot) \longrightarrow f(0, \cdot)$ uniformly on each level set of I and

$$\lim_{n \to \infty} \overline{\lim_{\epsilon \to 0}}\, \epsilon \log\left[\mu_\epsilon\left(\left\{q : \rho'\left(f_n(\epsilon, q), f(\epsilon, q)\right) \geq \delta\right\}\right)\right] = -\infty, \quad \delta \in (0, \infty).$$

Assuming that $\{\mu_\epsilon : \epsilon > 0\}$ is exponentially tight and that I governs the large deviations of $\{\mu_\epsilon : \epsilon > 0\}$, show that the function $I' : E' \longrightarrow [0, \infty]$ given by

$$I'(q') = \inf\{I(q) : q' = f(0, q)\}, \quad q' \in E',$$

is a good rate function and that it governs the large deviations of $\{\mu_\epsilon \circ f(\epsilon)^{-1} : \epsilon > 0\}$.

(iv) Again assume that $\{\mu_\epsilon : \epsilon > 0\}$ is an exponentially tight family whose large deviations are governed by I. Next, let X be a compact metric space and suppose that $f : [0,\infty) \times E \times X \longrightarrow E'$ is a measurable map with the property that there is a sequence $\{f_n\}_1^\infty \subseteq C\big([0,\infty) \times E \times X; E'\big)$ such that

$$\lim_{n\to\infty} \sup\Big\{\rho'\big(f_n(0,q,x), f(0,q,x)\big) : (q,x) \in E \times X \text{ and } I(q) \le L\Big\} = 0$$

for $L \in [0,\infty)$, and

$$\sup_{x\in X} \lim_{n\to\infty} \overline{\lim_{\epsilon\to 0}}\, \epsilon \log\Big[\mu_\epsilon\Big(\Big\{q : \rho'\big(f_n(\epsilon,q,x), f(\epsilon,q,x)\big) \ge \delta\Big\}\Big)\Big] = -\infty$$

for $\delta \in (0,\infty)$. Finally, define $I'_x : E' \longrightarrow [0,\infty]$ for $x \in X$ by

$$I'_x(q') = \inf\big\{I(q) : q' = f(0,q,x)\big\}, \quad q' \in E'.$$

Show that I'_x is a good rate function for each $x \in X$. In addition, show that if $x_\epsilon \longrightarrow x$ as $\epsilon \searrow 0$ and if $g(\epsilon,q) = f(\epsilon,q,x_\epsilon)$ for $(\epsilon,q) \in (0,\infty) \times E$, then I'_x governs the large deviations of $\big\{\mu_\epsilon \circ g(\epsilon)^{-1} : \epsilon > 0\big\}$.

Hint: By using the exponential tightness of $\{\mu_\epsilon : \epsilon > 0\}$, show that, for every $\delta \in (0,\infty)$,

$$\overline{\lim_{\epsilon\to\infty}}\, \epsilon \log\Big[\mu_\epsilon\Big(\Big\{q : \rho'\big(f_n(\epsilon,q,x_\epsilon), f_n(\epsilon,q,x)\big) \ge \delta\Big\}\Big)\Big] = -\infty.$$

(v) Refer to the setting of part **(iv)** above and suppose that $\Phi \in C\big(E'; \mathbb{R}\big)$ has the property that

$$\sup_{0<\epsilon\le 1} \sup_{x\in X} \left(\int_E \exp\Big[\frac{\alpha}{\epsilon}\,\Phi\big(f(\epsilon,q,x)\big)\Big]\, \mu_\epsilon(dq)\right)^\epsilon < \infty$$

for some $\alpha \in (1,\infty)$. Show that

$$\begin{aligned}\lim_{\epsilon\to 0} \epsilon \log&\left[\sup_{x\in X} \int_E \exp\Big[\frac{1}{\epsilon}\,\Phi\big(f(\epsilon,q,x)\big)\Big]\, \mu_\epsilon(dq)\right] \\ &= \sup\Big\{\Phi\big(f(0,q,x)\big) - I(q) : (q,x) \in E \times X\Big\}.\end{aligned}$$

2.1.21 Exercise.

The purpose of this exercise is to check that the full large deviation principle behaves in a functorial fashion under **projective limits**. To be precise, suppose that $\{E_n : n \in \mathbb{Z}^+\}$ is a sequence of Polish spaces and that, for each $n \in \mathbb{Z}^+$, ρ_n is a complete metric for E_n and $\pi_{n+1,n} : E_{n+1} \longrightarrow E_n$ is a mapping with the property that $\rho_n(\pi_{n+1,n}x_{n+1}, \pi_{n+1,n}y_{n+1}) \le \rho_{n+1}(x_{n+1}, y_{n+1})$ for all $x_{n+1}, y_{n+1} \in E_{n+1}$. Define $\mathbf{E}$ to be the set of $\mathbf{x} = (x_1, \dots, x_n, \dots) \in \prod_{n=1}^{\infty} E_n$ such that $x_n = \pi_{n+1,n}x_{n+1}$ for every $n \in \mathbb{Z}^+$, and let π_n denote the restriction to $\mathbf{E}$ of the natural projection map from $\prod_{n=1}^{\infty} E_n$ onto E_n. Give $\mathbf{E}$ the topology which it inherits from the product topology on $\prod_{n=1}^{\infty} E_n$, and define

$$\rho(\mathbf{x}, \mathbf{y}) = \sum_{n=1}^{\infty} \frac{1}{2^n} \frac{\rho_n(\pi_n\mathbf{x}, \pi_n\mathbf{y})}{1 + \rho_n(\pi_n\mathbf{x}, \pi_n\mathbf{y})}, \quad \mathbf{x}, \mathbf{y} \in \mathbf{E}. \tag{2.1.22}$$

(**i**) Show that $\mathbf{E}$ is a Polish space and, in fact, that ρ is a complete metric on $\mathbf{E}$. Further, check that $\mathbf{G}$ is an open subset of $\mathbf{E}$ if and only if $\mathbf{G} = \bigcup_{n=1}^{\infty} \pi_n^{-1} G_n$, where G_n is an open subset of E_n for each $n \in \mathbb{Z}^+$. Also, check that for each closed subset $\mathbf{F}$ of $\mathbf{E}$ and every $\delta > 0$, there is an $n \in \mathbb{Z}^+$ and a closed subset F_n of E_n such that $\mathbf{F} \subseteq \pi_n^{-1} F_n \subseteq \mathbf{F}^{(\delta)}$, where $\mathbf{F}^{(\delta)}$ is computed relative to the metric ρ. Finally, show that $\mathbf{K} \subset\subset \mathbf{E}$ if and only if $\mathbf{K} = \bigcap_{n=1}^{\infty} \pi_n^{-1} K_n$, where $K_n \subset\subset E_n$ for each $n \in \mathbb{Z}^+$. The metric space $(\mathbf{E}, \rho)$ is called the **projective limit** of the sequence $\{(E_n, \pi_{n+1,n}, \rho_n) : n \in \mathbb{Z}^+\}$.

(**ii**) For each $n \in \mathbb{Z}^+$ suppose that $I_n : E_n \longrightarrow [0, \infty]$ is a good rate function, and define

$$\mathbf{I}(\mathbf{x}) = \sup_{n \in \mathbb{Z}^+} I_n(\pi_n\mathbf{x}), \quad x \in \mathbf{E}. \tag{2.1.23}$$

Show that $\mathbf{I}$ is a good rate function and that

$$I_n(x_n) = \inf\{\mathbf{I}(\mathbf{x}) : x_n = \pi_n(\mathbf{x})\}, \quad n \in \mathbb{Z}^+ \text{ and } x_n \in E_n.$$

(**iii**) Again let I_n be a good rate function on E_n for each $n \in \mathbb{Z}^+$ and define $\mathbf{I}$ accordingly as in (2.1.23). Next, let $\{\mu_\epsilon : \epsilon > 0\} \subseteq \mathbf{M}_1(\mathbf{E})$; and, for each $n \in \mathbb{Z}^+$ set $\mu_{n,\epsilon} = \mu_\epsilon \circ \pi_n^{-1}$. If, for each $n \in \mathbb{Z}^+$,

$$\varliminf_{\epsilon \to 0} \epsilon \log\Big(\mu_{n,\epsilon}(G_n)\Big) \ge -\inf_{G_n} I_n$$

for every open G_n in E_n, show that

$$\varliminf_{\epsilon \to 0} \epsilon \log\Big(\mu_\epsilon(\mathbf{G})\Big) \ge -\inf_{\mathbf{G}} \mathbf{I}$$

for every open $\mathbf{G}$ in $\mathbf{E}$. Similarly, if, for each $n \in \mathbb{Z}^+$,

$$\varlimsup_{\epsilon \to 0} \epsilon \log\Big(\mu_{n,\epsilon}(F_n)\Big) \le -\inf_{F_n} I_n$$

for all closed F_n in E_n, show that

$$\varlimsup_{\epsilon \to 0} \epsilon \log\Big(\mu_\epsilon(\mathbf{F})\Big) \le -\inf_{\mathbf{F}} \mathbf{I}$$

for every closed $\mathbf{F}$ in $\mathbf{E}$.

2.1.24 Exercise.

Assume that $\{\mu_\epsilon : \epsilon > 0\}$ satisfies the full large deviation principle with respect to the good rate function I, and suppose that $\Phi \in C(E;\mathbb{R})$ satisfies the condition in (2.1.9).

(i) Show that

$$\varliminf_{\epsilon\to 0} \epsilon \log\left(\int_G \exp\bigl[\Phi(q)/\epsilon\bigr]\,\mu_\epsilon(dq)\right) \geq \sup_G\bigl(\Phi - I\bigr) \quad \text{for open } G \subseteq E$$

and that

$$\varlimsup_{\epsilon\to 0} \epsilon \log\left(\int_F \exp\bigl[\Phi(q)/\epsilon\bigr]\,\mu_\epsilon(dq)\right) \leq \sup_F\bigl(\Phi - I\bigr) \quad \text{for closed } F \subseteq E.$$

In particular, conclude that

$$\lim_{L\to\infty} \inf\bigl\{I(q) - \Phi(q) : q \in E \text{ with } \Phi(q) > L\bigr\} = \infty.$$

Hint: For $\Gamma \in \mathcal{B}_E$, set

$$\Phi_\Gamma(q) = \begin{cases} \Phi(q) & \text{if } q \in \Gamma \\ -\infty & \text{if } q \notin \Gamma; \end{cases}$$

and apply Lemma 2.1.7 and Lemma 2.1.8 to Φ_G and Φ_F, respectively.

(ii) For $\epsilon > 0$, define

$$\nu_\epsilon(\Gamma) = \frac{\int_\Gamma \exp\bigl[\Phi(q)/\epsilon\bigr]\,\mu_\epsilon(dq)}{\int_E \exp\bigl[\Phi(q)/\epsilon\bigr]\,\mu_\epsilon(dq)}, \quad \Gamma \in \mathcal{B}_E.$$

Next, set $J(q) = I(q) - \Phi(q) - \alpha$, $q \in E$, where $\alpha \equiv \inf_E\bigl(I - \Phi\bigr)$. Show that J is a good rate function and that it governs the large deviations of $\{\nu_\epsilon : \epsilon > 0\}$. Finally, check that when there is precisely one $p \in E$ at which J vanishes then the measures ν_ϵ converge to δ_p.

2.1.25 Exercise.

In Theorem 1.4.25, we made the assumption that the diffusion matrix $a : \mathbb{R}^d \longrightarrow \mathbb{R}^d \otimes \mathbb{R}^d$ was symmetric and that a together with the drift coefficient $b : \mathbb{R}^d \longrightarrow \mathbb{R}^d$ satisfy (1.4.10) and (1.4.11). Here we replace those assumptions by

$$0 \leq a(x) \leq M I_{\mathbb{R}^d} \quad \text{and} \quad |b(x)| \leq M, \quad x \in \mathbb{R}^d,$$

and, with $\sigma \equiv a^{1/2}$,

$$\|\sigma(y) - \sigma(x)\|_{\mathrm{H.S.}} \vee |b(y) - b(x)| \leq M|y - x|, \quad x, y \in \mathbb{R}^d.$$

Using Lemma 2.1.4, show that the function $I^{a,b}_{T,x} : \Omega_T \longrightarrow [0, \infty]$ given by

$$I^{a,b}_{T,x}(X) = \inf\Big\{ I_T(\psi) : \psi \in H^1 \text{ and, for } t \in [0,T],$$
$$X(t) = x + \int_0^t a^{1/2}\big(X(s)\big)\dot{\psi}(s)\,ds + \int_0^t b\big(X(s)\big)\,ds \Big\}$$

for $X \in \Omega_T$ (cf. (1.3.36) for the notation here) is a good rate function and that (1.4.26) continues to hold. Next (cf. Exercise 1.4.33) extend this result to cover the case when the preceding upper bounds on a and b are replaced by

$$0 \leq a(x) \leq M\big(1 + |x|^2\big)^{1/2} I_{\mathbb{R}^d} \quad \text{and} \quad |b(x)| \leq M\big(1 + |x|^2\big)^{1/2}, \quad x \in \mathbb{R}^d.$$

2.2 Large Deviations and Convex Analysis

As we saw in Chapter I, it is sometimes the case that the state space E is a separable BANACH space. Furthermore, even when E is not itself a vector space, it often turns out that it is a convex subset of one. For this reason we formulate the following somewhat cumbersome hypothesis about E.

(C) E is a closed convex subset of the locally convex, HAUSDORFF topological (real) vector space X, and E is a Polish space with respect to the topology that it inherits as a subset of X.

2.2.1 Remark.

The two examples which should be kept in mind are when E is itself a separable BANACH space (in which case we take $X = E$) and when $E = \mathbf{M}_1(\Sigma)$, where Σ is a Polish space. In the latter case, we take $X = \mathbf{M}(\Sigma)$ to be the space of all finite signed measures on Σ and endow $\mathbf{M}(\Sigma)$ with the topology generated by the sets

$$\left\{\beta \in \mathbf{M}(\Sigma) : \left|\int \phi\, d(\beta - \alpha)\right| < r\right\}, \tag{2.2.2}$$

where $\alpha \in \mathbf{M}(\Sigma)$, $\phi \in C_b(\Sigma; \mathbb{R})$, and $r > 0$. As is well known (cf. Lemma 3.2.2 below), the LÉVY metric on $\mathbf{M}_1(\Sigma)$ is a complete separable metric, which is consistent with the restriction of this topology to $\mathbf{M}_1(\Sigma)$.

Throughout the rest of this section we will be assuming, without further comment, that we are in the situation described in (**C**). In this connection, we will be using X^* to denote the (real) topological dual of X; and, for $\mu \in \mathbf{M}_1(E)$, we will define the **logarithmic moment generating function of** μ to be the map $\lambda \in X^* \longmapsto \Lambda_\mu(\lambda) \in [-\infty, \infty]$ given by

$$\Lambda_\mu(\lambda) = \log\left(\int_E \exp\left[{}_{X^*}\langle \lambda, x\rangle_X\right] \mu(dx)\right). \tag{2.2.3}$$

As we saw in Sections 1.2 and 1.3, when $\{\mu_\epsilon : \epsilon > 0\}$ is a family of measures on a separable BANACH space, the logarithmic moment generating functions of the μ_ϵ 's can play an important role in the analysis of the large deviations of $\{\mu_\epsilon : \epsilon > 0\}$. It should therefore come as no surprise that the same is true even when we are working with the more general situation described by (**C**). The reason for this is partially explained by the next result.

2.2.4 Theorem. *Let* $\{\mu_\epsilon : \epsilon > 0\} \subseteq \mathbf{M}_1(E)$ *and assume that*

$$\Lambda(\lambda) \equiv \lim_{\epsilon \to 0} \epsilon \Lambda_{\mu_\epsilon}(\lambda/\epsilon) \in [-\infty, \infty] \tag{2.2.5}$$

exists for every $\lambda \in X^*$. *Then* Λ *is a convex function on* X^*. *Moreover, if the* LEGENDRE *transform* $q \in E \longmapsto \Lambda^*(q)$ *of* Λ *is defined by*

$$\Lambda^*(q) = \sup\left\{{}_{X^*}\langle \lambda, q\rangle_X - \Lambda(\lambda) : \lambda \in X^*\right\}, \tag{2.2.6}$$

then Λ^* *is a non-negative, lower semi-continuous, convex function; and, for any* $F \subset\subset E$,

$$\varlimsup_{\epsilon \to 0} \epsilon \log\left(\mu_\epsilon(F)\right) \le -\inf_F \Lambda^*. \tag{2.2.7}$$

Finally, if in addition, $\{\mu_\epsilon : \epsilon > 0\}$ *is exponentially tight, then* (2.2.7) *continues to hold for all closed subsets* F *of* E.

PROOF: The convexity of Λ follows from that of the Λ_{μ_ϵ} 's, which in turn is a consequence of HÖLDER's inequality. To see that $\Lambda^*(q) \ge 0$ for every $q \in E$, simply note that $\Lambda(0) = 0$. Also, because Λ^* is the point-wise supremum over continuous affine functions on E, it is lower semi-continuous and convex.

The proof of (2.2.7) for compact F is little more than a re-run of the argument used to derive (1.3.16). (Note that when $\mu_\epsilon = \mathcal{W}_\epsilon$ one has that $\epsilon\Lambda_{\mu_\epsilon}(\lambda/\epsilon) = \Lambda_{\mathcal{W}}(\lambda)$ for all $\epsilon > 0$.) Namely, let $p \in E$ and $\delta \in (0, 1]$ be given and choose $\lambda \in X^*$ so that

$${}_{X^*}\langle \lambda, p\rangle_X - \Lambda(\lambda) \ge \begin{cases} 1 + \frac{1}{\delta} & \text{if } \Lambda^*(p) = \infty \\ \Lambda^*(p) - \frac{\delta}{2} & \text{if } \Lambda^*(p) < \infty. \end{cases}$$

Next choose $r > 0$ so that ${}_{X^*}\langle \lambda, p - q \rangle_X \leq \delta/2$ for $q \in \overline{B}(p, r)$. Since

$$\epsilon \log \left[\mu_\epsilon\big(\overline{B}(p,r)\big)\right] \leq -{}_{X^*}\langle \lambda, p \rangle_X + \epsilon \Lambda_{\mu_\epsilon}(\lambda/\epsilon) + \frac{\delta}{2},$$

we see that

$$\mu_\epsilon\big(\overline{B}(p,r)\big) \leq \begin{cases} \exp\left[-\frac{1}{\epsilon\delta}\right] & \text{if } \Lambda^*(p) = \infty \\ \exp\left[-\frac{\Lambda^*(p)-\delta}{\epsilon}\right] & \text{if } \Lambda^*(p) < \infty \end{cases} \tag{2.2.8}$$

for all sufficiently small ϵ and r. Once one has (2.2.8), (2.2.7) for compact F follows from the last part of (**iv**) in Exercise 2.1.14. Finally, the extension to all closed F when $\{\mu_\epsilon : \epsilon > 0\}$ is exponentially tight is precisely the same as the last part of the proof of Lemma 2.1.5. ∎

Although the preceding indicates that, when Λ exists, its LEGENDRE transform Λ^* is a good candidate for the rate function governing the large deviations of $\{\mu_\epsilon : \epsilon > 0\}$, we know that, in general, Λ^* will not be the correct rate function. Indeed, from Lemma 2.1.4 we know that when the μ_ϵ's come from pushing measures ν_ϵ forward under a continuous mapping f and if J governs the large deviations of $\{\nu_\epsilon :> 0\}$, then the large deviations of $\{\mu_\epsilon : \epsilon > 0\}$ will be governed by the rate function I given by $I(p) = \inf\{J(q) : p = f(q)\}$. Since it is extremely unlikely that such an I will be convex even if J is, we see that convexity of I will be more the exception than the rule.

With the preceding in mind and assuming that $\{\mu_\epsilon : \epsilon > 0\}$ satisfies the full large deviation principle with some rate function I, one might ask if convexity is the only obstruction to the identification of I with Λ^*. As we are about to see, the answer to this question is, apart from minor technicalities, "yes." There are two steps in the proof. The first one is the easy application of Theorem 2.1.10 alluded to above.

2.2.9 Lemma. *Let $\{\mu_\epsilon : \epsilon > 0\} \subseteq \mathbf{M}_1(E)$ satisfy the condition that*

$$\sup_{0<\epsilon\leq 1} \left(\int_E \exp\left[\frac{1}{\epsilon}\, {}_{X^*}\langle \lambda, x \rangle_X\right] \mu_\epsilon(dx) \right)^\epsilon < \infty, \quad \lambda \in X^*. \tag{2.2.10}$$

If $\{\mu_\epsilon : \epsilon > 0\}$ satisfies the full large deviation principle with the good rate function I, then the limit $\Lambda(\lambda)$ in (2.2.5) exists for every $\lambda \in X^*$ and satisfies

$$\Lambda(\lambda) = \sup\left\{ {}_{X^*}\langle \lambda, q \rangle_X - I(q) : q \in E \right\}, \quad \lambda \in X^*. \tag{2.2.11}$$

PROOF: Note that (2.2.10) guarantees that

$$\sup_{0<\epsilon\leq 1}\left(\int_E \exp\left[\frac{2}{\epsilon}\, {}_{X^*}\langle\lambda, x\rangle_X\right]\mu_\epsilon(dx)\right)^\epsilon < \infty$$

for each $\lambda \in X^*$. Hence, we can apply Theorem 2.1.10 to the function

$$q \longmapsto {}_{X^*}\langle\lambda, q\rangle_X,$$

and thereby conclude not only that $\Lambda(\lambda)$ exists but also that (2.2.11) holds. ∎

2.2.12 Remark.

Let everything be as in Lemma 2.2.9 and define

$$\tilde{I}(x) = \begin{cases} \infty & \text{if } x \in X \setminus E \\ I(x) & \text{if } x \in E. \end{cases} \tag{2.2.13}$$

Obviously, (2.2.11) is then equivalent to

$$\Lambda(\lambda) = \sup\left\{{}_{X^*}\langle\lambda, x\rangle_X - \tilde{I}(x) : x \in X\right\}, \quad \lambda \in X^*. \tag{2.2.14}$$

Moreover, $\tilde{I}$ is always lower semi-continuous on X. Finally, $\tilde{I}$ is convex on X if I is convex on E.

The second step in our program is contained in the following theorem about one of the basic properties of the LEGENDRE transform. If one looks carefully at the proof, one realizes that this property is an analytic statement of the geometric fact that at each point on the graph of a convex function there is a tangent line which never goes above the graph.

2.2.15 Theorem. *Let $f : X \longrightarrow (-\infty, \infty]$ be a lower semi-continuous, convex function and define $g : X^* \longrightarrow (-\infty, \infty]$ by*

$$g(\lambda) = \sup\{{}_{X^*}\langle\lambda, x\rangle_X - f(x) : x \in X\}.$$

If f is not identically equal to ∞, then

$$f(x) = \sup\{{}_{X^*}\langle\lambda, x\rangle_X - g(\lambda) : \lambda \in X^*\}, \quad x \in X. \tag{2.2.16}$$

PROOF: The first step in the proof is to develop the geometric picture alluded to above. To this end, define

$$\mathcal{E}(f) = \{(x, \alpha) \in X \times \mathbb{R} : \alpha \geq f(x)\}$$

and

$$\mathcal{E}^*(f) = \{(\lambda, \beta) \in X^* \times \mathbb{R} : f(x) \geq {}_{X^*}\langle \lambda, x \rangle_X - \beta \text{ for every } x \in X\}.$$

It is then an easy matter to check from our assumptions that $\mathcal{E}(f)$ is a non-empty, closed, convex subset of $X \times \mathbb{R}$. Indeed, the closedness and convexity of $\mathcal{E}(f)$ come from the lower semi-continuity and convexity of f; and it is clear that $(x_0, f(x_0)) \in \mathcal{E}(f)$, where x_0 is any element of X for which $f(x_0) < \infty$. On the other hand, although $\mathcal{E}^*(f)$ is obviously closed and convex, it is less obvious that it is non-empty. To see that $\mathcal{E}^*(f) \neq \emptyset$, choose $x_0 \in E$ as above and apply the HAHN–BANACH Theorem to find a $(\mu, \rho, \gamma) \in X^* \times \mathbb{R} \times \mathbb{R}$ with the properties that the closed affine half space

$$H(\mu, \rho, \gamma) \equiv \{(x, \xi) \in X \times \mathbb{R} : {}_{X^*}\langle \mu, x \rangle_X - \rho\xi \leq \gamma\} \tag{2.2.17}$$

contains the set $\mathcal{E}(f)$ but not the point $(x_0, f(x_0) - 1)$. Then, since

$${}_{X^*}\langle \mu, x_0 \rangle_X - \rho\xi \leq \gamma \text{ for } \xi \geq f(x_0)$$

while

$${}_{X^*}\langle \mu, x_0 \rangle_X - \rho(f(x_0) - 1) > \gamma,$$

we see that $\rho > 0$, and therefore that

$$(\lambda_0, \beta_0) \equiv \left(\frac{\mu}{\rho}, \frac{\gamma}{\rho}\right) \in \mathcal{E}^*(f). \tag{2.2.18}$$

Next, noting that $\beta \geq g(\lambda)$ for every $(\lambda, \beta) \in \mathcal{E}^*(f)$ and

$$(\lambda, g(\lambda)) \in \mathcal{E}^*(f) \text{ for any } \lambda \in X^* \text{ with } g(\lambda) < \infty,$$

one sees that

$$g(\lambda) = \inf\{\beta : (\lambda, \beta) \in \mathcal{E}^*(f)\},$$

and therefore that (2.2.16) is equivalent to

$$f(x) = \sup\{{}_{X^*}\langle \lambda, x \rangle_X - \beta : (\lambda, \beta) \in \mathcal{E}^*(f)\}, \quad x \in X. \tag{2.2.19}$$

Since it is clear that $f(x) \geq {}_{X^*}\langle \lambda, x \rangle_X - \beta$ for any $x \in X$ and $(\lambda, \beta) \in \mathcal{E}^*(f)$, we will have proved (2.2.19) as soon as we show that, for each $(x, \alpha) \notin \mathcal{E}(f)$, there is a $(\lambda, \beta) \in \mathcal{E}^*(f)$ such that

$${}_{X^*}\langle \lambda, x \rangle_X - \beta > \alpha. \tag{2.2.20}$$

In order to prove the existence of the pair $(\lambda, \beta) \in \mathcal{E}^*(f)$ in (2.2.20), suppose that $x \in X$ and that $\alpha < f(x)$ are given. Then, since $(x, \alpha) \notin \mathcal{E}(f)$, the HAHN–BANACH Theorem again provides the existence of $(\mu, \rho, \gamma) \in X^* \times \mathbb{R} \times \mathbb{R}$ so that the $H(\mu, \rho, \gamma)$ in (2.2.17) contains $\mathcal{E}(f)$ and $(x, \alpha) \notin H(\mu, \rho, \gamma)$. In particular, since ${}_{X^*}\langle \mu, x_0 \rangle_X - \rho\xi \leq \gamma$ for $\xi \geq f(x_0)$, we know that $\rho \geq 0$. Hence, for every $\delta > 0$,

$$(\lambda_\delta, \beta_\delta) \equiv \left(\frac{\mu + \delta\lambda_0}{\rho + \delta}, \frac{\gamma + \delta\beta_0}{\rho + \delta} \right) \in \mathcal{E}^*(f),$$

where (λ_0, β_0) is the element of $\mathcal{E}^*(f)$ described in (2.2.18). (The introduction of $\delta > 0$ here is to take care of the case when the tangent hyperplane is vertical and therefore $\rho = 0$.) At the same time, for sufficiently small $\delta > 0$ one has that

$${}_{X^*}\langle \lambda_\delta, x \rangle_X - \alpha = \frac{1}{\rho + \delta} \Big({}_{X^*}\langle \mu + \delta\lambda_0, x \rangle_X - (\rho + \delta)\alpha \Big) > \frac{\gamma + \delta\beta_0}{\rho + \delta} = \beta_\delta.$$

Hence, (2.1.20) holds with $(\lambda, \beta) = (\lambda_\delta, \beta_\delta)$ for any sufficiently small $\delta > 0$. ∎

By combining Lemma 2.2.10 and Remark 2.2.12 with Theorem 2.2.15, we arrive at the following useful algorithm for identifying convex rate functions.

2.2.21 Theorem. *Assume that* $\{\mu_\epsilon : \epsilon > 0\} \subseteq \mathbf{M}_1(E)$ *satisfies* (2.2.10) *and that* I *is a convex, good rate function which governs the large deviations of* $\{\mu_\epsilon : \epsilon > 0\}$. *Then, not only does the limit* $\Lambda(\lambda)$ *in* (2.2.5) *exist for every* $\lambda \in X^*$, *but also* (2.2.11) *holds and*

$$(2.2.22) \qquad I(q) = \Lambda^*(q) \equiv \sup\big\{ {}_{X^*}\langle \lambda, q \rangle_X - \Lambda(\lambda) : \lambda \in X^* \big\}, \quad q \in X.$$

2.2.23 Exercise.

Suppose that $\{\mu_\epsilon : \epsilon > 0\}$ satisfies the weak large deviation principle with rate function I. Further, assume that the limit $\Lambda(\lambda)$ in (2.2.5) exists for each $\lambda \in X^*$.

(i) Show that $\Lambda^* \leq I$.

(ii) If one knows, in addition, that Λ and I satisfy (2.2.11), show that $\Lambda^* \geq f$ for every lower semi-continuous convex $f : E \longrightarrow (-\infty, \infty]$ which satisfies $f \leq I$. In other words, Λ^* is the **lower semi-continuous, convex minorant of** I.

III General Cramèr Theory

3.1: Preliminary Formulation

We want in this section to extend the CRAMÈR Theorem (cf. Theorem 1.2.6) to a more general setting. In order to describe the setting which we have in mind, it will be necessary to introduce the following embellished form of the hypothesis (**C**) made at the beginning of Section 2.2.

($\hat{\mathbf{C}}$) E and X are the same as they were in (**C**). In addition there is a metric ρ on E which is compatible with the topology on E induced by the topology on X and a measurable norm $\|\cdot\|$ on X (which need not be compatible with the topology on X) such that: (E,ρ) is Polish; $\|\cdot\|$ is bounded on ρ-bounded subsets of E;

$$\rho\big(\alpha p_1 + (1-\alpha)p_2,\, \alpha q_1 + (1-\alpha)q_2\big) \le \rho\big(p_1, q_1\big) \vee \rho\big(p_2, q_2\big)$$

for all $\alpha \in [0,1]$ and all elements $p_1, p_2, q_1, q_2 \in E$; and

$$\rho(p,q) \le \|q-p\| \quad \text{for all } p, q \in E.$$

Without further mention, we will be working in this section with the situation which we now describe. E, X, ρ, and $\|\cdot\|$ are as in ($\hat{\mathbf{C}}$), and $\Omega \equiv E^{\mathbb{Z}^+}$ is given the product topology. Note that, since E is a separable metric space, the BOREL field $\mathcal{B}_\Omega$ over Ω coincides with the product σ-algebra $\big(\mathcal{B}_E\big)^{\mathbb{Z}^+}$. Next, for $n \in \mathbb{Z}^+$, we use $X_n : \Omega \longrightarrow E$ to denote the nth coordinate map (i.e., $X_n(\omega) = \omega_n$). In view of the preceding remark about $\mathcal{B}_\Omega$, one sees that not only is each of the maps X_n measurable from

$(\Omega, \mathcal{B}_\Omega)$ into $(X, \mathcal{B}_X)$ but so are linear combinations of these maps. Given $0 \le m \le n$, we will use $\mathbf{S}_n^m$ to denote $\sum_{\ell=m+1}^n X_\ell$ ($\equiv 0$ when $m = n$) and $\overline{\mathbf{S}}_n^m$ to stand for $\frac{\mathbf{S}_n^m}{n-m}$; and when $m = 0$, we will usually drop the superscript. Finally, $\mu \in \mathbf{M}_1(E)$, $P \equiv \mu^{\mathbb{Z}^+}$ (again using the remark about $\mathcal{B}_\Omega$, one sees that $P \in \mathbf{M}_1(\Omega)$) and $\mu_n \in \mathbf{M}_1(E)$ is the distribution of $\overline{\mathbf{S}}_n$ under P.

Our purpose will be to study the large deviation theory for the family $\{\mu_n : n \ge 1\}$. Obviously, to whatever extent we succeed, we will have generalized CRAMÈR's Theorem. Our approach is an amalgam of ideas coming from D. RUELLE via O. LANFORD and the results obtained in Section 2.2. In particular, we will first use LANFORD's argument to show, in complete generality, that $\{\mu_n : n \ge 1\}$ satisfies a weak large deviation principle with a convex rate function. We will then do our best to replace the weak principle with the full large deviation principle and to identify the governing rate function.

The main reason for our needing to make the assumptions in $(\hat{\mathbf{C}})$ is that we will want to use the technical facts proved in the following lemma.

3.1.1 Lemma. *Let A be a non-empty, open convex subset of E. Then for any $K \subset\subset A$, the closed convex hull $\hat{K}$ of K is also a compact subset of A. In particular, if $\nu \in \mathbf{M}_1(E)$, then, for each $\delta \in (0,1)$, there is a convex $K \subset\subset A$ such that $\nu(K) \ge (1-\delta)\nu(A)$.*

PROOF: Suppose that $K \subset\subset A$. Given $0 < \delta < \rho(K, A^c)$, choose

$$p_1, \dots, p_M \in K \quad \text{so that} \quad K \subseteq \bigcup_1^M B(p_m, \delta)$$

and denote by $\Gamma(\delta)$ the set of points $\sum_1^M \alpha_m q_m$, where $\{\alpha_m\}_1^M \subseteq [0,1]$ with $\sum_1^M \alpha_m = 1$ and $q_m \in \overline{B}(p_m, \delta)$, $1 \le m \le M$. Clearly, $\Gamma(\delta) \subseteq A$ and is closed in E. Moreover, because $(\hat{\mathbf{C}})$ implies that ρ-balls are convex, it is easy to show that $\Gamma(\delta)$ is convex. Hence, $\hat{K} \subseteq \Gamma(\delta)$. This not only proves that $\hat{K} \subseteq A$, but it also gives us an easy way to see that $\hat{K}$ is compact. Indeed, again using $(\hat{\mathbf{C}})$, one sees that

$$\Gamma(\delta) \subseteq \left(\{p_1, \dots, p_M\}^\frown\right)^{(2\delta)},$$

where $\{p_1, \dots, p_M\}^\frown$ is the convex hull of $\{p_1, \dots, p_M\}$ and, as such, is compact. Since $\hat{K} \subseteq \Gamma(\delta)$ and δ can be taken arbitrarily small, it follows immediately that $\hat{K}$ is totally bounded and therefore, since it is closed in E, compact.

Given the first part, the second part of the lemma is an immediate consequence of the well-known ULAM's Lemma which says that, because E and therefore A are Polish spaces, there is a $K \subset\subset A$ such that $\nu(K) \geq (1-\delta)\nu(A)$; and obviously, the first part says that we may as well take K to be convex. ∎

Our first application of Lemma 3.1.1 occurs already in the second part of the next key result.

3.1.2 Lemma. *For each convex $C \in \mathcal{B}_E$, $n \in \mathbb{Z}^+ \longmapsto \mu_n(C)$ is super-multiplicative. In addition, if A is an open convex subset of E, then either $\mu_n(A) = 0$ for all $n \in \mathbb{Z}^+$ or there exists an $N \in \mathbb{Z}^+$ such that $\mu_n(A) > 0$ for all $n \geq N$.*

PROOF: To prove the first assertion, observe that, by convexity,

$$\{\omega : \overline{\mathbf{S}}_{m+n}(\omega) \in C\} \supseteq \{\omega : \overline{\mathbf{S}}_m(\omega) \in C\} \cap \{\omega : \overline{\mathbf{S}}^m_{m+n}(\omega) \in C\};$$

and therefore, by shift invariance and independence,

$$\mu_{m+n}(C) \geq \mu_m(C)\mu_n(C).$$

We next turn to the second assertion. Suppose that $\mu_m(A) > 0$ for some $m \in \mathbb{Z}^+$, and, using Lemma 3.1.1, choose a convex $K \subset\subset A$ so that $\mu_m(K) > 0$. Let $0 < 2\delta < \rho(K, A^c)$, take $G = \{q \in E : \|q - K\| < \delta\}$, and set $M = \sup\{\,\|q\| : q \in K\}$. Then, for $n = sm + r$, where $0 \leq r < m$,

$$\begin{aligned} \mu_n(A) &\geq P\left(\left\{\omega : \frac{1}{n}\mathbf{S}_{sm}(\omega) \in G \text{ and } \frac{1}{n}\left\|\mathbf{S}^{sm}_n(\omega)\right\| < \delta\right\}\right) \\ &\geq \mu_{sm}(K)P(\{\omega : \|\mathbf{S}_r(\omega)\| < n\delta\}) \end{aligned}$$

as long as $mM < n\delta$. Thus, if we choose N so that $mM < N\delta$ and

$$\min_{1 \leq r < m} P(\{\omega : \|\mathbf{S}_r(\omega)\| < N\delta\}) \geq \frac{1}{2},$$

then, since K is convex, we have that

$$\mu_n(A) \geq \frac{1}{2}\left(\mu_m(K)\right)^{[n/m]} > 0$$

for all $n \geq N$. ∎

Before we can use Lemma 3.1.2, we recall the following simple fact about sub-additive functions.

3.1.3 Lemma. *Let $f : \mathbb{Z}^+ \longrightarrow [0,\infty]$ be a sub-additive function and assume that there is an $N \in \mathbb{Z}^+$ such that $f(n) < \infty$ for all $n \geq N$. Then*

$$\lim_{n\to\infty} \frac{f(n)}{n} = \inf_{n\geq N} \frac{f(n)}{n} \in [0,\infty).$$

PROOF: For $m \geq N$, set $M_m = \max\{f(n) : m \leq n \leq 2m\}$. For $n \geq m \geq N$,

$$\frac{f(n)}{n} \leq \frac{(s-1)f(m)}{n} + \frac{f(r)}{n} \leq \frac{(s-1)f(m)}{n} + \frac{M_m}{n},$$

where $s = [n/m]$ and $r = n - ms$. Hence,

$$\overline{\lim_{n\to\infty}} \frac{f(n)}{n} \leq \frac{f(m)}{m}, \qquad m \geq N. \quad \blacksquare$$

By combining Lemma 3.1.2 with Lemma 3.1.3, we know that if $\mathcal{C}^\circ$ denotes the collection of all non-empty, convex open sets A in E, then

$$\mathcal{L}(A) = \mathcal{L}_\mu(A) \equiv -\lim_{n\to\infty} \frac{1}{n} \log\left(\mu_n(A)\right) \in [0,\infty] \tag{3.1.4}$$

exists for every $A \in \mathcal{C}^\circ$.

Noting that if I is the rate function governing the large deviations of $\{\mu_n : n \geq 1\}$, then (cf. the proof of Lemma 2.1.1)

$$I(q) = I_\mu(q) \equiv \lim_{r\searrow 0} \mathcal{L}_\mu\big(B(q,r)\big) = \sup\{\mathcal{L}_\mu(A) : q \in A \in \mathcal{C}^\circ\}, \tag{3.1.5}$$

we see that there is no alternative to our adopting (3.1.5) as the definition of I. Of course, we still have to check that this I does indeed give rise to a large deviation principle.

3.1.6 Theorem. *The function I_μ in (3.1.5) is a convex rate function on E and $\{\mu_n : n \geq 1\}$ satisfies the weak large deviation principle with rate function I_μ. Furthermore, if G is a finite union of elements from $\mathcal{C}^\circ$, then*

$$\lim_{n\to\infty} \frac{1}{n} \log\left(\mu_n(G)\right) = -\inf_G I_\mu.$$

PROOF: The lower semi-continuity of I_μ is an immediate consequence of its definition. To prove that I_μ is convex, let $q_1, q_2 \in E$ be given, and set $q = \frac{q_1+q_2}{2}$. Given an $A \in \mathcal{C}^\circ$ containing q, choose $A_i \in \mathcal{C}^\circ$ so that $q_i \in A_i$

and $A \supseteq \frac{A_1+A_2}{2}$. Then

$$\begin{aligned}
\mathcal{L}(A) &= -\lim_{n\to\infty} \frac{1}{2n} \log\left(\mu_{2n}(A)\right) \\
&\leq -\underline{\lim}_{n\to\infty} \frac{1}{2n} \log\Big(P\big(\{\omega : \overline{\mathbf{S}}_n(\omega) \in A_1 \text{ and } \overline{\mathbf{S}}^n_{2n}(\omega) \in A_2\}\big)\Big) \\
&= \frac{1}{2}\left(-\lim_{n\to\infty} \frac{1}{n} \log\left[\mu_n(A_1)\right] - \lim_{n\to\infty} \frac{1}{n} \log\left[\mu_n(A_2)\right]\right) \\
&\leq \frac{I_\mu(q_1) + I_\mu(q_2)}{2};
\end{aligned}$$

and from this we conclude that $I_\mu(q) \leq \big(I_\mu(q_1) + I_\mu(q_2)\big)/2$. Because we already know that I_μ is lower semi-continuous, the convexity of I_μ is now proved by a familiar iteration argument followed by a passage to the limit.

The fact that $\underline{\lim}_{n\to\infty} \frac{1}{n} \log\left(\mu_n(G)\right) \geq -\inf_G I_\mu$ for arbitrary open G in E is built into the definition of I_μ. Next, suppose that $K \subset\subset E$ and let $\ell < \inf_K I_\mu$. Then, there is a finite cover $\{A_1, \ldots, A_M\} \subseteq \mathcal{C}^\circ$ of K with the property that $\mathcal{L}(A_m) > \ell$ for each $1 \leq m \leq M$. Hence,

$$\overline{\lim}_{n\to\infty} \frac{1}{n} \log\left(\mu_n(K)\right) \leq \overline{\lim}_{n\to\infty} \frac{1}{n} \log\left(\max_{1\leq m\leq M} \mu_n(A_m)\right) \leq -\ell;$$

and so we have proved that $\overline{\lim}_{n\to\infty} \frac{1}{n} \log\left(\mu_n(K)\right) \leq -\inf_K I_\mu$ and therefore that the weak large deviation principle holds.

To complete the proof, suppose that $G = \bigcup_1^M A_m$, where $\{A_m\}_1^M \subseteq \mathcal{C}^\circ$. Then an easy argument shows that

$$-\lim_{n\to\infty} \frac{1}{n} \log\left(\mu_n(G)\right) = \min_{1\leq m\leq M} \mathcal{L}(A_m).$$

Hence, it suffices for us to check that $\mathcal{L}(A) = \inf_A I_\mu$ for every $A \in \mathcal{C}^\circ$; and since we already know that $\mathcal{L}(A) \leq \inf_A I_\mu$, this comes down to checking that $\mathcal{L}(A) \geq \inf_A I_\mu$ when $\mathcal{L}(A) < \infty$. To this end, let $\delta \in (0,1)$ be given and choose N so that $-\frac{1}{n} \log\left(\mu_n(A)\right) \leq \mathcal{L}(A) + \delta$ for $n \geq N$. Next, we use Lemma 3.1.1 to find a convex $K \subset\subset A$ such that

$$\frac{1}{N} \log\left(\mu_N(A)\right) - \frac{1}{N} \log\left(\mu_N(K)\right) < \delta.$$

Then, by sub-additivity and the preceding paragraph,

$$\begin{aligned}
\inf_A I_\mu \leq \inf_K I_\mu &\leq \underline{\lim}_{n\to\infty} -\frac{1}{n} \log\left(\mu_n(K)\right) \\
&\leq \underline{\lim}_{n\to\infty} -\frac{1}{nN} \log\left(\mu_{nN}(K)\right) \leq \mathcal{L}(A) + 2\delta. \quad \blacksquare
\end{aligned}$$

3.1.7 Corollary. *If for each $L \geq 0$ there is a $K_L \subset\subset E$ such that*

$$\overline{\lim_{n\to\infty}} \frac{1}{n} \log \left(\mu_n\left(K_L^c\right)\right) \leq -L, \tag{3.1.8}$$

then I_μ is a good rate function and $\{\mu_n : n \geq 1\}$ satisfies the full large deviation principle with rate function I_μ. If, in addition,

$$\Lambda_\mu(\lambda) \equiv \log\left(\int_E \exp\left[{}_{X^*}\langle \lambda, q\rangle_X\right] \mu(dq)\right) < \infty$$

for every $\lambda \in X^*$, then

$$\Lambda_\mu(\lambda) = \sup\left\{{}_{X^*}\langle \lambda, q\rangle_X - I_\mu(q) : q \in E\right\}, \quad \lambda \in X^*, \tag{3.1.9}$$

and

$$I_\mu(q) = \sup\left\{{}_{X^*}\langle \lambda, q\rangle_X - \Lambda_\mu(\lambda) : \lambda \in X^*\right\}, \quad q \in E. \tag{3.1.10}$$

PROOF: The first assertion is no more than the conjunction of Theorem 3.1.6 and Lemma 2.1.5. The rest is an immediate consequence of the first part together with Theorem 2.2.21. ∎

3.1.11 Exercise.

In the case when $E = X$ is finite dimensional and $\rho(p, q) = \|q - p\|$, show that the whole of Corollary 3.1.7 applies as soon as $\Lambda_\mu(\lambda) < \infty$ for every $\lambda \in X^*$. This is, of course, the CRAMÈR Theorem in the general finite dimensional setting.

3.2 Sanov's Theorem

In this section we will specialize to the situation described in the second example of Remark 2.2.1. That is, $E = \mathbf{M}_1(\Sigma)$ and $X = \mathbf{M}(\Sigma)$, where Σ is a Polish space and $\mathbf{M}(\Sigma)$ is given the topology which is generated by the sets in (2.2.2). Clearly, the topology inherited by $\mathbf{M}_1(\Sigma)$ as a subset of $\mathbf{M}(\Sigma)$ is the **weak topology** (i.e., the topology corresponding to convergence against bounded continuous test functions). In order to show that $\mathbf{M}_1(\Sigma)$ and $\mathbf{M}(\Sigma)$ satisfy the hypothesis ($\hat{\mathbf{C}}$) at the beginning of Section 3.1, we must produce the metric ρ on $\mathbf{M}_1(\Sigma)$ and the norm $\|\cdot\|$ on $\mathbf{M}(\Sigma)$. The latter is easy; namely, we take $\|\alpha\|$ to be the total variation of $\|\alpha\|_{\text{var}}$ of $\alpha \in \mathbf{M}(\Sigma)$. Since

$$\|\alpha\|_{\text{var}} = \sup\left\{\int \phi \, d\alpha : \phi \in C_{\text{b}}(\Sigma; \mathbb{R}) \text{ with } \|\phi\|_{C_{\text{b}}} \le 1\right\},$$

we see that $\|\cdot\|_{\text{var}}$ is lower semi-continuous and therefore certainly measurable on $\mathbf{M}(\Sigma)$; and clearly $\|\cdot\|_{\text{var}}$ is bounded on $\mathbf{M}_1(\Sigma)$. We now turn to the metric for $\mathbf{M}_1(\Sigma)$. Following LÉVY and PROHOROV, define the **Lévy metric**

$$\begin{aligned}\rho(\alpha, \nu) = \inf\{\delta > 0 : \ & \alpha(F) \le \nu\big(F^{(\delta)}\big) + \delta \\ & \text{and } \nu(F) \le \alpha\big(F^{(\delta)}\big) + \delta \text{ for every closed } F \text{ in } \Sigma\}\end{aligned} \tag{3.2.1}$$

for $\alpha, \nu \in \mathbf{M}_1(\Sigma)$, where $F^{(\delta)}$ is defined relative to a complete metric on Σ. An easy argument shows that ρ is a metric and that it satisfies the convexity property required in ($\hat{\mathbf{C}}$). Since it is clear that $\rho(\alpha, \nu) \le \|\nu - \alpha\|$, all that remains is to show that ρ is compatible with the weak topology and that $(\mathbf{M}_1(\Sigma), \rho)$ is Polish. Before proving this, we will need to recall some elementary properties of the weak topology.

(i) The weak topology is second countable.

(ii) $\alpha_n \Longrightarrow \nu$ if and only if $\overline{\lim}_{n\to\infty} \alpha_n(F) \le \nu(F)$ for every closed F in Σ.

(iii) $\Gamma \subseteq \mathbf{M}_1(\Sigma)$ is relatively compact if and only if for each $\delta > 0$ there is a $K \subset\subset \Sigma$ such that $\alpha(K) \ge 1 - \delta$ for every $\alpha \in \Gamma$. (Such a subset $\Gamma$is said to be **tight**.)

(iv) If $\mathcal{F} \subseteq C_{\text{b}}(\Sigma; \mathbb{R})$ is uniformly bounded on all of Σ and is equi-continuous on each compact subset of Σ, then $\alpha_n \Longrightarrow \nu$ implies that

$$\sup\left\{\left|\int \phi \, d\alpha_n - \int \phi \, d\nu\right| : \phi \in \mathcal{F}\right\} \longrightarrow 0 \text{ as } n \longrightarrow \infty.$$

All of these facts are well-known, and their proofs can be found in any standard text in which the modern theory of weak convergence is discussed. We will now use them to check that the LÉVY metric possesses the properties which we want.

3.2.2 Lemma. (LÉVY & PROHOROV) *The metric* ρ *in* (3.2.1) *is compatible with the weak topology on* $\mathbf{M}_1(\Sigma)$, *and* $(\mathbf{M}_1(\Sigma), \rho)$ *is Polish.*

PROOF: In view of property (**ii**) above, it is obvious that $\alpha_n \Longrightarrow \nu$ if $\rho(\alpha_n, \nu) \longrightarrow 0$. To prove the opposite implication, let $\delta > 0$ be given and for each closed F in Σ define

$$\psi_F(\sigma) = \frac{\operatorname{dist}\left(\sigma, \left(F^{(\delta)}\right)^{\mathrm{c}}\right)}{\operatorname{dist}(\sigma, F) + \operatorname{dist}\left(\sigma, \left(F^{(\delta)}\right)^{\mathrm{c}}\right)}, \quad \sigma \in \Sigma,$$

where "dist" is measured with the same metric on Σ as the one used in the definition of $F^{(\delta)}$. It is then an easy matter to check that $\{\psi_F : F \text{ closed in } \Sigma\}$ is uniformly bounded and equi-continuous on Σ. Hence, by property (**iv**), if $\alpha_n \Longrightarrow \nu$, then $\int \psi_F \, d\alpha_n \longrightarrow \int \psi_F \, d\nu$ at a rate which is independent of F; and, since $\chi_F \le \psi_F \le \chi_{F^{(\delta)}}$, we conclude from this that $\rho(\alpha_n, \nu) \longrightarrow 0$ if $\alpha_n \Longrightarrow \nu$. (We use the notation χ_Γ to denote the indicator (or characteristic) function of a set Γ.) We have therefore proved that ρ is compatible with the weak topology on $\mathbf{M}_1(\Sigma)$.

To prove that ρ is a complete metric on $\mathbf{M}_1(\Sigma)$, suppose that

$$\sup_{n \ge m} \rho(\alpha_n, \alpha_m) \longrightarrow 0 \quad \text{as } m \longrightarrow \infty.$$

We must show that $\{\alpha_n\}_1^\infty$ is relatively compact. To this end, let $\delta > 0$ be given and, for $\ell \in \mathbb{Z}^+$, choose $m \in \mathbb{Z}^+$ so that $\sup_{n \ge m} \rho(\alpha_n, \alpha_m) \le \delta/2^\ell$, and then (using property (**iii**)) choose $K_\ell \subset\subset \Sigma$ so that $\alpha_k(K_\ell) \ge 1 - \delta/2^\ell$ for $k \le m$. Then

$$\alpha_n\left(K_\ell^{(\delta/2^\ell)}\right) \ge 1 - \delta/2^{\ell-1}$$

for all $n \in \mathbb{Z}^+$. Finally, set

$$K = \bigcap_{\ell=1}^{\infty} \overline{K_\ell^{(\delta/2^\ell)}},$$

note that K is closed and totally bounded with respect to a complete metric and is therefore compact, and check that $\alpha_n\left(K^{\mathrm{c}}\right) \le 2\delta$ for all $n \in \mathbb{Z}^+$. Thus, by property (**iii**), $\{\alpha_n\}_1^\infty$ is indeed relatively compact. ∎

Before getting down to the main business of this section, there is one more general fact about the space $\mathbf{M}(\Sigma)$ which it will be useful to have at our disposal. Namely, we want a good representation of $\mathbf{M}(\Sigma)^*$.

3.2.3 Lemma. *The duality relation*

$$(\phi, \alpha) \in C_{\mathrm{b}}(\Sigma; \mathbb{R}) \times \mathbf{M}(\Sigma) \longmapsto \int_\Sigma \phi \, d\alpha \tag{3.2.4}$$

determines a representation of $\mathbf{M}(\Sigma)^*$ *as* $C_{\mathrm{b}}(\Sigma;\mathbb{R})$.

PROOF: Clearly, for each $\phi \in C_{\mathrm{b}}(\Sigma;\mathbb{R})$, $\alpha \in \mathbf{M}(\Sigma) \longmapsto \int_\Sigma \phi\, d\alpha$ determines a unique element of $\mathbf{M}(\Sigma)^*$. Thus, all that we have to show is that every element of $\mathbf{M}(\Sigma)^*$ arises in this way. Let $\lambda \in \mathbf{M}(\Sigma)^*$ be given and define $\phi(\sigma) = \lambda(\delta_\sigma)$, $\sigma \in \Sigma$. Clearly, ϕ is continuous. Moreover, because of the way in which the topology on $\mathbf{M}(\Sigma)$ is defined, we can find a finite set $\{\psi_m\}_1^M \in C_{\mathrm{b}}(\Sigma;\mathbb{R})$ such that

$$\bigl|\lambda(\alpha)\bigr| \le \sum_1^M \left|\int_\Sigma \psi_m\, d\alpha\right|, \quad \alpha \in \mathbf{M}(\Sigma);$$

and from this it is clear that ϕ is bounded. Finally, it is obvious that $\lambda(\alpha) = \int_\Sigma \phi\, d\alpha$ if α is a linear combination of point masses; and, because such α's are dense in $\mathbf{M}(\Sigma)$, it follows that this equation holds for all $\alpha \in \mathbf{M}(\Sigma)$. ∎

Returning to the problem of large deviations, let $Q \in \mathbf{M}_1\bigl(\mathbf{M}_1(\Sigma)\bigr)$ be given and define $Q_n \in \mathbf{M}_1\bigl(\mathbf{M}_1(\Sigma)\bigr)$ to be the distribution of

$$\boldsymbol{\nu} = \bigl(\nu_1, \dots, \nu_n\bigr) \in \mathbf{M}_1(\Sigma)^n \longmapsto \frac{1}{n}\sum_{k=1}^n \nu_k \in \mathbf{M}_1(\Sigma)$$

under $Q^n \in \mathbf{M}_1\bigl(\mathbf{M}_1(\Sigma)^n\bigr)$. By the Weak Law of Large Numbers combined with the second countability of the weak topology on $\mathbf{M}_1(\Sigma)$, one can easily check that $Q_n \Longrightarrow \delta_{\mu_Q}$, where $\mu_Q \in \mathbf{M}_1(\Sigma)$ is defined by

$$\mu_Q(\Gamma) = \int_{\mathbf{M}_1(\Sigma)} \nu(\Gamma)\, Q(d\nu), \quad \Gamma \in \mathcal{B}_\Sigma.$$

Thus, it is reasonable to inquire about the large deviations of $\{Q_n : n \ge 1\}$. In fact, by the results which we proved in Section 3.1, we will know that the large deviations $\{Q_n : n \ge 1\}$ are governed by the rate function

$$\text{(3.2.5)} \qquad I_Q(\nu) = \Lambda_Q^*(\nu) \equiv \sup\left\{\int_\Sigma \phi\, d\nu - \Lambda_Q(\phi) : \phi \in C_{\mathrm{b}}(\Sigma;\mathbb{R})\right\}$$

for $\nu \in \mathbf{M}_1(\Sigma)$, where

$$\text{(3.2.6)} \quad \Lambda_Q(\phi) = \log\left(\int_{\mathbf{M}_1(\Sigma)} \exp\left[\int_\Sigma \phi\, d\nu\right] Q(d\nu)\right), \quad \phi \in C_{\mathrm{b}}(\Sigma;\mathbb{R}),$$

and that I_Q is good, as soon as we show that $\{Q_n : n \ge 1\}$ is exponentially tight. In order to do so, we employ the following remarkably useful general observation which will serve us well not only here but also later on.

3.2.7 Lemma. *Let $\mu \in \mathbf{M}_1(\Sigma)$ be a fixed and suppose that $\{V_m\}_{m=1}^{\infty}$ is a bounded sequence of non-negative, measurable functions on Σ which tends to 0 in μ-measure as $m \longrightarrow \infty$. Then, for each $M \in [1,\infty)$ and $\beta \in [1,\infty)$ there is a subsequence $\{V_{m_\ell}\}_{\ell=1}^{\infty}$ with the property that*

$$(3.2.8) \qquad R_\epsilon\left(\left\{\nu \in \mathbf{M}_1(\Sigma) : \sup_{\ell \geq L} \ell \int_\Sigma V_{m_\ell}\, d\nu \geq 1\right\}\right) \leq \exp[-L/\epsilon]$$

for $0 < \epsilon \leq 1$ and $L \in [1,\infty)$, whenever $\{R_\epsilon : \epsilon > 0\} \subseteq \mathbf{M}_1(\mathbf{M}_1(\Sigma))$ is a family which satisfies

$$(3.2.9) \qquad \sup_{0<\epsilon\leq 1}\left(\int_{\mathbf{M}_1(\Sigma)} \exp\left[\frac{1}{\epsilon}\int_\Sigma V\, d\nu\right] R_\epsilon(d\nu)\right)^\epsilon \leq M \int_\Sigma \exp[\beta V]\, d\mu$$

for every bounded measurable $V : \Sigma \longrightarrow [0,\infty]$. In particular, for each $L \in [1,\infty)$ there is a $C_L \subset\subset \mathbf{M}_1(\Sigma)$ such that

$$(3.2.10) \qquad R_\epsilon(C_L^{\mathrm{c}}) \leq \exp[-L/\epsilon], \quad 0 < \epsilon \leq 1,$$

for any $\{R_\epsilon : \epsilon > 0\}$ satisfying (3.2.9).

PROOF: Obviously, for any $\delta > 0$ and measurable $V : E \longrightarrow [0,\infty)$,

$$R_\epsilon\left(\left\{\nu \in \mathbf{M}_1(\Sigma) : \int_\Sigma V\, d\nu \geq \delta\right\}\right) \leq \left(Me^{-T\delta}\int_\Sigma \exp[\beta T V]\, d\mu\right)^{1/\epsilon}$$

for every $0 < \epsilon \leq 1$ and $T \in (0,\infty)$. Now let $\ell \in \mathbb{Z}^+$ be given, take $\delta = 1/\ell$, $T = \ell(\ell + \log(2M) + 1)$, and choose $m_\ell \in \mathbb{Z}^+$ so that

$$\int_\Sigma \exp[\beta T V_{m_\ell}]\, d\mu \leq 2.$$

One then has that

$$R_\epsilon\left(\left\{\nu \in \mathbf{M}_1(\Sigma) : \ell\int_\Sigma V_{m_\ell}\, d\nu \geq 1\right\}\right) \leq \exp[-(\ell+1)/\epsilon]$$

for $0 < \epsilon \leq 1$ and $\ell \in \mathbb{Z}^+$; and (3.2.8) is an immediate consequence of this.

To get (3.2.10), choose $\{K_m\}$ to be a sequence of compact subsets of Σ for which $\mu(K_m^{\mathrm{c}}) \longrightarrow 0$ as $m \longrightarrow \infty$. Next, apply the preceding (with $V_m = \chi_{K_m}$) to see that there is a subsequence $\{K_{m_\ell}\}_{\ell=1}^{\infty}$ for which (3.2.10) holds with

$$C_L \equiv \bigcap_{\ell \geq L}\left\{\nu : \nu(K_{m_\ell}) \geq 1 - \frac{1}{\ell}\right\}. \quad \blacksquare$$

To see how the exponential tightness of $\{Q_n : n \geq 1\}$ follows from Lemma 3.2.7, set $\mu_Q = \int_{\mathbf{M}_1(\Sigma)} \nu \, Q(d\nu)$ and note that

$$\int_{\mathbf{M}_1(\Sigma)} \exp\left[n \int_\Sigma V \, d\nu\right] Q_n(d\nu) = \int_{(\mathbf{M}_1(\Sigma))^n} \exp\left[\sum_{k=1}^n \int_\Sigma V \, d\nu_k\right] Q^n(d\boldsymbol{\nu})$$
$$= \left(\int_{\mathbf{M}_1(\Sigma)} \exp\left[\int_\Sigma V \, d\nu\right] Q(d\nu)\right)^n \leq \left(\int_\Sigma \exp[V] \, d\mu_Q\right)^n,$$

where we have used JENSEN's inequality at the final step. Obviously, exponential tightness for $\{Q_n : n \geq 1\}$ is now a trivial consequence of (3.2.10); and, as we mentioned just before Lemma 3.2.7, this is all that we needed in order to know that the I_Q in (3.2.5) is good and that it governs the large deviations of $\{Q_n : n \geq 1\}$.

A particularly interesting case of this general result is the one in which Q is the distribution $\tilde{\mu}$ of $\sigma \in \Sigma \longmapsto \delta_\sigma \in \mathbf{M}_1(\Sigma)$ under some $\mu \in \mathbf{M}_1(\Sigma)$. In this case, Q_n is the distribution $\tilde{\mu}_n$ of the **empirical distribution functional**

$$\boldsymbol{\sigma} = (\sigma_1, \ldots, \sigma_n) \in \Sigma^n \longmapsto \mathbf{L}_n(\boldsymbol{\sigma}) \equiv \frac{1}{n} \sum_{m=1}^n \delta_{\sigma_m} \tag{3.2.11}$$

under μ^n and the measure μ_Q introduced above coincides with μ. Specializing the preceding to this case, we see that

$$\Lambda_{\tilde{\mu}}(\phi) = \log\left[\int_\Sigma \exp[\phi] \, d\mu\right]$$

and therefore that the large deviations of $\{\tilde{\mu}_n : n \geq 1\}$ are governed by the good rate function

$$\begin{aligned} I_{\tilde{\mu}}(\nu) &= \Lambda^*_{\tilde{\mu}}(\nu) \\ &= \sup\left\{\int_\Sigma \phi \, d\nu - \log\left(\int_\Sigma \exp[\phi] \, d\mu\right) : \phi \in C_{\mathrm{b}}(\Sigma; \mathbb{R})\right\} \end{aligned} \tag{3.2.12}$$

for $\nu \in \mathbf{M}_1(\Sigma)$. However, before stating this as a theorem, we want to develop a more tractable expression for $I_{\tilde{\mu}}$.

3.2.13 Lemma. *For $\nu \in \mathbf{M}_1(\Sigma)$, define*

$$\mathbf{H}(\nu|\mu) = \begin{cases} \int_\Sigma f \log f \, d\mu & \text{if } \nu << \mu \text{ and } f = \frac{d\nu}{d\mu} \\ \infty & \text{otherwise.} \end{cases} \tag{3.2.14}$$

Then the $I_{\tilde{\mu}}$ in (3.2.12) is equal to $\mathbf{H}(\cdot\,|\mu)$.

PROOF: We first show that if $\nu << \mu$ and $\nu_\theta \equiv \theta\mu + (1-\theta)\nu$ for $\theta \in [0,1]$, then $\mathbf{H}(\nu|\mu) = \lim_{\theta\searrow 0} \mathbf{H}(\nu_\theta|\mu)$. To this end, set $f = \frac{d\nu}{d\mu}$ and $f_\theta = \theta + (1-\theta)f$. Since $x \in [0,\infty) \longmapsto x\log x$ is convex, JENSEN's inequality says that

$$\mathbf{H}(\nu_\theta|\mu) = \int_\Sigma f_\theta \log f_\theta \, d\mu \leq (1-\theta)\int_\Sigma f \log f \, d\mu = (1-\theta)\mathbf{H}(\nu|\mu).$$

At the same time, since $x \in [0,\infty) \longmapsto \log x$ is non-decreasing and concave, $\log f_\theta \geq \big(\log\theta\big) \vee \big((1-\theta)\log f\big)$; and therefore

$$\mathbf{H}(\nu_\theta|\mu) = \theta\int_\Sigma \log f_\theta \, d\mu + (1-\theta)\int_\Sigma f\log f_\theta \, d\mu \geq \theta\log\theta + (1-\theta)^2\mathbf{H}(\nu|\mu).$$

After combining these two, one clearly gets the asserted convergence.

We next show that if $\nu << \mu$, then $I_{\tilde{\mu}}(\nu) \leq \mathbf{H}(\nu|\mu)$. In view of the preceding and the obvious fact that $\nu \in \mathbf{M}_1(\Sigma) \longmapsto \Lambda^*_{\tilde{\mu}}(\nu)$ is lower semi-continuous, we may and will assume that $f = \frac{d\nu}{d\mu} \geq \theta$ for some $\theta \in (0,1)$. In particular, by JENSEN's inequality, we then have

$$\begin{aligned}\exp\left[\int_\Sigma \phi\,d\nu - \mathbf{H}(\nu|\mu)\right] &= \exp\left[\int_\Sigma (\phi - \log f)\,d\nu\right] \\ &\leq \int_\Sigma \frac{\exp[\phi]}{f}\,d\nu = \int_\Sigma \exp[\phi]\,d\mu;\end{aligned}$$

from which it is clear that $I_{\tilde{\mu}}(\nu) \leq \mathbf{H}(\nu|\mu)$.

As a consequence of the preceding, all that remains is to show that if $I_{\tilde{\mu}}(\nu) < \infty$, then $d\nu = f\,d\mu$ and

$$I_{\tilde{\mu}}(\nu) \geq \int_\Sigma f\log f\,d\mu. \tag{3.2.15}$$

Given ν with $I_{\tilde{\mu}}(\nu) < \infty$, one has

$$\int_\Sigma \phi\,d\nu - \log\left(\int_\Sigma \exp[\phi]\,d\mu\right) \leq I_{\tilde{\mu}}(\nu) < \infty \tag{3.2.16}$$

for every bounded continuous ϕ. Since the class of ϕ's for which (3.2.16) holds is closed under bounded point-wise convergence, (3.2.16) continues to be true for every bounded $\mathcal{B}_\Sigma$-measurable ϕ. In particular, we can now show that $\nu << \mu$. Indeed, suppose that $\Gamma \in \mathcal{B}_\Sigma$ with $\mu(\Gamma) = 0$. Then, by (3.2.16) with $\phi = r\,\chi_\Gamma$, $r\nu(\Gamma) \leq I_{\tilde{\mu}}(\nu)$, $r > 0$; and therefore $\nu(\Gamma) = 0$. Knowing that $\nu << \mu$, set $f = \frac{d\nu}{d\mu}$. If f is uniformly positive and uniformly bounded, then (3.2.15) is an immediate consequence of (3.2.16) with $\phi = \log f$. If f is uniformly positive but not necessarily uniformly bounded, set $f_n = f \wedge n$ and use (3.2.16) together with FATOU's Lemma to justify

$$\int_\Sigma f \log f \, d\mu = \int_\Sigma \log f \, d\nu$$
$$\leq \varliminf_{n\to\infty} \int_\Sigma \log f_n \, d\nu \leq I_{\tilde{\mu}}(\nu) + \varlimsup_{n\to\infty} \log\left(\int_\Sigma f \wedge n \, d\mu\right) = I_{\tilde{\mu}}(\nu).$$

Finally, to treat the general case, define ν_θ and $f_\theta = \theta + (1-\theta)f$ for $\theta \in [0,1]$ as in the first paragraph of this proof. By the preceding, $\int f_\theta \log f_\theta \, d\mu \leq I_{\tilde{\mu}}(\nu_\theta)$ as long as $\theta \in (0,1)$. Moreover, since $\theta \in [0,1] \longmapsto I_{\tilde{\mu}}(\nu_\theta)$ is bounded, lower semi-continuous, and convex on $[0,1]$, it is continuous there. In conjunction with the result obtained in the first paragraph, this now completes the proof. ∎

The quantity $\mathbf{H}(\nu|\mu)$ in (3.2.14) is called the **relative entropy of** ν **with respect to** μ. As we will see in the sequel, the relative entropy functional plays a central role in the theory of large deviations.

We have now proved the following theorem which, at least when $\Sigma = \mathbb{R}$, was originally derived by SANOV.

3.2.17 Theorem. (SANOV) *Let μ be a probability measure on the Polish space Σ and let $\tilde{\mu}_n \in \mathbf{M}_1(\mathbf{M}_1(\Sigma))$ be the distribution under μ^n of the function $\mathbf{L}_n$ in (3.2.11). Also, define $\mathbf{H}(\cdot\,|\mu)$ as in (3.2.14). Then $\mathbf{H}(\cdot\,|\mu)$ is a good, convex rate function on $\mathbf{M}_1(\Sigma)$ and $\{\tilde{\mu}_n : n \geq 1\}$ satisfies the full large deviation principle with rate function $\mathbf{H}(\cdot\,|\mu)$.*

Before dropping this topic, it seems appropriate to address a deficiency in the preceding result. Namely, the Weak Law of Large Numbers tells us that

$$\tilde{\mu}_n\left(\left\{\nu \in \mathbf{M}_1(\Sigma) : \left|\int_\Sigma \phi \, d\nu - \int_\Sigma \phi \, d\mu\right| \geq \epsilon\right\}\right) \longrightarrow 0, \quad \epsilon > 0,$$

for every bounded measurable $\phi : \Sigma \longrightarrow \mathbb{R}$; not just for bounded continuous ϕ's. Thus, $\{\tilde{\mu}_n\}_1^\infty$ actually tends to δ_μ in the strong topology on $\mathbf{M}_1(\Sigma)$ and not just the weak topology. It is therefore reasonable to ask whether one cannot also develop the corresponding large deviation theory relative to the strong topology. As we are about to see, not only is it possible to do so, but it is even a rather easy exercise to transform what we already have into a statement about the strong topology.

The **strong topology** (or **τ-topology**) on $\mathbf{M}_1(\Sigma)$ is the topology $\mathfrak{U}$ generated by the sets

$$U(\alpha;\phi;\epsilon) \equiv \left\{\beta \in \mathbf{M}_1(\Sigma) : \left|\int_\Sigma \phi\, d\beta - \int_\Sigma \phi\, d\alpha\right| < \epsilon\right\}, \quad \epsilon > 0,$$

as ϕ runs over the space $B(\Sigma;\mathbb{R})$ of bounded measurable functions on Σ into $\mathbb{R}$. Obviously, the strong topology is stronger than the weak one. In addition, it is clear that, for each $\alpha \in \mathbf{M}_1(\Sigma)$, the sets

$$U\big(\alpha;\, \Delta_1,\dots,\Delta_N;\, \epsilon\big) = \bigcap_{k=1}^{N} U\big(\alpha;\, \chi_{\Delta_k};\, \epsilon\big) \tag{3.2.18}$$

for $N \in \mathbb{Z}^+$, $\Delta_1,\dots,\Delta_N \in \mathcal{B}_\Sigma$, and $\epsilon > 0$ constitute a $\mathfrak{U}$-neighborhood basis at α. In particular, $\mathbf{M}_1(\Sigma)$ with the strong topology is usually not even first countable!

The key which will allow us to transform the SANOV Theorem to the strong topological setting is contained in the following, whose proof turns on another application of Lemma 3.2.7.

3.2.19 Lemma. *Let $\mu \in \mathbf{M}_1(\Sigma)$ and assume that*

$$\{R_\epsilon : \epsilon > 0\} \subseteq \mathbf{M}_1\big(\mathbf{M}_1(\Sigma)\big)$$

satisfies (3.2.9) for some β, $M \in [1,\infty)$ and all measurable $V : \Sigma \longrightarrow [0,\infty]$. Further, assume that $\{R_\epsilon : \epsilon > 0\}$ satisfies the weak large deviation principle with the rate function I. Then,

$$\mathbf{H}(\nu|\mu) \le \beta\big(I(\nu) + \log(2M)\big), \quad \nu \in \mathbf{M}_1(\Sigma). \tag{3.2.20}$$

Moreover, given $N \in \mathbb{Z}^+$ and $\Delta_1,\dots,\Delta_N \in \mathcal{B}_\Sigma$, define $f : \mathbf{M}_1(\Sigma) \longrightarrow \mathbb{R}^N$ by

$$f(\nu) = \Big(\nu\big(\Delta_1\big),\dots,\nu\big(\Delta_N\big)\Big), \quad \nu \in \mathbf{M}_1(\Sigma).$$

Then f is measurable and

$$\begin{aligned} -\inf\big\{I(\nu) : f(\nu) \in A^\circ\big\} &\le \varliminf_{\epsilon\to 0} \epsilon \log\Big(R_\epsilon\big(f^{-1}(A)\big)\Big) \\ &\le \varlimsup_{\epsilon\to 0} \epsilon \log\Big(R_\epsilon\big(f^{-1}(A)\big)\Big) \\ &\le -\inf\big\{I(\nu) : f(\nu) \in \overline{A}\big\} \end{aligned}$$

for every $A \in \mathcal{B}_{\mathbb{R}^N}$.

PROOF: First observe that, by Lemmas 2.1.5 and 3.2.7, our hypotheses imply that I must be a good rate function and that $\{R_\epsilon : \epsilon > 0\}$ satisfies the full large deviation principle with rate I.

To prove (3.2.20), use Theorem 2.1.10 and (3.2.9) to obtain

$$\begin{aligned}\int_\Sigma V\,d\nu - I(\nu) &\le \Lambda(V)\\ &\equiv \lim_{\epsilon\to 0}\epsilon\log\left(\int_{\mathbf{M}_1(\Sigma)}\exp\left[\frac{1}{\epsilon}\int_\Sigma V\,d\nu\right]R_\epsilon(d\nu)\right)\\ &\le \log\left(\int_\Sigma \exp\big[\beta|V|\big]\,d\mu\right)+\log M\end{aligned}$$

for all $\nu\in\mathbf{M}_1(\Sigma)$ and $V\in C_b(\Sigma;\mathbb{R})$; and so, for each $\nu\in\mathbf{M}_1(\Sigma)$,

$$I(\nu)\ge\int_\Sigma V\,d\nu-\log\left(\int_\Sigma\exp\big[\beta V\big]\,d\mu\right)-\log M$$

for every bounded measurable $V:\Sigma\longrightarrow[0,\infty)$. In particular, just as in the proof of Lemma 3.2.13, $I(\nu)=\infty$ if ν is not absolutely continuous with respect to μ. On the other hand, if $\nu<<\mu$ and $f=\frac{d\nu}{d\mu}$, take $V_n=\frac{1}{\beta}\log\big(1+f\wedge n\big)$ and conclude that

$$I(\nu)\ge\frac{1}{\beta}\int_\Sigma\log\big(1+f\wedge n\big)\,d\nu-\log(2M).$$

Finally, let $n\longrightarrow\infty$ and thereby arrive at (3.2.20).

We now turn to the proof of the second assertion. In view of Lemma 2.1.4, all that we have to do is produce a sequence $\{f_\ell\}_1^\infty\subseteq C\big(\mathbf{M}_1(\Sigma);\mathbb{R}^N\big)$ with the properties that

$$\varlimsup_{\ell\to\infty}\sup\Big\{\big|f(\nu)-f_\ell(\nu)\big|:\,I(\nu)\le L\Big\}=0,\quad L\in(0,\infty)$$

and

$$\lim_{\ell\to\infty}\varlimsup_{\epsilon\to 0}\epsilon\log\Big[R_\epsilon\Big(\big\{\nu:\,\big|f(\nu)-f_\ell(\nu)\big|\ge\delta\big\}\Big)\Big]=-\infty,\qquad\delta\in(0,\infty).$$

To this end, choose for each $1\le k\le N$ a sequence

$$\{\psi_{m,k}\}_{m=1}^\infty\subseteq C\big(\Sigma;[0,1]\big)$$

so that

$$V_m\equiv\left(\sum_{k=1}^N\big|\chi_{\Delta_k}-\psi_{m,k}\big|^2\right)^{1/2}\longrightarrow 0$$

in μ-measure as $m \longrightarrow \infty$. Next, apply Lemma 3.2.7 to find a subsequence $\{V_{m_\ell}\}_{\ell=1}^{\infty}$ for which

$$R_\epsilon\left(\left\{\nu : \sup_{\ell \geq L} \ell \int_\Sigma V_{m_\ell}\, d\nu \geq 1\right\}\right) \leq \exp\big[-L/\epsilon\big]$$

for $0 < \epsilon \leq 1$ and $L \in [1, \infty)$.

Finally, set

$$f_\ell(\nu) = \left(\int_\Sigma \psi_{m_\ell,1}\, d\nu, \dots, \int_\Sigma \psi_{m_\ell,N}\, d\nu\right), \quad \nu \in \mathbf{M}_1(\Sigma).$$

Clearly, $f_\ell \in C\big(\mathbf{M}_1(\Sigma); \mathbb{R}^N\big)$ for each $\ell \in \mathbb{Z}^+$. Moreover, since (cf. part (**i**) of Exercise 3.2.23 below) $\{\nu : \mathbf{H}(\nu|\mu) \leq K\}$ is uniformly absolutely continuous with respect to μ for every $K \in (0, \infty)$, one sees from (3.2.20) that

$$\int_\Sigma V_m\, d\nu \longrightarrow 0$$

uniformly over ν's with $I(\nu) \leq L$; and therefore $f_\ell(\nu) \longrightarrow f(\nu)$ uniformly for ν's in such sets. It is therefore clear that these f_ℓ's will serve. ∎

3.2.21 Theorem. *Let $\{R_\epsilon : \epsilon > 0\}$ and I be as in* Lemma *3.2.19 above. For each $L \geq 0$ the set $\{\nu \in \mathbf{M}_1(\Sigma) : I(\nu) \leq L\}$ is compact in the strong topology. Moreover, if Γ is a measurable subset of $\mathbf{M}_1(\Sigma)$ and $\overline{\Gamma}^{\tau}$ and $\Gamma^{\circ\tau}$ denote, respectively, the interior and closure of Γ in the strong topology, then*

$$-\inf_{\nu \in \Gamma^{\circ\tau}} I(\nu) \leq \varliminf_{\epsilon \to 0} \epsilon \log\left(R_\epsilon(\Gamma)\right) \leq \varlimsup_{\epsilon \to 0} \epsilon \log\left(R_\epsilon(\Gamma)\right) \leq -\inf_{\nu \in \overline{\Gamma}^{\tau}} I(\nu).$$

In particular, for any $\mu \in \mathbf{M}_1(\Sigma)$ and all $\Gamma \in \mathcal{B}_\Sigma$,

$$-\inf_{\Gamma^{\circ\tau}} \mathbf{H}(\,\cdot\,|\mu) \leq \varliminf_{n \to \infty} \frac{1}{n} \log\left(\tilde{\mu}_n(\Gamma)\right) \leq \varlimsup_{n \to \infty} \frac{1}{n} \log\left(\tilde{\mu}_n(\Gamma)\right) \leq -\inf_{\overline{\Gamma}^{\tau}} \mathbf{H}(\,\cdot\,|\mu).$$

PROOF: To see that $\{\nu : I(\nu) \leq L\}$ is strongly compact, simply observe that the strong and weak topologies coincide on subsets of $\mathbf{M}_1(\Sigma)$ consisting of measures which are uniformly absolutely continuous with respect to a fixed element of $\mathbf{M}_1(\Sigma)$ and use (3.2.20) together with part (**i**) of Exercise 3.2.23 below.

To prove the lower bound, note that if $\alpha \in \Gamma^{\circ\tau}$, then there exist $N \in \mathbb{Z}^+$,

$$\Delta_1, \dots, \Delta_N \in \mathcal{B}_\Sigma,$$

and an open set A in $\mathbb{R}^N$ such that

$$\alpha \in \{\nu : (\nu(\Delta_1), \dots, \nu(\Delta_N)) \in A\} \subseteq \Gamma;$$

and therefore Lemma 3.2.19 shows that

$$\underline{\lim}_{\epsilon \to 0} \epsilon \log(R_\epsilon(\Gamma)) \geq -I(\alpha).$$

To prove the upper bound, let $\Gamma \in \mathcal{B}_{\mathbf{M}_1(E)}$ be given and suppose that $F \supseteq \Gamma$ is a strongly closed subset of $\mathbf{M}_1(\Sigma)$. Using $\mathcal{P}$ to denote the collection of all finite partitions P of Σ into measurable sets and given $P = \{\Delta_1, \dots, \Delta_N\} \in \mathcal{P}$, let $A(P)$ be the closure in $\mathbb{R}^N$ of

$$\{(\nu(\Delta_1), \dots, \nu(\Delta_N)) : \nu \in F\}$$

and set $F(P) = \{\nu : (\nu(\Delta_1), \dots, \nu(\Delta_N)) \in A(P)\}$. Then, by Lemma 3.2.19, we know that

$$\overline{\lim}_{\epsilon \to 0} \epsilon \log(R_\epsilon(\Gamma)) \leq \overline{\lim}_{\epsilon \to 0} \epsilon \log\left[R_\epsilon(F(P))\right] \leq - \inf_{\nu \in F(P)} I(\nu).$$

Thus, we will have the upper bound once we show that

$$\inf_{\nu \in F} I(\nu) = \sup_{P \in \mathcal{P}} \inf_{\nu \in F(P)} I(\nu).$$

In proving this, we may and will assume that there is an $L \in (0, \infty)$ such that

$$\inf_{\nu \in F(P)} I(\nu) \leq L \quad \text{for every } P \in \mathcal{P}.$$

Noting that (because the level sets of I are strongly compact) there is, for each $P \in \mathcal{P}$, a $\nu_P \in F(P)$ such that $I(\nu_P) = \inf_{\nu \in F(P)} I(\nu)$ and using the fact that $\{\nu_P : P \in \mathcal{P}\} \subseteq \{\nu : I(\nu) \leq L\}$, choose any subnet $\{\nu_P : P \in \mathcal{P}'\}$ of $\{\nu_P : P \in \mathcal{P}\}$ (think of $\mathcal{P}$ being partially ordered by "refinement") so that $\{\nu_P : P \in \mathcal{P}'\}$ converges strongly to some α. Clearly, all that we have to do is check that $\alpha \in F$. To this end, let $N \in \mathbb{Z}^+$, $\Delta_1, \dots, \Delta_N \in \mathcal{B}_\Sigma$, and $\epsilon > 0$ be given; and denote by U the corresponding set defined by (3.2.18). Next, choose $P \in \mathcal{P}'$ so that $\{\Delta_1, \dots, \Delta_N\}$ is contained in the algebra over Σ which is generated by P and $\sum_{\Delta \in P} |\nu_P(\Delta) - \alpha(\Delta)| < \epsilon$. Then, if $\beta_P \in F$ is chosen so that $\beta_P(\Delta_k) = \nu_P(\Delta_k)$, $1 \leq k \leq N$, one sees that $\beta_P \in U$. In other words, $U \cap F \neq \emptyset$ for any U of the form in (3.2.18), and therefore $\alpha \in F$. ∎

3.2.22 Exercise.

(i) On the basis of the reader's previous information about the weak topology, derive the four properties listed at beginning of this section.

(ii) Suppose that $\{(\Sigma_n, \rho_n)\}_1^\infty$ is a sequence of complete separable metric spaces and that, for each $n \in \mathbb{Z}^+$, $\pi_{n+1,n} : \Sigma_{n+1} \longrightarrow \Sigma_n$ is a mapping with the property that

$$\rho_n\big(\pi_{n+1,n}\sigma_{n+1}, \pi_{n+1,n}\sigma'_{n+1}\big) \leq \rho_{n+1}(\sigma_{n+1}, \sigma'_{n+1}), \quad \sigma_{n+1}, \sigma'_{n+1} \in \Sigma_{n+1}.$$

Let (Σ, ρ) be the projective limit of $\{(\Sigma_n, \pi_{n+1,n}, \rho_n) : n \in \mathbb{Z}^+\}$, and show that $\mathbf{M}_1(\Sigma)$ is homeomorphic to the projective limit of $\{\mathbf{M}_1(\Sigma_n)\}_1^\infty$ when $\mu_{n+1} \longmapsto \mu_{n+1} \circ \pi_{n+1,n}^{-1}$ is the mapping from $\mathbf{M}_1(\Sigma_{n+1})$ into $\mathbf{M}_1(\Sigma_n)$ for each $n \in \mathbb{Z}^+$.

Hint: The only difficult step here is to check that if $\mu_n \in \mathbf{M}_1(\Sigma_n)$ and $\mu_n = \mu_{n+1} \circ \pi_{n+1,n}^{-1}$ for each $n \in \mathbb{Z}^+$, then there is a $\mu \in \mathbf{M}_1(\Sigma)$ such that $\mu_n = \mu \circ \pi_n^{-1}$, $n \in \mathbb{Z}^+$. However, the existence of such a μ can be seen as a consequence of the KOLMOGOROV Extension Theorem as presented in Theorem 1.1.10 of [**104**].

3.2.23 Exercise.

In this exercise we outline an approach to the SANOV Theorem which avoids the estimate in Lemma 3.2.7 and resembles, to a much greater extent, the ideas behind our proof of the classical CRAMÈR Theorem. Even though the proof avoids Lemma 3.2.7, it nonetheless uses the equation $\Lambda^*_{\tilde{\mu}} = \mathbf{H}(\cdot|\mu)$ provided by Lemma 3.2.13.

(i) It turns out (cf. Corollary 5.1.11) that both the goodness of $\Lambda^*_{\tilde{\mu}}$ as well as the upper bound (in terms of $\Lambda^*_{\tilde{\mu}}$) for all closed sets $C \subseteq \mathbf{M}_1(\Sigma)$ follow once one knows that $\Lambda_{\tilde{\mu}}$ has the property: for each $M \in (0, \infty)$, there is a $K(M) \subset\subset \Sigma$ such that $\Lambda_{\tilde{\mu}}(V) \leq 1$ whenever $V \in C_b(\Sigma; \mathbb{R})$ vanishes on $K(M)$ and is bounded by M. Prove that $\Lambda_{\tilde{\mu}}$ has this property.

Although it is true that the goodness of $\Lambda^*_{\tilde{\mu}}$ follows from the general principles which will be derived in Section 5.1, it is actually very easy to check that the level sets $\{\nu : \mathbf{H}(\nu|\mu) \leq L\}$ are not only weakly but even strongly compact. Indeed, it is clear that the set $\{f \in L^1(\mu)^+ : \int_\Sigma f \log f \, d\mu \leq L\}$ is uniformly μ-integrable and is therefore weakly compact as a subset of $L^1(\mu)$. Since weak convergence in $L^1(\mu)$ gives rise to strong convergence in $\mathbf{M}(\Sigma)$ of the associated measures, it follows that $\{\nu : \mathbf{H}(\nu|\mu) \leq L\}$ is strongly compact.

(ii) We now outline a proof of the lower bound in Theorem 3.2.21 which is based the same principle as the one used to prove the lower bound in the classical CRAMÈR Theorem. Since this lower bound is better than the lower bound in SANOV's Theorem, this will complete the program for this exercise.

Let $\mathbf{H}(\nu|\mu) < \infty$ and suppose that $G \in \mathcal{B}_{\mathbf{M}_1(\Sigma)}$ is a strongly open neighborhood of ν. Set $f = \frac{d\nu}{d\mu}$, define $F_n(\sigma) = \prod_{m=1}^n f_m(\sigma_m)$ for $\sigma \in \Sigma^n$, and let

$$A_n = \{\sigma \in \Sigma^n : \mathbf{L}_n(\sigma) \in G \text{ and } F_n(\sigma) > 0\}.$$

Using the Law of Large Numbers, check that $\nu^n(A_n) \longrightarrow 1$ as $n \longrightarrow \infty$. Next, using JENSEN's inequality and the fact that $x \log x \geq -e^{-1}$, $x \in [0, \infty)$, verify the following steps:

$$\begin{aligned}
\log\big(\tilde{\mu}_n(G)\big) &\geq \log\left(\int_{A_n} \frac{1}{F_n(\sigma)}\, \nu^n(d\sigma)\right) \\
&\geq \log\big(\nu^n(A_n)\big) - \frac{1}{\nu^n(A_n)} \int_{A_n} \log\big(F_n(\sigma)\big)\, \nu^n(d\sigma) \\
&\geq \log\big(\nu^n(A_n)\big) - \frac{1}{e\nu^n(A_n)} - \frac{1}{\nu^n(A_n)} \int_{\Sigma^n} \log\big(F_n(\sigma)\big)\, \nu^n(d\sigma) \\
&= \log\big(\nu^n(A_n)\big) - \frac{1}{e\nu^n(A_n)} - n\frac{\mathbf{H}(\nu|\mu)}{\nu^n(A_n)}
\end{aligned}$$

as long as $\nu^n(A_n) > 0$. Combining this with $\nu^n(A_n) \longrightarrow 1$, arrive at the lower bound in Theorem 3.2.21.

3.2.24 Exercise.

Prove that

$$\|\nu - \mu\|_{\text{var}}^2 \leq 2\mathbf{H}(\nu|\mu), \quad \mu, \nu \in \mathbf{M}_1(\Sigma). \tag{3.2.25}$$

A proof of (3.2.25) can be based on the observation that

$$3(x-1)^2 \leq (4+2x)(x \log x - x + 1), \quad x \in [0, \infty),$$

the fact that $\|\nu - \mu\|_{\text{var}} = \|f - 1\|_{L^1(\mu)}$ if $\nu << \mu$ and $f = \frac{d\nu}{d\mu}$, and SCHWARTZ's inequality

3.2.26 Exercise.

For $R \in \mathbf{M}_1\big(\mathbf{M}_1(\Sigma)\big)$, define $\mu_R \in \mathbf{M}_1(\Sigma)$ by

$$\mu_R(\Delta) = \int_{\mathbf{M}_1(\Sigma)} \nu(\Delta)\, R(d\nu), \quad \Delta \in \mathcal{B}_\Sigma,$$

and show that $R \in \mathbf{M}_1(\mathbf{M}_1(\Sigma)) \longmapsto \mu_R \in \mathbf{M}_1(\Sigma)$ is a continuous mapping. Next, let $Q \in \mathbf{M}_1(\mathbf{M}_1(\Sigma))$ be given, and show that

$$I_Q(\nu) = \Lambda_Q^*(\nu) = \inf\{\mathbf{H}(R|Q) : \mu_R = \nu\}, \quad \nu \in \mathbf{M}_1(\Sigma).$$

(**Hint:** Either use the variational formula for relative entropy, or combine the results of the present section with Lemma 2.1.4.) Finally, apply JENSEN's inequality to check that

$$I_Q(\nu) \geq \mathbf{H}(\nu|\mu_Q), \quad \nu \in \mathbf{M}_1(\Sigma).$$

Conclude that

$$\mathbf{H}(\mu_R|\mu_Q) \leq I_Q(\mu_R) \leq \mathbf{H}(R|Q), \quad R \in \mathbf{M}_1(\mathbf{M}_1(\Sigma)), \tag{3.2.27}$$

and therefore that $\nu << \mu_Q$ if $I_Q(\nu) < \infty$.

3.2.28 Exercise.

Let $\{Q_\epsilon : \epsilon > 0\}$ be a family of probability measures on $\mathbf{M}_1(\Sigma)$ and $I : \mathbf{M}_1(\Sigma) \longrightarrow [0, \infty]$ a rate function with the property that $\{\nu : I(\nu) \leq L\}$ is compact in the strong topology for each $L > 0$. Assume, in addition, that

$$-\inf_{\Gamma^{\circ\tau}} I \leq \varliminf_{\epsilon \to 0} \epsilon \log\big(Q_\epsilon(\Gamma)\big) \leq \varlimsup_{\epsilon \to 0} \epsilon \log\big(Q_\epsilon(\Gamma)\big) \leq -\inf_{\overline{\Gamma}^{\tau}} I$$

(see Theorem 3.2.21 for the notation here) for all $\Gamma \in \mathcal{B}_{\mathbf{M}_1(\Sigma)}$. Given a $\mathcal{B}_{\mathbf{M}_1(\Sigma)}$-measurable function $\Phi : \mathbf{M}_1(\Sigma) \longrightarrow \mathbb{R}$ which satisfies

$$\sup_{0<\epsilon\leq 1} \left(\int_{\mathbf{M}_1(\Sigma)} \exp\big[\alpha\Phi/\epsilon\big]\, dQ_\epsilon \right)^\epsilon < \infty$$

for some $\alpha \in (1, \infty)$, show that

$$\lim_{\epsilon \to 0} \epsilon \log \left(\int_{\mathbf{M}_1(\Sigma)} \exp\big[\Phi/\epsilon\big]\, dQ_\epsilon \right) = \sup\{\Phi(\nu) - I(\nu) : \nu \in \mathbf{M}_1(\Sigma)\}$$

if Φ is continuous with respect to the strong topology.

3.3 Cramèr's Theorem for Banach Spaces

In this section, we will be assuming that $E = X$ is a separable real BANACH space with norm $\|\cdot\|_E$, and we will be attempting to prove the analogue of CRAMÈR's Theorem (cf. Theorem 1.2.6) in this setting. That is, we want to show that if μ is a probability measure on $(E, \mathcal{B}_E)$ for which

$$\int_E \exp\bigl[\alpha\|x\|_E\bigr]\,\mu(dx) < \infty, \quad \alpha \in [0,\infty), \tag{3.3.1}$$

and if μ_n denotes the distribution of $\overline{\mathbf{S}}_n$ under μ^n (cf. the second paragraph of Section 3.1), then $\{\mu_n : n \geq 1\}$ satisfies the full large deviation principle with rate function Λ_μ^*, where

$$\Lambda_\mu^*(q) = \sup\left\{{}_{E^*}\langle\lambda, q\rangle_E - \Lambda_\mu(\lambda) : \lambda \in E^*\right\}, \quad q \in E \tag{3.3.2}$$

and

$$\Lambda_\mu(\lambda) = \log\left[\int_E \exp\left[{}_{E^*}\langle\lambda, x\rangle_E\right]\,\mu(dx)\right], \quad \lambda \in E^*. \tag{3.3.3}$$

However, before getting into the details, it might be helpful to know what it is that "non-deviant behavior" means in this situation. For this reason, we begin with a proof of the Strong Law of Large Numbers for E-valued random variables.

3.3.4 Theorem. (RANGA RAO) *Let $\{X_n\}_1^\infty$ be a sequence of independent, identically distributed, E-valued random variables on some probability space $(\Omega, \mathcal{M}, P)$; and let μ be the distribution of X_1. If $\int_E \|x\|_E\,\mu(dx) < \infty$, then there is an $m(\mu) \in E$ such that*

$$\frac{1}{n}\sum_1^n X_k(\omega) \longrightarrow m(\mu)$$

for P-almost every $\omega \in \Omega$. Moreover, $m(\mu)$ is the unique element y of E with the property that

$$ {}_{E^*}\langle\lambda, y\rangle_E = \int_E {}_{E^*}\langle\lambda, x\rangle_E\,\mu(dx), \quad \lambda \in E^*. \tag{3.3.5}$$

PROOF: First, by KOLMOGOROV's Zero–One Law, if $\frac{1}{n}\sum_1^n X_k(\omega)$ has a limit for P-almost every $\omega \in \Omega$, then that limit is, P-almost surely, independent of $\omega \in \Omega$. Knowing this and using the Classical Strong Law of Large Numbers, one can easily check that, if it exists, then the P-almost sure limit of $\frac{1}{n}\sum_1^n X_k$ must satisfy (3.3.5). Hence, all that we have to do is show that convergence takes place.

Note that the Classical Strong Law together with the second countability of the weak topology on $\mathbf{M}_1(E)$ lead to the conclusion that

$$\text{(3.3.6)} \qquad \frac{1}{n}\sum_1^n \delta_{X_k(\omega)} \Longrightarrow \mu$$

for P-almost every $\omega \in \Omega$. As we are about to see, this turns to be surprisingly close to the result which we are seeking. Indeed, suppose, for the moment, that there is an $R \in (0,\infty)$ such that $\|X_n(\omega)\|_E \leq R$ for all $n \in \mathbb{Z}^+$ and $\omega \in \Omega$. For $\lambda \in E^*$, define

$$\text{(3.3.7)} \qquad g_\lambda^{(R)}(x) = \eta_R(x)\, {}_{E^*}\langle \lambda, x\rangle_E, \quad x \in E$$

where $\eta_R \in C(E;[0,1])$ satisfies

$$\eta_R(x) = \begin{cases} 1 & \text{if } \|x\|_E \leq R \\ 0 & \text{if } \|x\|_E \geq R+1. \end{cases}$$

Then $\{g_\lambda^{(R)} : \|\lambda\|_{E^*} \leq 1\}$ is uniformly bounded and equi-continuous. Hence, by property **iv**) of the weak topology (cf. the first paragraph of Section 3.2) and (3.3.6)

$$\sup_{\|\lambda\|_{E^*}\leq 1} \left| \frac{1}{n}\sum_1^n g_\lambda^{(R)}(X_k(\omega)) - \int_E g_\lambda^{(R)}(x)\,\mu(dx)\right| \longrightarrow 0$$

and therefore

$$\sup_{\|\lambda\|_{E^*}\leq 1} \left| {}_{E^*}\left\langle \lambda, \frac{1}{n}\sum_1^n X_k(\omega)\right\rangle_E - \int_E {}_{E^*}\langle \lambda, x\rangle_E\,\mu(dx)\right| \longrightarrow 0$$

for P-almost every $\omega \in \Omega$. Thus, for P-almost every $\omega \in \Omega$,

$$\left\{\frac{1}{n}\sum_1^n X_k(\omega)\right\}_{n=1}^\infty$$

is a CAUCHY sequence in E; and, as we pointed out in the preceding paragraph, that is all that we need. In other words, our result has been proved in the case of bounded random variables.

To handle the general case, define, for $R \in (0,\infty)$,

$$X_n^{(R)}(\omega) = \chi_{[0,R]}\big(\|X_n(\omega)\|_E\big)X_n(\omega)$$

and

$$Y_n^{(R)}(\omega) = \|X_n(\omega) - X_n^{(R)}(\omega)\|_E, \quad \omega \in \Omega.$$

Given $\epsilon > 0$, choose $R \in (0, \infty)$ so that

$$\int_{\|x\|_E \geq R} \|x\|_E \, \mu(dx) < \frac{\epsilon}{4};$$

and use the preceding result applied to $\{X_n^{(R)}\}_1^\infty$ together with the Classical Strong Law applied to $\{Y_n^{(R)}\}_1^\infty$ to conclude that

$$\begin{aligned}
\overline{\lim_{N\to\infty}} \, P\left(\left\{\omega : \sup_{n\geq N} \left\| \frac{1}{n}\sum_1^n X_k(\omega) - \frac{1}{N}\sum_1^N X_k(\omega)\right\|_E \geq \epsilon \right\}\right) \\
\leq \overline{\lim_{N\to\infty}} \, P\left(\left\{\omega : \sup_{n\geq N} \frac{1}{n}\sum_1^n Y_k^{(R)}(\omega) \geq \frac{\epsilon}{2}\right\}\right) = 0.
\end{aligned}$$

Since this clearly shows that $\{\frac{1}{n}\sum_1^n X_k(\omega)\}_1^\infty$ is CAUCHY in E for P-almost every $\omega \in \Omega$, the proof is complete. ∎

The quantity $m(\mu)$ in (3.3.5) is called the **mean of the measure** μ. Obviously, a consequence of Theorem 3.3.4 is the fact that $\mu_n \Longrightarrow \delta_{m(\mu)}$. In order to study the large deviations from this, we again want to use Corollary 3.1.7. Thus, we must show that there are compact K_L 's for which (3.1.8) holds. Actually, most of the work required for their construction has already been done in Lemma 3.2.7; all that we need in addition is the following relatively simple observation.

3.3.8 Lemma. *If* $\Gamma \subseteq \mathbf{M}_1(E)$ *satisfies*

$$\lim_{R\to\infty} \sup_{\nu\in\Gamma} \int_{\|x\|_E > R} \|x\|_E \, \nu(dx) = 0,$$

then $\nu \in \overline{\Gamma} \longmapsto m(\nu)$ *is a continuous map. In particular, if* $f : E \longrightarrow [0, \infty]$ *is a lower semi-continuous function satisfying*

$$\lim_{\|x\|_E\to\infty} \frac{f(x)}{\|x\|_E} = \infty$$

and if

$$\Gamma(f, L) \equiv \left\{\nu \in \mathbf{M}_1(E) : \int_E f(x)\, \nu(dx) \leq L\right\}, \quad L \in [0, \infty),$$

then, for each $L \in [0,\infty)$, $\Gamma(f,L)$ is closed and $\nu \in \Gamma(f,L) \longmapsto m(\nu)$ is continuous. Thus,

$$\Gamma(f) \equiv \left\{\nu : \int_E f\, d\nu < \infty\right\} \tag{3.3.9}$$

is a measurable subset of $\mathbf{M}_1(E)$ on which $\nu \longmapsto m(\nu)$ is a measurable mapping.

PROOF: The second assertion is just an application of the first assertion once one notes that, for $\nu \in \Gamma(f,L)$,

$$\int_{\|x\|_E > R} \|x\|_E\, \nu(dx) \le \sup_{\|x\|_E > R} \frac{\|x\|_E}{f(x)} \int_E f(x)\, \nu(dx) \le L \sup_{\|x\|_E > R} \frac{\|x\|_E}{f(x)}.$$

Thus, we need only prove the first assertion. Moreover, since $\nu \longmapsto \int_{\|x\|_E > R} \|x\|_E\, \nu(dx)$ is lower semi-continuous for each $R \in (0,\infty)$, we may and will assume from the outset that the set Γ is closed in $\mathbf{M}_1(E)$. Now, suppose that $\{\nu_n\}_1^\infty \subseteq \Gamma$ and that $\nu_n \Longrightarrow \nu_\infty$. Then

$$\begin{aligned}
&\|m(\nu_n) - m(\nu_\infty)\|_E = \sup\left\{ {}_{E^*}\langle \lambda, m(\nu_n) - m(\nu_\infty)\rangle_E : \|\lambda\|_{E^*} \le 1\right\} \\
&\le 2 \sup_{\nu\in\Gamma} \int_{\|x\|_E > R} \|x\|_E\, \nu(dx) \\
&\qquad + \sup\left\{ \left| \int_E g_\lambda^{(R)}(x)\, \nu_n(dx) - \int_E g_\lambda^{(R)}(x)\, \nu_\infty(dx) \right| : \|\lambda\|_{E^*} \le 1 \right\},
\end{aligned}$$

where the functions $g_\lambda^{(R)}$ are the ones defined in (3.3.7). Hence, for each $R \in (0,\infty)$,

$$\varlimsup_{n\to\infty} \|m(\nu_n) - m(\nu_\infty)\|_E \le 2 \sup_{\nu\in\Gamma} \int_{\|x\|_E > R} \|x\|_E\, \nu(dx);$$

from which the desired result is clear. ∎

The preceding enables us to prove the following variant of Lemma 3.2.7.

3.3.10 Lemma. *Let $\mu \in \mathbf{M}_1(E)$ be given, assume that*

$$g(\alpha) \equiv \log\left(\int_E \exp\left[\alpha\|x\|\right] \mu(dx) \right) < \infty, \quad \alpha \in [0,\infty),$$

and set

$$f(x) = \sup\left\{\alpha\|x\| - g(\alpha) : \alpha \in [0,\infty)\right\}, \quad x \in E.$$

Then $f(0) = 0$; $x \in E \longmapsto f(x) \in [0,\infty]$ *is lower semi-continuous;*

$$\lim_{\|x\|\to\infty} \frac{f(x)}{\|x\|} = \infty;$$

and

$$\int_E \exp\bigl[\delta f\bigr]\, d\mu \le \frac{1}{1-\delta}, \quad \delta \in [0,1).$$

Finally, if $\{R_\epsilon : \epsilon > 0\} \subseteq \mathbf{M}_1(\mathbf{M}_1(E))$ *satisfies*

$$\sup_{0<\epsilon\le 1} \left(\int_{\mathbf{M}_1(E)} \exp\left[\frac{1}{\epsilon}\int_E V\, d\nu\right] R_\epsilon(d\nu)\right)^{\epsilon} \le M \int_E \exp\bigl[\beta V\bigr]\, d\mu$$

for some β, $M \in [1,\infty)$ *and all measurable* $V : E \longrightarrow [0,\infty]$, *then for each* $L \in [1,\infty)$ *there is a* $K_L \subset\subset E$ *(whose choice depends only on* μ, β, *and* M *and is otherwise independent of* $\{R_\epsilon : \epsilon > 0\}$*) such that*

$$\rho_\epsilon\bigl(K_L^{\mathrm{c}}\bigr) \le \exp\bigl[-L/\epsilon\bigr], \quad 0 < \epsilon \le 1,$$

where $\rho_\epsilon \in \mathbf{M}_1(E)$ *denotes the distribution of* $\nu \in \Gamma(f) \longmapsto m(\nu) \in E$ *under* R_ϵ *and* $\Gamma(f)$ *is the set in* (3.3.9).

PROOF: Define

$$h(r) = \sup\bigl\{\alpha r - g(\alpha) : \alpha \in [0,\infty)\bigr\}, \quad r \in [0,\infty).$$

It is then clear that $h(0) = 0$ and that $r \in [0,\infty) \longmapsto h(r) \in [0,\infty]$ is a lower semi-continuous, non-decreasing function for which

$$\lim_{r\to\infty} \frac{h(r)}{r} = \infty.$$

Furthermore, just as in the proof of Lemma 1.2.5 (only even more easily), one sees that

$$\mu\bigl(\{x \in E : \|x\| \ge r\}\bigr) \le e^{-h(r)}, \quad r \in [0,\infty);$$

from which it is easy to show that

$$\int_E \exp\bigl[\delta f\bigr]\, d\mu \le \frac{1}{1-\delta}, \quad \delta \in [0,1).$$

Now let $\{R_\epsilon : \epsilon > 0\}$ satisfying our hypothesis be given. By Lemma 3.2.7, we know that, for each $L \in [1,\infty)$, there is a $C_L \subset\subset \mathbf{M}_1(E)$ (depending only on μ, β, and M) such that

$$R_\epsilon\bigl(C_L^{\mathrm{c}}\bigr) \le e^{-L/\epsilon}, \quad 0 < \epsilon \le 1.$$

In addition, since

$$\left(\int_{\mathbf{M}_1(E)} \exp\left[\frac{1}{2\beta\epsilon}\int_E f\,d\nu\right] R_\epsilon(d\nu)\right)^\epsilon \le 2M, \quad 0 < \epsilon \le 1,$$

we have, when $r_L \equiv 2\beta\big(L + \log(4M)\big)$, that

$$R_\epsilon\Big(\Gamma(f, r_L)^c\Big)$$
$$\le \left(\frac{e^{-L}}{4M}\right)^{1/\epsilon} \int_{\mathbf{M}_1(E)} \exp\left[\frac{1}{2\beta\epsilon}\int_E f\,d\nu\right] R_\epsilon(d\nu) \le \frac{e^{-L/\epsilon}}{2}$$

for all $L \in [1, \infty)$ and $0 < \epsilon \le 1$. Hence, if

$$K_L \equiv \Big\{m(\nu) : \nu \in \Gamma(f, r_L) \cap C_{L+\log 2}\Big\},$$

then, as the continuous image of a compact set, $K_L \subset\subset E$ and

$$\rho_\epsilon(K_L^c) \le R_\epsilon(C_{L+\log 2}^c) + R_\epsilon\Big(\Gamma(f, r_L)^c\Big) \le e^{-L/\epsilon}$$

for all $L \in [1, \infty)$ and $0 < \epsilon \le 1$. ∎

We now have all the machinery which we need to prove the following extension of the CRAMÈR Theorem.

3.3.11 Theorem. (DONSKER & VARADHAN) *Assume that* (3.3.1) *holds. Then, for each* $L \in [0, \infty)$, *there is a* $K_L \subset\subset E$ *such that*

$$\mu_n(K_L^c) \le e^{-nL}, \quad n \in \mathbb{Z}^+ \text{ and } L \in [1, \infty).$$

In particular, $\{\mu_n : n \in \mathbb{Z}^+\}$ *is exponentially tight and the function* Λ_μ^* *in* (3.3.2) *is a good rate function which governs the large deviations of* $\{\mu_n : n \ge 1\}$.

PROOF: In view of Corollary 3.1.7, all that we need to check is the first assertion. To this end, define $\tilde{\mu}_n \in \mathbf{M}_1\big(\mathbf{M}_1(E)\big)$, as in Section 3.2, to be the distribution of

$$\mathbf{L}_n \equiv \frac{1}{n}\sum_{m=1}^{n} \delta_{x_m}$$

under μ^n. It is then clear that the measures $R_{1/n} \equiv \tilde{\mu}_n$ satisfy the hypotheses of Lemma 3.3.10 with respect to μ; and therefore the existence of the required K_L's is an immediate consequence of that lemma. ∎

3.3.12 Exercise.

It is amusing and instructive to prove the large deviation result of this section as a corollary of the SANOV Theorem. To this end, let $f : E \longrightarrow [0,\infty]$ be the function in the proof of Lemma 3.3.10 and define the sets $\Gamma(f, L)$, $L \geq 0$ accordingly. Noting that the mean value functional m is continuous on each $\Gamma(f, L)$, use part (**i**) of Exercise 2.1.20 together with the SANOV Theorem to show that $I : E \longrightarrow [0,\infty]$ given by

$$I(x) \equiv \inf\left\{\mathbf{H}(\nu|\mu) : \int_E \|y\|_E \, \nu(dy) < \infty \text{ and } m(\nu) = x\right\} \tag{3.3.13}$$

is a good rate function on E and that it governs the large deviations of $\{\mu_n : n \geq 1\}$. Conclude, in particular, that $I = \Lambda_\mu^*$.

A direct proof of the equality in (3.3.13) is not easy, even in special cases. For example, suppose that $E = \Theta$ and that μ is WIENER's measure $\mathcal{W}$ as in Section 1.3. As we pointed in the discussion (1.3.7), the large deviation principle proved in SCHILDER's Theorem can be thought of as an example of CRAMÈR's Theorem; and so (3.3.13) gives us another variational formula for $I_{\mathcal{W}} = \Lambda_{\mathcal{W}}^*$. To give a direct proof of (3.3.13) in this case, one can use the fact that $P \in \mathbf{M}_1(\Theta)$ is absolutely continuous with respect to $\mathcal{W}$ if and only if there is a $\{\mathcal{B}_t : t \in [0,\infty)\}$-progressively measurable map $b : [0,\infty) \times \Theta \longrightarrow \mathbb{R}^d$ such that

$$\int_{[0,\infty)} |b(t,\theta)|^2 \, dt < \infty \quad (\text{a.s., } \mathcal{W})$$

and

$$P(d\theta) = \exp\left[\int_0^\infty b(s,\theta)\, d\theta(s) - \frac{1}{2}\int_0^\infty |b(s,\theta)|^2 \, ds\right] \mathcal{W}(d\theta);$$

in which case the distribution of

$$\theta \in \Theta \longrightarrow \theta - \int_0^{\cdot} b(s,\theta)\, ds$$

under P is $\mathcal{W}$. Using this, one can check that, when $P << \mathcal{W}$,

$$\mathbf{H}(P|\mathcal{W}) = \frac{1}{2}\int_\Theta \left[\int_0^\infty |b(s,\theta)|^2 \, ds\right] \mathcal{W}(d\theta).$$

Finally, for a given $\psi \in H^1$, show that, among $P \in \mathbf{M}_1(\Theta)$ satisfying $\int_\Theta \|\theta\|_\Theta \, P(d\theta) < \infty$ and $m(P) = \psi$, the one which minimizes $\mathbf{H}(P|\mathcal{W})$ is $\mathcal{W}^\psi$.

3.4 Large Deviations for Gaussian Measures

In this section, we will generalize SCHILDER's Theorem to cover all **centered Gaussian measures** on a separable, real BANACH space. That is, we will be assuming that $(E, \|\cdot\|_E)$ is a separable, real BANACH space and that μ is a probability measure on $(E, \mathcal{B}_E)$ with the property that

$$\text{(3.4.1)} \quad \int_E \exp\left[\sqrt{-1}\ {}_{E^*}\langle \lambda, x\rangle_E\right] \mu(dx) = \exp\left[-Q_\mu(\lambda, \lambda)/2\right], \quad \lambda \in E^*$$

for some symmetric, bilinear map $Q_\mu : E^* \times E^* \longrightarrow [0, \infty)$. The bilinear map Q_μ is called the **covariance of** μ.

At least when E is infinite dimensional, the archetype for this sort of measure is WIENER's measure $\mathcal{W}$ on the space Θ (cf. Section 1.3), in which case

$$Q_{\mathcal{W}}(\lambda, \lambda) = \int_{[0,\infty)} \lambda\big((t, \infty)\big)^2 \, dt = \int_E \int_E s \wedge t \ \lambda(ds) \cdot \lambda(dt)$$

and SCHILDER's Theorem gave us a large deviation principle for the family $\{\mathcal{W}_\epsilon : \epsilon > 0\}$. Our aim here will be to come as close as possible to duplicating SCHILDER's result in general.

3.4.2 Lemma. *There exists an $\alpha \in (0, \infty)$ such that*

$$\text{(3.4.3)} \qquad \int_E \exp\left[\alpha \|x\|_E^2\right] \mu(dx) < \infty.$$

In particular,

$$\text{(3.4.4)} \qquad 2\Lambda_\mu(\lambda) = Q_\mu(\lambda, \lambda) = \int_E {}_{E^*}\langle \lambda, x\rangle_E^2 \, \mu(dx) \leq B\|\lambda\|_{E^*}^2$$

for $\lambda \in E^*$, where $B \equiv \int_E \|x\|_E^2 \, \mu(dx) \in (0, \infty)$. Finally, Λ_μ^* is a good rate function; and, for all $x \in E$ and $\xi \in \mathbb{R}$, $\Lambda_\mu^*(\xi x) = \xi^2 \Lambda_\mu^*(x)$. (See Exercise *3.4.15 below for a little more information.)*

PROOF: The existence of an $\alpha > 0$ for which (3.4.3) holds is a consequence of FERNIQUE's Theorem (cf. Theorem 1.3.24). Furthermore, the equalities in (3.4.4) are all obtained from consideration of the $\mathbb{R}$-valued, centered GAUSSian random variable $x \in E \longmapsto {}_{E^*}\langle \lambda, x\rangle_E$; the inequality is trivial; and the finiteness of B follows trivially from (3.4.3). Finally, given (3.4.3), the fact that Λ_μ^* is a good rate function is covered in the statement of Theorem 3.3.11; and the homogeneity of Λ_μ^* is an immediate consequence of the homogeneity of Λ_μ. ∎

Following the pattern in SCHILDER's Theorem, we now define μ_ϵ to be the distribution of $x \in E \longmapsto \epsilon^{1/2} x \in E$ under μ; and, as a first approximation to his result, we present the following.

3.4.5 Theorem. *The family $\{\mu_\epsilon : \epsilon > 0\}$ satisfies the full large deviation principle with the good rate function Λ_μ^*.*

PROOF: We have already pointed out that, as a consequence of Theorem 3.3.11, Λ_μ^* is a good rate function. Furthermore, since $\mu_{1/n}$ is the distribution under μ^n of $\mathbf{x} \in E^n \longmapsto \frac{1}{n}\sum_1^n x_k$ (i.e., $\mu_{1/n}$ here is the same measure as the one which we denoted by μ_n in Section 3.3), Theorem 3.3.11 allows us to also conclude from (3.4.3) that $\{\mu_{1/n} : n \geq 1\}$ satisfies the full large deviation principle with rate Λ_μ^*. In order to pass from this statement to the desired one, set $n(\epsilon) = [1/\epsilon] \vee 1$ and $\gamma(\epsilon) = \epsilon n(\epsilon)$ for $\epsilon > 0$. It is then clear that $x \longmapsto \gamma(\epsilon)^{1/2}x$ under $\mu_{1/n(\epsilon)}$ has distribution μ_ϵ and that $\gamma(\epsilon) \in [1-\epsilon, 1]$ for $0 < \epsilon < 1$. Now suppose that F is a closed subset of E and set $\tilde{F} = \{\gamma^{-1/2}x : \gamma \in \left[\frac{1}{2}, 1\right] \text{ and } x \in F\}$. Then $\tilde{F}$ is also closed, and so

$$\begin{aligned}\overline{\lim_{\epsilon\to 0}}\, \epsilon \log\left(\mu_\epsilon\left(\gamma(\epsilon)^{-1/2}F\right)\right) &= \overline{\lim_{\epsilon\to 0}}\, \frac{\gamma(\epsilon)}{n(\epsilon)} \log\left(\mu_{n(\epsilon)}(F)\right) \\ &\leq \overline{\lim_{n\to\infty}}\, \frac{1}{n} \log\left(\mu_{1/n}(\tilde{F})\right) \leq -\inf_{\tilde{F}} \Lambda_\mu^*.\end{aligned}$$

Since

$$\inf_{\tilde{F}} \Lambda_\mu^* = \inf_{\gamma\in[1/2,1]}\ \inf_{x\in F} \Lambda_\mu^*\left(\gamma^{1/2}x\right) = \inf_{\gamma\in[1/2,1]} \gamma^{-1} \inf_F \Lambda_\mu^* = \inf_F \Lambda_\mu^*,$$

this proves the upper bound in the large deviation principle. To prove the lower bound, let G be an open set in E and suppose that $x \in G$. Then we can find an open neighborhood U of x and an $\epsilon_0 \in (0, 1/2]$ such that $U \subseteq \gamma(\epsilon)^{-1/2}G$ for all $0 < \epsilon < \epsilon_0$. Hence

$$\begin{aligned}\underline{\lim_{\epsilon\to 0}}\, \epsilon \log\left(\mu_\epsilon(G)\right) &= \underline{\lim_{\epsilon\to 0}}\, \frac{\gamma(\epsilon)}{n(\epsilon)} \log\left(\mu_{n(\epsilon)}\left(\gamma(\epsilon)^{-1/2}G\right)\right) \\ &\geq \underline{\lim_{n\to\infty}}\, \frac{1}{n} \log\left(\mu_{1/n}(U)\right) \geq -\inf_U \Lambda_\mu^* \geq -\Lambda_\mu^*(x).\end{aligned}$$

Thus, the lower bound is also proved. ∎

As a dividend of Theorem 3.4.5, we get the following sharpening of the estimate in (3.4.3).

3.4.6 Corollary. (DONSKER & VARADHAN) *Set*

$$a = \inf\left\{\Lambda_\mu^*(x) : \|x\|_E = 1\right\} \text{ and } b = \sup\left\{Q_\mu(\lambda, \lambda) : \|\lambda\|_{E^*} = 1\right\}. \tag{3.4.7}$$

Then

$$\lim_{R\to\infty} \frac{1}{R^2} \log\left[\mu\left(\{x \in E : \|x\|_E \geq R\}\right)\right] = -a = -\frac{1}{2b}. \tag{3.4.8}$$

In particular, $\frac{1}{2a} \le \int_E \|x\|_E^2 \,\mu(dx) < \infty$, (3.4.3) *holds for* $\alpha \in (0, a)$, *and it fails for* $\alpha \in (a, \infty)$.

PROOF: It suffices to prove (3.4.8). To prove the first equality, set $B = B(0,1)$ and note that

$$\inf_{B^c} \Lambda_\mu^* = \inf_{r \ge 1} \inf_{x \in \partial B} \Lambda^*(rx) = a \inf_{r \ge 1} r^2 = a$$

and similarly that $\inf_{\overline{B}^c} \Lambda_\mu^* = a$. Hence, by Theorem 3.4.5, we see that

$$\lim_{R \to \infty} \frac{1}{R^2} \log\left[\mu\big(\overline{B}(0,R)^c\big)\right] = \lim_{\epsilon \to 0} \epsilon \log\left[\mu_\epsilon\big(\overline{B}^c\big)\right] = -\inf_{\overline{B}^c} \Lambda_\mu^* = -a.$$

To prove the second equality in (3.4.8), first observe that

$$\begin{aligned}\Lambda_\mu^*(x) &= \sup\left\{\xi\, {}_{E^*}\langle \lambda, x\rangle_E - \xi^2 \Lambda_\mu(\lambda,\lambda) : \xi \in \mathbb{R} \text{ and } \|\lambda\|_{E^*} = 1\right\} \\ &= \sup\left\{\Lambda_{\mu^\lambda}^*\left({}_{E^*}\langle \lambda, x\rangle_E\right) : \|\lambda\|_{E^*} = 1\right\} \\ &= \sup\left\{\frac{{}_{E^*}\langle \lambda, x\rangle_E^2}{2Q_\mu(\lambda,\lambda)} : \|\lambda\|_{E^*} = 1\right\},\end{aligned}$$

where μ^λ denotes the distribution under μ of $x \in E \longmapsto {}_{E^*}\langle \lambda, x\rangle_E \in \mathbb{R}$, and we have used (**iv**) of Exercise 1.2.11 to see that

$$\Lambda_{\mu^\lambda}^*(\xi) = \frac{\xi^2}{2Q_\mu(\lambda,\lambda)}, \quad \xi \in \mathbb{R}. \tag{3.4.9}$$

In particular, if $\|x\|_E = 1$, then

$$\Lambda_\mu^*(x) \ge \sup\left\{\frac{{}_{E^*}\langle \lambda, x\rangle_E^2}{2b} : \|\lambda\|_{E^*} = 1\right\} = \frac{1}{2b}.$$

To prove the opposite inequality, suppose $\|\lambda\|_{E^*} = 1$ and note that

$$\begin{aligned}-a &= \lim_{R \to \infty} \frac{1}{R^2} \log\left[\mu\big(\overline{B}(0,R)^c\big)\right] \\ &\ge \lim_{R \to \infty} \frac{1}{R^2} \log\left[\mu\Big(\{x \in E : {}_{E^*}\langle \lambda, x\rangle_E > R\}\Big)\right] \\ &= -\left[\Lambda_{\mu^\lambda}^*(1) \wedge \Lambda_{\mu^\lambda}^*(-1)\right] = -\frac{1}{2Q_\mu(\lambda,\lambda)},\end{aligned}$$

where we have applied the first equality in (3.4.8) to both μ and to μ^λ, and we again used (3.4.9) to get the final equality. Since this shows that $a \le \frac{1}{2Q_\mu(\lambda,\lambda)}$ whenever $\|\lambda\|_{E^*} = 1$, we have now shown that $a = \frac{1}{2b}$. ∎

Before leaving the topic of centered GAUSSian measures μ, we want to show that one can always develop a representation of Λ_μ^* analogous to the one for $\Lambda_{\mathcal{W}}^*$ given in (1.3.12). For this purpose, it will be convenient to introduce a new notion. Namely, we will say that (E, H, S, μ) is a **Wiener quadruple** if

(i) E is a separable, real BANACH space,

(ii) H is a separable, real HILBERT space,

(iii) S is a continuous, linear injection from H into E,

(iv) μ is a probability measure on $(E, \mathcal{B}_E)$ with the property that

$$\int_E \exp\left[\sqrt{-1}\; {}_{E^*}\langle \lambda, x\rangle_E\right] \mu(dx) = \exp\left[-\frac{1}{2}\|S^*\lambda\|_H^2\right], \quad \lambda \in E^*, \tag{3.4.10}$$

where $S^* : E^* \longrightarrow H$ is the adjoint map to S.

Obviously, if (E, H, S, μ) is a WIENER quadruple, then μ is a centered GAUSSian measure on E and the covariance of μ is

$$Q_\mu(\lambda, \lambda') = (S^*\lambda, S^*\lambda')_H.$$

In particular, S^* is bounded from E^* to H with operator norm given by

$$\|S^*\|_{E^*\to H}^2 \le \int_E \|x\|_E^2\, \mu(dx).$$

Hence, the norm of S also satisfies

$$\|S\|_{H\to E} \le \left(\int_E \|x\|_E^2\, \mu(dx)\right)^{1/2}. \tag{3.4.11}$$

3.4.12 Theorem. *If μ is a centered GAUSSian measure on the separable, real BANACH space E, then there exist a separable, real HILBERT space H and a continuous, linear injection $S : H \longrightarrow E$ such that (E, H, S, μ) is a WIENER quadruple. Moreover, if (E, H, S, μ) is any WIENER quadruple, then S is a compact map, S satisfies (3.4.11), and*

$$\Lambda_\mu^*(x) = \begin{cases} \frac{1}{2}\|S^{-1}x\|_H^2 & \text{if } x \in SH \\ \infty & \text{if } x \in E \setminus SH. \end{cases} \tag{3.4.13}$$

PROOF: To prove the first statement, let H denote the closure in $L^2(\mu)$ of the subspace spanned by the functions ${}_{E^*}\langle \lambda, \cdot\rangle_E$, $\lambda \in E^*$; and set $\|h\|_H = \|h\|_{L^2(\mu)}$ for $h \in H$. In order to define S, we must first define (cf. Theorem

3.3.4) "$m(f\mu) \in E$" for $f \in L^2(\mu)$. To this end, assume that $f \in L^2(\mu)$ is non-negative and that $\int_E f\,d\mu = 1$. Then, since

$$\int_E \|x\|_E f(x)\,\mu(dx) \le \left(\int_E \|x\|_E^2\,\mu(dx)\right)^{1/2} \|f\|_H < \infty,$$

we can define $m(f\mu)$, where $f\mu$ is the probability measure ν given by $\nu(dx) = f(x)\,\mu(dx)$. Now, extend $f \longmapsto m(f\mu)$ to the whole of $L^2(\mu)$ by linearity; and let S be the restriction of this map to H. Note that

$$(3.4.14) \qquad {}_{E^*}\langle \lambda, Sh\rangle_E = \left({}_{E^*}\langle \lambda, \cdot\rangle_E, h\right)_H, \quad \lambda \in E^* \text{ and } h \in H.$$

In particular, if $h \in H$ and $Sh = 0$, then $h \perp_H \{{}_{E^*}\langle \lambda, \cdot\rangle_E : \lambda \in E^*\}$, and therefore $h = 0$. That is, S is an injection. Finally, to complete the proof that (E, H, S, μ) is a WIENER quadruple, let $\lambda \in E^*$ be given and, using (3.4.14), check that $S^*\lambda = {}_{E^*}\langle \lambda, \cdot\rangle_E$ and therefore that $\|S^*\lambda\|_H^2 = Q_\mu(\lambda, \lambda)$.

Now let (E, H, S, μ) be any WIENER quadruple. We have already seen that (3.4.11) holds. Moreover, since Λ_μ^* is a good rate function, the compactness of S will follow as soon as we show that (3.4.13) is true. To prove (3.4.13), first suppose that $x = Sh$ for some $h \in H$. Then

$$\begin{aligned}\Lambda_\mu^*(x) &= \sup\left\{{}_{E^*}\langle \lambda, Sh\rangle_E - \frac{1}{2}\|S^*\lambda\|_H^2 : \lambda \in E^*\right\}\\ &= \sup\left\{{}_{E^*}\langle S^*\lambda, h\rangle_E - \frac{1}{2}\|S^*\lambda\|_H^2 : \lambda \in E^*\right\}\\ &= \sup\left\{{}_{E^*}\langle h', h\rangle_E - \frac{1}{2}\|h'\|_H^2 : h' \in H\right\} = \frac{1}{2}\|h\|_H^2\end{aligned}$$

since S is an injection and therefore S^*E^* is dense in H. Conversely, suppose that $x \in E$ and that $\Lambda_\mu^*(x) < \infty$. Then, since

$$\Lambda_\mu^*(x) = \sup\left\{\frac{{}_{E^*}\langle \lambda, x\rangle_E^2}{2} : \lambda \in E^* \text{ and } \|S^*\lambda\|_H = 1\right\},$$

we have that

$$\left|{}_{E^*}\langle \lambda, x\rangle_E\right| \le \left(2\Lambda_\mu^*(x)\right)^{1/2}\|S^*\lambda\|_H, \quad \lambda \in E^*.$$

Hence, because S^*E^* is dense in H, there is a unique, continuous, linear functional F on H such that $F(S^*\lambda) = {}_{E^*}\langle \lambda, x\rangle_E$, $\lambda \in E^*$; and therefore, by the RIESZ Representation Theorem, we know that there is a unique $h \in H$ with the property that

$${}_{E^*}\langle \lambda, Sh\rangle_E = \left(S^*\lambda, h\right)_H = {}_{E^*}\langle \lambda, x\rangle_E, \ \lambda \in E^*.$$

Thus, we conclude that $x = Sh$ and therefore that (3.4.13) holds. ∎

3.4.15 Exercise.

Let μ be a centered GAUSSian measure on the separable, real BANACH space E. Show that $(\lambda, \lambda') \in E^* \times E^* \longmapsto Q_\mu(\lambda, \lambda')$ is continuous with respect to the weak* topology; and conclude from this that there is a $\lambda_0 \in E^*$ with $\|\lambda_0\|_{E^*} = 1$ and $a = \frac{1}{2Q_\mu(\lambda_0, \lambda_0)}$, where a is defined as in (3.4.7). In particular, use this to show that

$$\int_E \exp\left[a\|x\|_E^2\right] \mu(dx) = \infty.$$

3.4.16 Exercise.

Let $E = C(\Sigma; \mathbb{R})$, where Σ is a compact metric space and we think of E as a BANACH space with the uniform norm. Given a centered GAUSSian measure μ on E, define $q_\mu(s,t) = Q_\mu(\delta_s, \delta_t)$ for s, $t \in \Sigma$. Show that $q_\mu \in C(\Sigma^2; [0, \infty))$ and that

$$Q_\mu(\alpha, \beta) = \int_\Sigma \int_\Sigma q_\mu(s,t)\, \alpha(ds)\beta(dt), \quad \alpha,\ \beta \in E^*.$$

Next, show that $q_\mu(s,t)^2 \le q_\mu(s,s) q_\mu(t,t)$ for all s, $t \in \Sigma$, and use this to conclude that $b = \sup_{s \in \Sigma} q_\mu(s,s)$, where b is defined as in (3.4.7).

IV Uniform Large Deviations

4.1 Markov Chains

In this section, we present a theory which generalizes the results in Chapter III in an important direction. Namely, we will see how to replace the independence which we assumed there with the MARKOV property here and still end up with a SAVOV-type result for the empirical distribution functional. Of course, we will have to impose strong ergodicity conditions in order to assure that there is a "typical behavior" from which large deviations may occur (cf. Example 4.1.1 below).

Throughout, Σ will denote a Polish space and E, X, ρ, and $\|\cdot\|$ will be as in $(\hat{\mathbf{C}})$ at the beginning of Section 3.2. Set $\hat{\Sigma} = \Sigma \times E$, $\hat{\Omega} = \hat{\Sigma}^{\mathbb{N}}$; and, for $n \in \mathbb{N}$, let $\hat{\omega} \longmapsto \hat{\Sigma}_n(\hat{\omega}) = \big((\Sigma_n(\hat{\omega}), X_n(\hat{\omega})\big) \in \Sigma \times E$ denote the nth coordinate map on $\hat{\Omega}$, and set $\hat{\mathfrak{M}}_n = \sigma\big(\hat{\Sigma}_m : 0 \le m \le n\big)$. Next, let

$$\hat{\sigma} \in \hat{\Sigma} \longmapsto \hat{\Pi}(\hat{\sigma}, \cdot) \in \mathbf{M}_1(\hat{\Sigma})$$

be a **transition probability function on** $\hat{\Sigma}$ (i.e., $\hat{\sigma} \in \hat{\Sigma} \longmapsto \hat{\Pi}(\hat{\sigma}, \hat{\Gamma})$ is measurable for every $\hat{\Gamma} \in \mathcal{B}_{\hat{\Sigma}}$); and for each $\hat{\sigma} \in \hat{\Sigma}$ denote by $\hat{P}_{\hat{\sigma}}$ the unique probability measure on $\big(\hat{\Omega}, \mathcal{B}_{\hat{\Omega}}\big)$ with the properties that

$$\hat{P}_{\hat{\sigma}}\big(\{\hat{\omega} : \hat{\Sigma}_0(\hat{\omega}) = \hat{\sigma}\}\big) = 1$$

and

$$\hat{P}_{\hat{\sigma}}\big(\{\hat{\omega} : \hat{\Sigma}_{n+1}(\hat{\omega}) \in \hat{\Gamma}\}\big|\hat{\mathfrak{M}}_n\big) = \hat{\Pi}\big(\hat{\Sigma}_n(\,\cdot\,), \hat{\Gamma}\big) \quad (\text{a.s.}, \hat{P}_{\hat{\sigma}})$$

for each $n \in \mathbb{N}$ and $\hat{\Gamma} \in \mathcal{B}_{\hat{\Sigma}}$. That is, $\hat{P}_{\hat{\sigma}}$ is the distribution of the **Markov chain on $\hat{\Sigma}$ starting from $\hat{\sigma}$ with transition function** $\hat{\Pi}$. Finally, define

$$\mathbf{S}_n(\hat{\omega}) = \sum_{k=1}^{n} X_k(\hat{\omega}) \quad \text{and} \quad \overline{\mathbf{S}}_n(\hat{\omega}) = \frac{1}{n}\mathbf{S}_n(\hat{\omega}) \quad \text{for } n \in \mathbb{Z}^+,$$

and let $\mu_{\hat{\sigma},n} \in \mathbf{M}_1(E)$ be the distribution of $\overline{\mathbf{S}}_n$ under $\hat{P}_{\hat{\sigma}}$. What we want to do is study the large deviation theory for the families $\{\mu_{\hat{\sigma},n} : n \geq 1\}$, $\hat{\sigma} \in \hat{\Sigma}$; and, in so far as possible, our treatment will be based on the ideas introduced in Chapter III.

4.1.1 Example.

The example to be kept in mind is the one in which $E = \mathbf{M}_1(\Sigma)$ and

$$(4.1.2) \quad \hat{\Pi}\big((\sigma,\nu),\hat{\Gamma}\big) = \Pi\big(\sigma, \{\tau \in \Sigma : (\tau, \delta_\tau) \in \hat{\Gamma}\}\big), \quad (\sigma,\nu) \in \Sigma \times \mathbf{M}_1(\Sigma),$$

where $\sigma \in \Sigma \longmapsto \Pi(\sigma,\cdot)$ is a transition probability function on Σ. In this case, it is unnecessary to deal with $\hat{\Omega}$ at all. Instead, one should set $\Omega = \Sigma^{\mathbb{N}}$, let $\omega \in \Omega \longmapsto \Sigma_n(\omega) \in \Sigma$ be the nth coordinate map on Ω, and take $P_\sigma \in \mathbf{M}_1(\Omega)$ to be measure defined by the conditions

$$P_\sigma\big(\{\omega : \Sigma_0(\omega) = \sigma\}\big) = 1$$

and

$$P_\sigma\big(\{\omega : \Sigma_{n+1}(\omega) \in \Gamma\}\big|\mathfrak{M}_n\big) = \Pi\big(\Sigma_n(\,\cdot\,),\Gamma\big) \quad (\text{a.s.}, P_\sigma)$$

for all $n \in \mathbb{N}$ and $\Gamma \in \mathcal{B}_\Sigma$. (In other words, P_σ is the distribution of the Markov chain on Σ starting at σ and having transition function Π.) It is then an easy matter to check that for any $\hat{\sigma} = (\sigma,\nu)$, $\mu_{\hat{\sigma},n}$ is the distribution of the empirical distribution functional

$$(4.1.3) \quad \omega \in \Omega \longmapsto \mathbf{L}_n(\omega) \equiv \frac{1}{n}\sum_{k=1}^{n} \delta_{\Sigma_k}(\omega)$$

under P_σ. In particular, $\mu_{(\sigma,\nu),n}$ is independent of $\nu \in \mathbf{M}_1(\Sigma)$, and therefore, in this case, we will use the notation $\mu_{\sigma,n}$ instead.

In order to explain why it is that one might suspect that the measures $\mu_{\sigma,n}$, $n \in \mathbb{Z}^+$, are candidates for a large deviation theory, assume for the moment that the Markov chain determined by Π is sufficiently ergodic to allow one to conclude that there is a $\mu \in \mathbf{M}_1(\Sigma)$ with the property that $\mathbf{L}_n(\omega) \Longrightarrow \mu$ (a.e., P_σ) for each $\sigma \in \Sigma$. One would then have that $\mu_{\sigma,n} \Longrightarrow \delta_\mu$ for every $\sigma \in \Sigma$; and, obviously, a large deviation theory for $\{\mu_{\sigma,n} : n \in \mathbb{Z}^+\}$ would then be the precise analogue for Markov chains of what Sanov did for sums of independent random variables: the difference being that here we have to rely on ergodicity, whereas there we had the Strong Law working for us.

It will be convenient to have the notation

$$\mathbf{S}_n^m(\hat{\omega}) \equiv \sum_{k=m+1}^{n} X_k(\hat{\omega}) \quad \text{and} \quad \overline{\mathbf{S}}_n^m(\hat{\omega}) \equiv \frac{1}{n-m}\mathbf{S}_n^m(\hat{\omega}) \quad \text{for } 0 \le m < n.$$

We now present the analogue of Lemma 3.1.2.

4.1.4 Lemma. *For $n \geq 1$ define*

$$\mathcal{P}_n(\Gamma) = \inf_{\hat{\sigma}\in\hat{\Sigma}} \mu_{\hat{\sigma},n}(\Gamma), \quad \Gamma \in \mathcal{B}_E.$$

Then, for each convex $\Gamma \in \mathcal{B}_E$, $n \in \mathbb{Z}^+ \longmapsto \mathcal{P}_n(\Gamma)$ is super-multiplicative. In addition, if

$$\lim_{R\to\infty} \sup_{\hat{\sigma}\in\hat{\Sigma}} \hat{P}_{\hat{\sigma}}\left(\{\hat{\omega} : \|\mathbf{S}_n(\hat{\omega})\| \geq R\}\right) = 0, \quad n \in \mathbb{Z}^+, \tag{4.1.5}$$

then for every ρ-bounded $\Gamma \in \mathcal{B}_E$ which is convex, either $\mathcal{P}_n(\Gamma) = 0$ for all every $n \in \mathbb{Z}^+$, or, for each $\delta > 0$, there is an $m \in \mathbb{Z}^+$ such that $\mathcal{P}_n(\Gamma^{(\delta)}) > 0$, $n \geq m$. (Throughout, $\Gamma^{(\delta)}$ is defined relative to the metric ρ on E.)

PROOF: Suppose that $\Gamma \in \mathcal{B}_E$ is convex. Then, for $\hat{\sigma} \in \hat{\Sigma}$ and $m, n \in \mathbb{Z}^+$,

$$\begin{aligned}\mu_{\hat{\sigma},m+n}(\Gamma) &= \hat{P}_{\hat{\sigma}}\left(\{\hat{\omega} : \overline{\mathbf{S}}_{m+n}(\hat{\omega}) \in \Gamma\}\right) \\ &\geq \hat{P}_{\hat{\sigma}}\left(\left\{\hat{\omega} : \overline{\mathbf{S}}_{m+n}^m(\hat{\omega}) \in \Gamma \text{ and } \overline{\mathbf{S}}_m(\hat{\omega}) \in \Gamma\right\}\right) \\ &= \int_{\{\hat{\omega}:\overline{\mathbf{S}}_m(\hat{\omega})\in\Gamma\}} \mu_{\hat{\Sigma}_m(\hat{\omega}),n}(\Gamma)\, \hat{P}_{\hat{\sigma}}(d\hat{\omega}) \geq \mu_{\hat{\sigma},m}(\Gamma)\mathcal{P}_n(\Gamma).\end{aligned}$$

Hence, $\mathcal{P}_{m+n}(\Gamma) \geq \mathcal{P}_m(\Gamma)\mathcal{P}_n(\Gamma)$.

To prove the second statement, assume that $\Gamma \in \mathcal{B}_E$ is ρ-bounded and convex; and suppose that $\mathcal{P}_m(\Gamma) > 0$ for some $m \in \mathbb{Z}^+$. For $n > m$, set $q_n = \left[\frac{n}{m}\right]$ and $r_n = n - q_n m$. Then, since, by ($\hat{\mathbf{C}}$), Γ is $\|\cdot\|$-bounded,

$$\begin{aligned}\mu_{\hat{\sigma},n}\left(\Gamma^{(\delta)}\right) &= \hat{P}_{\hat{\sigma}}\left(\{\hat{\omega} : \overline{\mathbf{S}}_n(\hat{\omega}) \in \Gamma^{(\delta)}\}\right) \\ &\geq \hat{P}_{\hat{\sigma}}\left(\{\hat{\omega} : \overline{\mathbf{S}}_n^{r_n}(\hat{\omega}) \in \Gamma \text{ and } \|\overline{\mathbf{S}}_n(\hat{\omega}) - \overline{\mathbf{S}}_n^{r_n}(\hat{\omega})\| < \delta\}\right) \\ &\geq \hat{P}_{\hat{\sigma}}\left(\{\hat{\omega} : \left\|\mathbf{S}_{r_n}(\hat{\omega})\right\| < n\delta/2 \text{ and } \overline{\mathbf{S}}_n^{r_n}(\hat{\omega}) \in \Gamma\}\right) \\ &\geq \hat{P}_{\hat{\sigma}}\left(\{\hat{\omega} : \|\mathbf{S}_{r_n}(\hat{\omega})\| < n\delta/2\}\right)\mathcal{P}_{q_n m}(\Gamma) \\ &\geq \hat{P}_{\hat{\sigma}}\left(\{\hat{\omega} : \|\mathbf{S}_{r_n}(\hat{\omega})\| < n\delta/2\}\right)\mathcal{P}_m(\Gamma)^{q_n}\end{aligned}$$

for all sufficiently large n's. Since

$$\inf_{\hat{\sigma}\in\hat{\Sigma}} \min_{1\le r<m} \hat{P}_{\hat{\sigma}}\left(\{\hat{\omega} : \|\mathbf{S}_r(\hat{\omega})\| < n\delta/2\}\right) \longrightarrow 1 \quad \text{as } n \longrightarrow \infty,$$

we now see that $\mathcal{P}_n\big(\Gamma^{(\delta)}\big) > 0$ for all large enough n's. ∎

Assuming that (4.1.5) holds and proceeding as in Section 3.1, we first use Lemma 4.1.4 together with Lemma 3.1.3 to define

$$\mathcal{L}(q,r) = \begin{cases} \infty & \text{if } \sup_{n\in\mathbb{Z}^+} \mathcal{P}_n\big(B(q,r/2)\big) = 0 \\ -\lim_{n\to\infty} \frac{1}{n}\log \mathcal{P}_n\big(B(q,r)\big) & \text{otherwise} \end{cases}$$

for $(q,r) \in E \times (0,\infty)$; and then take

$$I_{\hat{\Pi}}(q) = \sup_{r>0} \mathcal{L}(q,r), \quad q \in E. \tag{4.1.6}$$

4.1.7 Lemma. *Assume that* (4.1.5) *holds. Then the function* $I_{\hat{\Pi}}$ *in* (4.1.6) *is lower semi-continuous and convex. In addition, for every open* G *in* E,

$$\varliminf_{n\to\infty} \frac{1}{n}\log \mathcal{P}_n(G) \ge -\inf_G I_{\hat{\Pi}}. \tag{4.1.8}$$

PROOF: The proof is, more or less, the same as that of Theorem 3.1.6.

To see that $I_{\hat{\Pi}}$ is lower semi-continuous, suppose that $\ell < I_{\hat{\Pi}}(p)$. Then $\mathcal{L}(p,r) \ge \ell$ for some $r > 0$; and therefore $I_{\hat{\Pi}}(q) \ge \mathcal{L}(q,r/2) \ge \mathcal{L}(p,r) \ge \ell$ for all $q \in B(p,r/4)$. To prove the convexity, let $p,\ q \in E$ with $I_{\hat{\Pi}}(p) \vee I_{\hat{\Pi}}(q) < \infty$ be given. For $r > 0$, choose $\delta > 0$ so that

$$\frac{B(p,\delta) + B(q,\delta)}{2} \subseteq B\left(\frac{p+q}{2}, r\right).$$

Then

$$\begin{aligned}
\mu_{\hat{\sigma},2n}\left(B\left(\frac{p+q}{2},r\right)\right) &\ge \hat{P}_{\hat{\sigma}}\left(\{\hat{\omega} : \overline{\mathbf{S}}_n(\hat{\omega}) \in B(p,\delta) \ \&\ \overline{\mathbf{S}}^n_{2n}(\hat{\omega}) \in B(q,\delta)\}\right) \\
&= \int_{\{\hat{\omega}:\overline{\mathbf{S}}_n(\hat{\omega})\in B(p,\delta)\}} \mu_{\hat{\Sigma}_n(\hat{\omega}),n}\big(B(q,\delta)\big)\,\hat{P}_{\hat{\sigma}}(d\hat{\omega}) \\
&\ge \mathcal{P}_n\big(B(q,\delta)\big)\mu_{\hat{\sigma},n}\big(B(p,\delta)\big) \\
&\ge \mathcal{P}_n\big(B(q,\delta)\big)\mathcal{P}_n\big(B(p,\delta)\big) > 0
\end{aligned}$$

for all large enough $n \in \mathbb{Z}^+$. Hence,

$$\mathcal{L}\left(\frac{p+q}{2},r\right) \le \frac{1}{2}\big(\mathcal{L}(p,\delta) + \mathcal{L}(q,\delta)\big) \le \frac{1}{2}\big(I_{\hat{\Pi}}(p) + I_{\hat{\Pi}}(q)\big), \quad r > 0;$$

and so $I_{\hat\Pi}\left(\frac{p+q}{2}\right) \le \left(I_{\hat\Pi}(p) + I_{\hat\Pi}(q)\right)/2$. Since $I_{\hat\Pi}$ is lower semi-continuous, it follows from this that it is also convex.

Next, suppose that G is open in E and that $p \in G$ with $I_{\hat\Pi}(p) < \infty$. Then, for $r > 0$ with $B(p,r) \subseteq G$,

$$\varliminf_{n\to\infty} \frac{1}{n} \log \mathcal{P}_n(G) \ge \varliminf_{n\to\infty} \frac{1}{n} \log \mathcal{P}_n\big(B(p,r)\big) \ge -\mathcal{L}(p,r);$$

and therefore $\varliminf_{n\to\infty} \frac{1}{n} \log \mathcal{P}_n(G) \ge -I_{\hat\Pi}(p)$. ∎

We now introduce an assumption which, among other things, guarantees that our MARKOV chain is uniformly ergodic (cf. Exercise 4.1.48 below). Namely, we will assume that there exist ℓ, $N \in \mathbb{Z}^+$ and $M \in [1,\infty)$ with $\ell \le N$ such that

$$(\hat{\mathbf{U}}) \qquad \begin{cases} \hat\Pi^\ell(\hat\sigma, \cdot\,) \le \frac{M}{N} \sum_{m=1}^N \hat\Pi^m(\hat\tau, \cdot\,) & \text{for all } \hat\sigma,\, \hat\tau \in \hat\Sigma \\ \sup_{\hat\sigma\in\hat\Sigma} \int_E \exp\big[\alpha\|x\|\big]\hat\Pi_E(\hat\sigma, dx) < \infty & \text{for } \alpha \in (0,\infty), \end{cases}$$

where

$$\hat\Pi^{m+1}(\hat\sigma, \cdot\,) \equiv \int_{\hat\Sigma} \hat\Pi^m(\hat\xi, \cdot\,)\, \hat\Pi(\hat\sigma, d\hat\xi), \quad m \ge 1,$$

and $\hat\Pi_E(\hat\sigma,\Gamma) = \hat\Pi(\hat\sigma, \Sigma\times\Gamma)$ for $\Gamma \in \mathcal{B}_E$.

The next lemma contains some important preliminary consequences of $(\hat{\mathbf{U}})$.

4.1.9 Lemma. *Assume that* $(\hat{\mathbf{U}})$ *holds. If*

$$H_\alpha \equiv \log\left(\sup_{\hat\sigma\in\hat\Sigma} \int_E \exp\big[\alpha\|x\|\big]\, \hat\Pi_E(\hat\sigma, dx)\right), \quad \alpha \in (0,\infty),$$

then, for all $m \in \mathbb{Z}^+$, $\delta > 0$, $n \in \mathbb{N}$, *and* $\alpha \in (0,\infty)$:

$$(4.1.10) \qquad \sup_{\hat\sigma\in\hat\Sigma} \hat P_{\hat\sigma}\Big(\{\hat\omega : \|\mathbf{S}^n_{m+n}(\hat\omega)\| \ge \delta\}\Big) \le \exp\big[-\alpha\delta + mH_\alpha\big].$$

In particular, this means that (4.1.5) is satisfied. Furthermore, if $\hat\nu_0 \in \mathbf{M}_1(\hat\Sigma)$ *and* $Q \in \mathbf{M}_1(E)$ *is defined by*

$$Q(\Gamma) = \frac{1}{N}\sum_{k=1}^N \int_{\hat\Sigma} \hat\Pi^k(\hat\sigma, \Sigma\times\Gamma)\, \hat\nu_0(d\hat\sigma), \quad \Gamma \in \mathcal{B}_E,$$

then

$$(4.1.11) \qquad \int_E \exp\big[\alpha\|x\|\big]\, Q(dx) \le \exp\big[H_\alpha\big], \quad \alpha \in (0,\infty),$$

and

$$\text{(4.1.12)} \quad \int_{\hat{\Omega}} F\big(X_{\ell+m}(\hat{\omega}),\dots,X_{n\ell+m}(\hat{\omega})\big)\hat{P}_{\hat{\sigma}}(d\hat{\omega}) \le M^n \int_{E^n} F(\mathbf{x})\, Q^n(d\mathbf{x})$$

for all $m \in \mathbb{N}$, $n \in \mathbb{Z}^+$, *and all measurable* $F : E^n \longrightarrow [0,\infty)$. *Finally, if either* E *is the space of probability measures on some Polish space or* $(X, \|\cdot\|)$ *is a separable* BANACH *space, then, for each* $L \ge 0$, *there is a* $K_L \subset\subset E$ *for which*

$$\text{(4.1.13)} \qquad \overline{\lim_{n\to\infty}} \frac{1}{n} \log\left(\sup_{\hat{\sigma}\in\hat{\Sigma}} \hat{P}_{\hat{\sigma}}\left(\{\hat{\omega} : \overline{\mathbf{S}}^{\ell}_{(n+1)\ell}(\hat{\omega}) \notin K_L\}\right)\right) \le -L.$$

PROOF: Noting that

$$\int_{\hat{\Omega}} \exp\Big[\alpha\|\mathbf{S}^n_{m+n}(\hat{\omega})\|\Big]\, \hat{P}_{\hat{\sigma}}(d\hat{\omega}) = \int_{\hat{\Sigma}} \left(\int_{\hat{\Omega}} \exp\Big[\alpha\|\mathbf{S}_m(\hat{\omega})\|\Big]\, \hat{P}_{\hat{\xi}}(d\hat{\omega})\right) \hat{\Pi}^n(\hat{\sigma}, d\hat{\xi}),$$

we see that (4.1.10) will be proved as soon as we show that

$$\int_{\hat{\Omega}} \exp\left[\alpha\|\mathbf{S}_m(\hat{\omega})\|\right]\, \hat{P}_{\hat{\sigma}}(d\hat{\omega}) \le \exp\big[mH_\alpha\big], \quad \hat{\sigma} \in \hat{\Sigma}.$$

But

$$\int_{\hat{\Omega}} \exp\left[\alpha\|\mathbf{S}_{m+1}(\hat{\omega})\|\right]\, \hat{P}_{\hat{\sigma}}(d\hat{\omega}) \le \int_{\hat{\Omega}} \exp\left[\alpha\|\mathbf{S}_m(\hat{\omega})\|\right] \left(\int_E \exp\left[\alpha\|x\|\right] \hat{\Pi}_E\big(\hat{\Sigma}_m(\hat{\omega}), d\xi\big)\right) \hat{P}_{\hat{\sigma}}(d\hat{\omega}),$$

and so the required estimate follows by induction on m.

Next let $\hat{\nu}_0$ be given and define Q accordingly. Then, since

$$\hat{\Pi}^{m+1}(\hat{\sigma}, \cdot\,) = \int_{\hat{\Sigma}} \hat{\Pi}(\hat{\xi}, \cdot\,)\hat{\Pi}^m(\hat{\sigma}, d\hat{\xi}) \le \sup_{\hat{\xi}\in\hat{\Sigma}} \hat{\Pi}(\hat{\xi}, \cdot\,),$$

we see that Q satisfies (4.1.11). In addition, since, for any $m \in \mathbb{N}$, $n \in \mathbb{Z}^+$, and measurable $F : E^n \longrightarrow [0,\infty)$

$$\begin{aligned}
&\int_{\hat{\Omega}} F\big(X_{\ell+m}(\hat{\omega}),\dots,X_{n\ell+m}(\hat{\omega})\,\big)\hat{P}_{\hat{\sigma}}(d\hat{\omega}) \\
&= \int_{\hat{\Sigma}} \left(\int_{\hat{\Omega}} F\big(X_\ell(\hat{\omega}),\dots,X_{n\ell}(\hat{\omega})\big)\, \hat{P}_{\hat{\xi}}(d\hat{\omega})\right) \hat{\Pi}^m(\hat{\sigma}, d\hat{\xi}) \\
&\le \sup_{\hat{\xi}\in\hat{\Sigma}} \int_{\hat{\Omega}} F\big(X_\ell(\hat{\omega}),\dots,X_{n\ell}(\hat{\omega})\big)\, \hat{P}_{\hat{\xi}}(d\hat{\omega}),
\end{aligned}$$

we see that it suffices to prove (4.1.12) when $m = 0$. But, by **($\hat{\mathbf{U}}$)**,

$$\begin{aligned}\int_{\hat{\Omega}} F\big(X_\ell(\hat{\omega}), \ldots, X_{(n+1)\ell}(\hat{\omega})\big)\, \hat{P}_{\hat{\sigma}}(d\hat{\omega}) &= \int_{\hat{\Omega}} \left(\int_E F\big(X_\ell(\hat{\omega}), \ldots, X_{n\ell}(\hat{\omega}), x\big)\, \hat{\Pi}_E^\ell\big(\hat{\Sigma}_{n\ell}(\hat{\omega}), dx\big) \right) \hat{P}_{\hat{\sigma}}(d\hat{\omega}) \\ &\leq M \int_{\hat{\Omega}} \left(\int_E F\big(X_\ell(\hat{\omega}), \ldots, X_{n\ell}(\hat{\omega}), x\big)\, Q(dx) \right) \hat{P}_{\hat{\sigma}}(d\hat{\omega});\end{aligned}$$

and therefore the desired result follows easily by induction on n.

Given (4.1.10) and (4.1.12), the proof of (4.1.13) can be accomplished as follows. Using Lemma 3.2.7 in the case when E is the space of probability measures on a Polish space and Theorem 3.3.11 when $(X, \|\cdot\|)$ is a separable BANACH space, we can find compact sets K_L in E such that $\overline{\lim}_{n\to\infty} \frac{1}{n} \log\left(Q_n(K_L^c)\right) \leq -(M + L)$. Furthermore, by Lemma 3.1.1, we may assume that these K_L 's are convex. Hence,

$$\hat{P}_{\hat{\sigma}}\left(\{\hat{\omega} : \overline{\mathbf{S}}^\ell_{(n+1)\ell}(\hat{\omega}) \notin K_L\}\right) \leq \sum_{k=1}^{\ell} \hat{P}_{\hat{\sigma}}\left(\left\{ \hat{\omega} : \frac{1}{n} \sum_{m=1}^{n} X_{m\ell+k}(\hat{\omega}) \notin K_L \right\} \right);$$

and so, by (4.1.12), the required estimate follows. ∎

We are now ready to prove the basic large deviation result of this section.

4.1.14 Theorem. *Assume that* ($\hat{\mathbf{U}}$) *holds and let* $I_{\hat{\Pi}}$ *be the function defined in* (4.1.6). *Then, for every* $K \subset\subset E$,

$$\overline{\lim_{n\to\infty}} \frac{1}{n} \log\left(\sup_{\hat{\sigma}\in\hat{\Sigma}} \mu_{\hat{\sigma},n}(K) \right) \leq -\inf_K I_{\hat{\Pi}}. \tag{4.1.15}$$

Furthermore, if either E *is the space of probability measures on some Polish space or* $(X, \|\cdot\|)$ *is a separable* BANACH *space, then* $I_{\hat{\Pi}}$ *is a good rate function and, for all* $\Gamma \in \mathcal{B}_E$,

$$\begin{aligned}-\inf_{\Gamma^\circ} I_{\hat{\Pi}} &\leq \underline{\lim_{n\to\infty}} \frac{1}{n} \log\left(\inf_{\hat{\sigma}\in\hat{\Sigma}} \mu_{\hat{\sigma},n}(\Gamma) \right) \\ &\leq \overline{\lim_{n\to\infty}} \frac{1}{n} \log\left(\sup_{\hat{\sigma}\in\hat{\Sigma}} \mu_{\hat{\sigma},n}(\Gamma) \right) \leq -\inf_{\overline{\Gamma}} I_{\hat{\Pi}}.\end{aligned} \tag{4.1.16}$$

PROOF: Set $\mu^\ell_{\hat{\sigma},n} = \hat{P}_{\hat{\sigma}} \circ \left(\overline{\mathbf{S}}^\ell_{(n+1)\ell}\right)^{-1}$, and note that

$$\mu^\ell_{\hat{\sigma},n} = \int_{\hat{\Sigma}} \mu_{\hat{\xi},n\ell} \Pi^\ell(\hat{\sigma}, d\hat{\xi}).$$

In particular,

$$\inf_{\hat\tau}\mu_{\hat\tau,n\ell}(\Gamma)\le\mu^{\ell}_{\hat\sigma,n}(\Gamma)\le\sup_{\hat\tau}\mu_{\hat\tau,n\ell}(\Gamma),\quad\Gamma\in\mathcal{B}_E.\tag{4.1.17}$$

What we will show first is that everything holds when $\mu_{\hat\sigma,n}$ is replaced by $\mu^{\ell}_{\hat\sigma,n}$ and $I_{\hat\Pi}$ is replaced by $\ell I_{\hat\Pi}$. It will then be a relatively simple matter to pass to the desired statements.

We begin by showing that

$$\varlimsup_{r\to0}\varlimsup_{n\to\infty}\frac{1}{n}\log\left(\sup_{\hat\sigma\in\Sigma}\mu^{\ell}_{\hat\sigma,n}\big(B(p,r)\big)\right)\le-\ell I_{\hat\Pi}(p)\tag{4.1.18}$$

for every $p\in E$. To this end, suppose that $0<\alpha<I_{\hat\Pi}(p)$ and choose $\delta>0$ so that $\mathcal{L}(p,4\delta)>\alpha$. Then, by $(\hat{\mathbf{U}})$, for all $\hat\sigma,\hat\tau\in\hat\Sigma$,

$$\begin{aligned}\mu^{\ell}_{\hat\sigma,n}\big(B(p,\delta)\big)&=\int_{\hat\Sigma}\mu_{\hat\xi,n\ell}\big(B(p,\delta)\big)\,\hat\Pi^{\ell}(\hat\sigma,d\hat\xi)\\&\le\frac{M}{N}\sum_{m=1}^{N}\int_{\hat\Sigma}\mu_{\hat\xi,n\ell}\big(B(p,\delta)\big)\,\hat\Pi^{m}(\hat\tau,d\hat\xi).\end{aligned}$$

At the same time, for each $1\le m\le N$,

$$\begin{aligned}\int_{\hat\Sigma}\mu_{\hat\xi,n\ell}\big(B(p,\delta)\big)\,\hat\Pi^{m}(\hat\tau,d\hat\xi)&=\hat P_{\hat\tau}\Big(\big\{\hat\omega:\overline{\mathbf{S}}^{m}_{m+n\ell}\in B(p,\delta)\big\}\Big)\\&\le\mu_{\hat\tau,n\ell}\big(B(p,2\delta)\big)+\hat P_{\hat\tau}\Big(\big\{\hat\omega:\big\|\overline{\mathbf{S}}^{m}_{m+n\ell}-\overline{\mathbf{S}}_{n\ell}\big\|\ge\delta\big\}\Big)\\&\le\mu_{\hat\tau,n\ell}\big(B(p,2\delta)\big)+2\exp\big[-n\ell\alpha+NH_\beta\big],\end{aligned}$$

where $\beta=\alpha/\delta$ and we have used (4.1.10). Hence, we now see that

$$\sup_{\hat\sigma\in\hat\Sigma}\mu^{\ell}_{\hat\sigma,n}\big(B(p,\delta)\big)\le M\mathcal{P}_{n\ell}\big(B(p,2\delta)\big)+2\exp\big[-n\ell\alpha+NH_\beta\big],\quad n\in\mathbb{Z}^+.$$

Since $\mathcal{P}_n\big(B(p,2\delta)\big)\le\exp\big[-n\alpha\big]$ for sufficiently large $n\ge1$, we conclude from the above that

$$\varlimsup_{n\to\infty}\frac{1}{n}\log\left(\sup_{\hat\sigma\in\hat\Sigma}\mu^{\ell}_{\hat\sigma,n}\big(B(p,\delta)\big)\right)\le-\ell\alpha,$$

and therefore that (4.1.18) holds.

Given (4.1.18), one can proceed in exactly the same way as one did in part (**iv**) of Exercise 2.1.14 to show that

$$\lim_{\delta\to0}\varlimsup_{n\to\infty}\frac{1}{n}\log\left[\sup_{\hat\sigma\in\hat\Sigma}\mu^{\ell}_{\hat\sigma,n}\big(K^{(\delta)}\big)\right]\le-\ell\inf_{K}I_{\hat\Pi},\quad K\subset\subset E.\tag{4.1.19}$$

In particular, we now know that (4.1.15) holds when $\mu_{\hat{\sigma},n}$ and $I_{\hat{\Pi}}$ are replaced by $\mu^{\ell}_{\hat{\sigma},n}$ and $\ell I_{\hat{\Pi}}$, respectively. Furthermore, from (4.1.8) and (4.1.17), we see that

$$\underline{\lim_{n\to\infty}} \frac{1}{n} \log\left(\inf_{\hat{\sigma}\in\hat{\Sigma}} \mu^{\ell}_{\hat{\sigma},n}(G)\right) \geq -\ell \inf_{G} I_{\hat{\Pi}}$$

for every open G in E. Finally, suppose that E is either the space of probability measures on some Polish space or that $(X, \|\cdot\|)$ is a separable BANACH space. By (4.1.13), we know that there is a family $\{K_L : L \geq 0\}$ of compact, convex E such that

$$\overline{\lim_{n\to\infty}} \frac{1}{n} \log\left(\sup_{\hat{\sigma}\in\hat{\Sigma}} \mu^{\ell}_{\hat{\sigma},n}(K_L^{\mathrm{c}})\right) \leq -L, \quad L > 0.$$

Hence, just as in the proof of Lemma 2.1.5, we conclude not only that $I_{\hat{\Pi}}$ is good but also that (4.1.16) holds with $\mu^{\ell}_{\hat{\sigma},n}$ and $\ell I_{\hat{\Pi}}$ in place of $\mu_{\hat{\sigma},n}$ and $I_{\hat{\Pi}}$.

In order to complete the proof, note that, from (4.1.10),

$$\overline{\lim_{n\to\infty}} \frac{1}{n} \log\left(\sup_{\hat{\sigma}\in\hat{\Sigma}} \mu_{\hat{\sigma},n}(\Gamma)\right) \leq \overline{\lim_{n\to\infty}} \frac{1}{n\ell} \log\left(\sup_{\hat{\sigma}\in\hat{\Sigma}} \mu^{\ell}_{\hat{\sigma},n}(\Gamma^{(\delta)})\right), \quad \delta > 0.$$

for every $\Gamma \in \mathcal{B}_E$. Hence, (4.1.15) follows from (4.1.19); and, when E is either the space of probability measures on some Polish space or a separable BANACH space, the right hand side of (4.1.16) is an easy consequence of the fact that it holds when $I_{\hat{\Pi}}$ and $\mu_{\hat{\sigma},n}$ are replaced by $\ell I_{\hat{\Pi}}$ and $\mu^{\ell}_{\hat{\sigma},n}$, respectively. Since the left hand side of (3.1.16) is precisely (3.1.8), the proof is now complete. ∎

4.1.20 Corollary. *Assume that* $(\hat{\mathbf{U}})$ *holds and that either* E *is the space of probability measures on some Polish space* Σ' *or that* $(X, \|\cdot\|)$ *is a separable real* BANACH *space. Then, for every* $\Phi \in C(E;\mathbb{R})$ *which satisfies*

$$\sup_{n\in\mathbb{Z}^+} \sup_{\hat{\sigma}\in\hat{\Sigma}} \left(\int_E \exp\big[n\alpha\Phi\big]\, d\mu_{\hat{\sigma},n}\right)^{1/n} < \infty \tag{4.1.21}$$

for some $\alpha \in (1,\infty)$, *one has that*

$$\lim_{n\to\infty} \sup_{\hat{\sigma}\in\hat{\Sigma}} \left| \frac{1}{n} \log\left(\int_E \exp[\Phi]\, d\mu_{\hat{\sigma},n}\right) - \sup\{\Phi(q) - I_{\hat{\Pi}}(q) : q \in E\} \right| = 0. \tag{4.1.22}$$

In particular, if

$$(4.1.23)\qquad \Lambda_{\hat{\Pi}}(\lambda) \equiv \overline{\lim_{n\to\infty}} \frac{1}{n} \sup_{\hat{\sigma}\in\hat{\Sigma}} \Lambda_{\mu_{\hat{\sigma},n}}(n\lambda), \quad \lambda \in X^*,$$

then $\Lambda_{\hat{\Pi}}(\lambda) \in \mathbb{R}$, $\lambda \in X^*$,

$$(4.1.24)\qquad \begin{aligned} \Lambda_{\hat{\Pi}}(\lambda) &= \sup\left\{{}_{X^*}\langle \lambda, q\rangle_X - I_{\hat{\Pi}}(q) : q \in E\right\}, \quad \lambda \in X^*, \\ I_{\hat{\Pi}}(q) &= \sup\left\{{}_{X^*}\langle \lambda, q\rangle_X - \Lambda_{\hat{\Pi}}(\lambda) : \lambda \in X^*\right\}, \quad q \in E, \end{aligned}$$

and

$$(4.1.25)\qquad \lim_{n\to\infty} \sup_{\hat{\sigma}\in\hat{\Sigma}} \left| \frac{1}{n}\Lambda_{\mu_{\hat{\sigma},n}}(n\lambda) - \Lambda_{\hat{\Pi}}(\lambda)\right| = 0, \quad \lambda \in X^*.$$

(Remember that, when $E = \mathbf{M}_1(\Sigma')$, $X^* = C_b(\Sigma';\mathbb{R})$ *and*

$${}_{X^*}\langle \lambda, q\rangle_X = \int_{\Sigma'} \lambda(\sigma')\, q(d\sigma')$$

for $\lambda \in C_b(\Sigma';\mathbb{R})$ *and* $q \in \mathbf{M}_1(\Sigma')$*.)*

PROOF: The first assertion is an immediate consequence of (4.1.16) combined with Exercise 2.1.15. Once one has (4.1.22), (4.1.24) follows from the estimate (4.1.10) together with Theorem 2.2.21. Finally, (4.1.25) is just a special case of (4.1.22). ∎

4.1.26 Remark.

It should be clear that Theorem 4.1.14 and Corollary 4.1.20 applied to the case when $\hat{\Pi}(\hat{\sigma},\cdot) = \mu \in \mathbf{M}_1(E)$, $\hat{\sigma} \in \hat{\Sigma}$, can be used to recover both the SANOV as well as the CRAMÈR Theorems.

What we want to do now is turn our attention to the situation described in Example 4.1.1. In other words: $E = \mathbf{M}_1(\Sigma)$; Π is a transition probability function on Σ; for each $\sigma \in \Sigma$, P_σ on $\Omega = \Sigma^{\mathbb{N}}$ is the MARKOV chain starting at σ with transition probability Π; and $\{\mu_{\sigma,n} : n \geq 1\}$ is the distribution of the empirical distribution functional $\omega \longmapsto \mathbf{L}_n(\omega)$ in (4.1.3). As a consequence of the preceding, we know that if there exist $\ell, N \in \mathbb{Z}^+$ with $1 \leq \ell \leq N$and $M \in [1,\infty)$ such that

$$(\mathrm{U})\qquad \Pi^\ell(\sigma,\,\cdot\,) \leq \frac{M}{N}\sum_{m=1}^{N} \Pi^m(\tau,\,\cdot\,) \text{ for } \sigma,\, \tau \in \Sigma,$$

(the second part of $(\hat{\mathbf{U}})$ is trivially satisfied in this case) then for every $\Gamma \in \mathcal{B}_{\mathbf{M}_1(\Sigma)}$

(4.1.27)
$$
\begin{aligned}
-\inf_{\Gamma^\circ} \Lambda_\Pi^* &\le \underline{\lim}_{n\to\infty} \frac{1}{n}\log\left(\inf_{\sigma\in\Sigma} \mu_{\sigma,n}(\Gamma)\right) \\
&\le \overline{\lim}_{n\to\infty} \frac{1}{n}\log\left(\sup_{\sigma\in\Sigma} \mu_{\sigma,n}(\Gamma)\right) \le -\inf_{\overline{\Gamma}} \Lambda_\Pi^*,
\end{aligned}
$$

where
(4.1.28)
$$\Lambda_\Pi^*(\nu) = \sup\left\{\int_\Sigma V\,d\nu - \Lambda_\Pi(V) : V \in C_b(\Sigma;\mathbb{R})\right\}, \quad \nu \in \mathbf{M}_1(\Sigma),$$

and

(4.1.29)
$$\Lambda_\Pi(V) \equiv \overline{\lim}_{n\to\infty} \frac{1}{n}\log\left(\sup_{\sigma\in\Sigma}\int_\Omega \exp\left[\sum_{k=1}^n V\big(\Sigma_k(\omega)\big)\right] P_\sigma(d\omega)\right)$$

for $V \in C_b(\Sigma;\mathbb{R})$. Of course, these functions Λ_Π and Λ_Π^* make perfectly good sense even when one does not assume that $(\mathbf{U})$ holds; and, as we will see below, the program on which we are about to embark makes no use of $(\mathbf{U})$.

Let $\big(B(\Sigma;\mathbb{R}), \|\cdot\|_B\big)$ denote the BANACH space of bounded, measurable real-functions on Σ with $\|\cdot\|_B$ being the uniform norm; and, again using the expression in (4.1.29), extend Λ_Π to the whole of $B(\Sigma;\mathbb{R})$. Clearly,

(4.1.30)
$$\big|\Lambda_\Pi(V) - \Lambda_\Pi(W)\big| \le \|V - W\|_B, \quad V, W \in B(\Sigma,\mathbb{R}).$$

Also, as a consequence of HÖLDER's inequality, note that Λ_Π is a convex function on $B(\Sigma;\mathbb{R})$.

Our aim is to find alternative expressions for Λ_Π and Λ_Π^*. In particular, we want to give an expression for Λ_Π^* which is more directly related to Π itself. In doing so, we will need to introduce the operators $\Pi_V : B(\Sigma;\mathbb{R}) \longrightarrow B(\Sigma;\mathbb{R})$, $V \in B(\Sigma;\mathbb{R})$, defined by

(4.1.31)
$$\big[\Pi_V\phi\big](\sigma) = \exp\big[V(\sigma)\big]\int_\Sigma \phi(\tau)\,\Pi(\sigma,d\tau)$$

for $\sigma \in \Sigma$ and $\phi \in B(\Sigma;\mathbb{R})$. When $V \equiv 0$, we will use Π to denote the operator Π_V. Also, it will be useful to recall the concept of the **logarithmic spectral radius** of a bounded linear operator $L : B(\Sigma;\mathbb{R}) \longrightarrow B(\Sigma;\mathbb{R})$. Namely, the logarithmic spectral radius $\rho(L)$ of L is the number given by

(4.1.32)
$$\rho(L) = \lim_{n\to\infty} \frac{1}{n}\log\big(\|L^n\|_{op}\big),$$

where $\|L\|_{op} \equiv \sup\{\|L\phi\|_B : \|\phi\|_B \le 1\}$ and $\|\cdot\|_B$ is the uniform norm on $B(\Sigma;\mathbb{R})$. (Note that $n \in \mathbb{Z}^+ \longmapsto \log(\|L^n\|_{op}) \in \mathbb{R}$ is sub-additive and therefore that the limit in (4.1.31) necessarily exits.)

The first step in our program is taken in the following trivial version of the FEYNMAN–KAC formula.

4.1.33 Lemma. *For any $V \in B(\Sigma;\mathbb{R})$,*

$$\int_\Omega \exp\left[\sum_{k=1}^n V(\Sigma_k(\omega))\right] \phi(\Sigma_{n+1}(\omega))\, P_\sigma(d\omega) = \exp[-V(\sigma)]\,[\Pi_V^{n+1}\phi](\sigma), \quad (n,\sigma) \in \mathbb{Z}^+ \times \Sigma, \tag{4.1.34}$$

for all $\phi \in B(\Sigma,\mathbb{R})$. In particular,

$$\Lambda_\Pi(V) = \rho(\Pi_V) = \lim_{n\to\infty} \frac{1}{n} \log\left[\left\| [\Pi_V^n 1] \right\|_B\right], \quad V \in B(\Sigma;\mathbb{R}). \tag{4.1.35}$$

PROOF: To prove (4.1.34), note that

$$\begin{aligned}
&\int_\Omega \exp\left[\sum_{k=1}^n V(\Sigma_k(\omega))\right] \phi(\Sigma_{n+1}(\omega))\, P_\sigma(d\omega) \\
&= \int_\Omega \exp\left[\sum_{k=1}^{n-1} V(\Sigma_k(\omega))\right] \left(\int_\Sigma \phi(\tau)\,\Pi_V(\Sigma_n(\omega), d\tau)\right) P_\sigma(d\omega) \\
&= \int_\Omega \exp\left[\sum_{k=1}^{n-1} V(\Sigma_k(\omega))\right] [\Pi_V\phi](\Sigma_n(\omega))\, P_\sigma(d\omega)
\end{aligned}$$

and that

$$\int_\Omega \phi(\Sigma_1(\omega))\, P_\sigma(d\omega) = \exp[-V(\sigma)]\,[\Pi_V\phi](\sigma).$$

Hence, (4.1.34) follows by induction.

Once one has (4.1.34), (4.1.35) is obvious. ∎

4.1.36 Lemma. *Let Π be a transition probability function on Σ and suppose that $\mathcal{A}$ is a closed subalgebra of $B(\Sigma;\mathbb{R})$ with the properties that $1 \in \mathcal{A}$ and $f \circ \phi \in \mathcal{A}$ whenever $f \in C^\infty(\mathbb{R};\mathbb{R})$ and $\phi \in \mathcal{A}$. If $\mathcal{A}$ is invariant under the operator Π, then*

$$\begin{aligned}
&\sup\left\{\int_\Sigma V\,d\nu - \Lambda_\Pi(V) : V \in \mathcal{A}\right\} \\
&= \sup\left\{-\int_\Sigma \log\frac{[\Pi u]}{u}\,d\nu : u \in \mathcal{A} \text{ and } u \ge 1\right\}, \quad \nu \in \mathbf{M}_1(\Sigma).
\end{aligned} \tag{4.1.37}$$

In particular, if Π *is* FELLER *continuous (i.e.,* $C_b(\Sigma;\mathbb{R})$ *is invariant under* Π*), then* $\Lambda_\Pi^* = J_\Pi$*, where*

$$J_\Pi(\nu) \equiv \sup\left\{-\int_\Sigma \log\frac{[\Pi u]}{u}\,d\nu : u \in C_b\big(\Sigma;[1,\infty)\big)\right\} \tag{4.1.38}$$

for $\nu \in \mathbf{M}_1(\Sigma)$*. Finally, in any case,*

$$\begin{aligned} J_\Pi(\nu) &= \sup\left\{-\int_\Sigma \log\frac{[\Pi u]}{u}\,d\nu : u \in B\big(\Sigma;[1,\infty)\big)\right\} \\ &= \sup\left\{\int_\Sigma V\,d\nu - \Lambda_\Pi(V) : V \in B(\Sigma;\mathbb{R})\right\} \end{aligned} \tag{4.1.39}$$

for every $\nu \in \mathbf{M}_1(\Sigma)$.

PROOF: Let $\Lambda_\Pi^{*\mathcal{A}}$ and $J_\Pi^{\mathcal{A}}$ denote the left and right hand sides of (4.1.37), respectively. Given $u \in \mathcal{A}$ with $u \geq 1$, set $V = \log\frac{u}{[\Pi u]}$. Then, since $[\Pi_V\, u] = u$, $\Lambda_\Pi(V) = 0$, and so

$$\Lambda_\Pi^{*\mathcal{A}}(\nu) \geq \int_\Sigma \log\frac{u}{[\Pi u]}\,d\nu, \quad \nu \in \mathbf{M}_1(\Sigma).$$

Thus, we now know that $\Lambda_\Pi^{*\mathcal{A}} \geq J_\Pi^{\mathcal{A}}$. To prove the opposite inequality, suppose that $\alpha > \Lambda_\Pi(V)$. Then, by (4.1.35), $\sum_{n=0}^\infty \exp[-n\alpha]\,[\Pi_V^n 1]$ converges uniformly to an element $u_V \in \mathcal{A}$ which satisfies $u_V \geq 1$. In addition, since $[\Pi_V\, u_V] = e^\alpha\big(u_V - 1\big)$,

$$\int_\Sigma V\,d\nu + \int_\Sigma \log\frac{[\Pi u_V]}{u_V}\,d\nu = \int_\Sigma \log\frac{[\Pi_V\, u_V]}{u_V}\,d\nu \leq \alpha, \quad \nu \in \mathbf{M}_1(\Sigma).$$

Hence,

$$\int_\Sigma V\,d\nu - \Lambda_\Pi(V) \leq J_\Pi^{\mathcal{A}}(\nu), \quad V \in \mathcal{A} \text{ and } \nu \in \mathbf{M}_1(\Sigma);$$

and clearly this is more than enough to conclude that $\Lambda_\Pi^{*\mathcal{A}} \leq J_\Pi^{\mathcal{A}}$.

Finally, note that $J_\Pi(\nu)$ is obviously dominated by the right hand side of (4.1.39). At the same time, for fixed $\nu \in \mathbf{M}_1(\Sigma)$, the set of $u \in B\big(\Sigma;[1,\infty)\big)$ for which $-\int_\Sigma \log\frac{[\Pi u]}{u}\,d\nu \leq J_\Pi(\nu)$ contains $C_b\big(\Sigma;[1,\infty)\big)$, is closed under bounded point-wise convergence, and therefore contains $B\big(\Sigma;[1,\infty)\big)$. ∎

We now want to show that one can say more about the topic in Lemma 4.1.36 when (**U**) holds.

4.1.40 Lemma. *Assume that* (**U**) *holds; and, given* $\nu \in \mathbf{M}_1(\Sigma)$, *define* $\mu \in \mathbf{M}_1(\Sigma)$ *by*

$$\mu = \frac{1}{N}\sum_{k=1}^{N}\int_{\Sigma}\Pi^k(\sigma,\cdot)\,\nu(d\sigma).$$

Then, for each $V \in B(\Sigma;\mathbb{R})$,

$$\int_{\Omega}\exp\left[\sum_{k=1}^{(n+1)\ell} V\big(\Sigma_k(\omega)\big)\right] P_\sigma(d\omega)$$
$$\leq \exp\big[\ell\|V\|_B\big]\left(M\int_{\Sigma}\exp\big[\ell V(\sigma)\big]\,\mu(d\sigma)\right)^n, \quad V \in B(\Sigma;\mathbb{R}),$$

and therefore

$$\Lambda_\Pi(V) \leq \frac{1}{\ell}\Big(\Lambda_{\tilde{\mu}}(\ell V) + \log M\Big), \quad V \in B(\Sigma;\mathbb{R}), \tag{4.1.41}$$

(*See discussion preceding* (3.2.11) *for the notation* $\tilde{\mu}$, *and use* (3.2.6) *with* $Q = \tilde{\mu}$ *to define* $\Lambda_{\tilde{\mu}}$.) *In addition, if* $V \in B(\Sigma;\mathbb{R})$ *and* $\{V_n\}_1^\infty$ *is a bounded sequence in* $B(\Sigma;\mathbb{R})$ *such that* $V_n(\sigma) \longrightarrow V(\sigma)$ *for* μ-*almost every* $\sigma \in \Sigma$, *then* $\Lambda_\Pi(V_n) \longrightarrow \Lambda_\Pi(V)$. *In particular,*

$$\Lambda_\Pi^* = J_\Pi \tag{4.1.42}$$

even when Π *is not necessarily* FELLER *continuous.*

PROOF: Choose and fix $\sigma_0 \in \Sigma$ and set $\mu = \frac{1}{N}\sum_{m=1}^{N}\Pi^m(\sigma_0,\,\cdot\,)$. Using (4.1.12) (note that the Q there is the distribution of $\sigma \in \Sigma \longmapsto \delta_\sigma \in \mathbf{M}_1(\Sigma)$ under the μ here) and HÖLDER's inequality, note that

$$\int_{\Omega}\exp\left[\sum_{k=1}^{(n+1)\ell} V\big(\Sigma_k(\omega)\big)\right] P_\sigma(d\omega)$$
$$\leq \exp\big[\ell\|V\|_B\big]\int_{\Omega}\prod_{k=1}^{\ell}\exp\left[\sum_{m=1}^{n} V\big(\Sigma_{m\ell+k}(\omega)\big)\right] P_\sigma(d\omega)$$
$$\leq \exp\big[\ell\|V\|_B\big]\prod_{k=1}^{\ell}\left(\int_{\Omega}\exp\left[\sum_{m=1}^{n} \ell V\big(\Sigma_{m\ell+k}(\omega)\big)\right] P_\sigma(d\omega)\right)^{1/\ell};$$

from which (4.1.41) follows.

To prove the asserted convergence result, let $\{V_n\}_1^\infty$ and V be given. Then, by convexity, (4.1.34), and the preceding,

$$\Lambda_\Pi(V) \leq \frac{1}{p}\Lambda_\Pi(pV_n) + \frac{1}{p'}\Lambda_\Pi\big(p'(V - V_n)\big)$$
$$\leq \frac{1}{p}\Lambda_\Pi(pV_n) + \frac{1}{\ell p'}\Big(\log M + \Lambda_{\tilde{\mu}}\big(p'\ell(V - V_n)\big)\Big)$$

for every $p \in (1,\infty)$. Since $\lim_{n\to\infty} \Lambda_{\tilde{\mu}}\big(p'\ell(V - V_n)\big) = 0$ for every $p \in (1,\infty)$, one concludes that $\Lambda_\Pi(V) \le \underline{\lim}_{n\to\infty} \Lambda_\Pi(V_n)$ by letting $p \searrow 1$ in the preceding. Because the same argument leads to $\overline{\lim}_{n\to\infty} \Lambda_\Pi(V_n) \le \Lambda_\Pi(V)$, we now have that $\Lambda_\Pi(V) = \lim_{n\to\infty} \Lambda_\Pi(V_n)$.

With the preceding in hand, we know that $V \in B(\Sigma;\mathbb{R}) \longmapsto \Lambda_\Pi(V) \in \mathbb{R}$ is continuous under bounded point-wise convergence; and therefore, it is an easy matter to check that, for each $\nu \in \mathbf{M}_1(\Sigma)$,

$$\Lambda_\Pi^*(\nu) = \sup\left\{\int_\Sigma V\,d\nu - \Lambda_\Pi(V) : V \in B(\Sigma;\mathbb{R})\right\}.$$

Thus, by (4.1.39), $J_\Pi = \Lambda_\Pi^*$. ∎

By combining the above considerations with Theorem 4.1.14, we now arrive at the following version of a theorem proved originally by DONSKER and VARADHAN [**30**].

4.1.43 Theorem. (DONSKER & VARADHAN) *Assume that* (**U**) *holds and define* J_Π *as in* (4.1.38). *Then* J_Π *is a good convex rate function and for every* $\Gamma \in \mathcal{B}_{\mathbf{M}_1(\Sigma)}$

$$(4.1.44)\qquad \begin{aligned} -\inf_{\Gamma^\circ} J_\Pi &\le \underline{\lim}_{n\to\infty} \frac{1}{n}\log\left(\inf_{\sigma\in\Sigma} P_\sigma\big(\{\omega : \mathbf{L}_n(\omega)\in\Gamma\}\big)\right) \\ &\le \overline{\lim}_{n\to\infty} \frac{1}{n}\log\left(\sup_{\sigma\in\Sigma} P_\sigma\big(\{\omega : \mathbf{L}_n(\omega)\in\Gamma\}\big)\right) \le -\inf_{\overline{\Gamma}} J_\Pi. \end{aligned}$$

Having gone to the trouble of replacing Λ_Π^* in Theorem 4.1.14 by J_Π, it is only reasonable to ask whether the effort was worthwhile. A partial answer is provided by the following sharpened version of a result due to DONSKER and VARADHAN.

4.1.45 Lemma. *For any transition probability functions* Π,

$$(4.1.46)\qquad \|\nu - \nu\Pi\|_{\text{var}}^2 \le 2J_\Pi(\nu), \quad \nu \in \mathbf{M}_1(\Sigma),$$

where

$$(4.1.47)\qquad \nu\Pi \equiv \int_\Sigma \Pi(\sigma,\cdot\,)\,\nu(d\sigma), \quad \nu \in \mathbf{M}_1(\Sigma).$$

In particular, $J_\Pi(\mu) = 0$ *if and only if* $\mu = \mu\Pi$.

PROOF: By JENSEN's inequality and Lemma 3.2.13,

$$\begin{aligned} J_\Pi(\nu) &= \sup\left\{-\int_\Sigma \log\frac{[\Pi e^\psi]}{e^\psi}\,d\nu : \psi \in C_b(\Sigma;\mathbb{R})\right\} \\ &= \sup\left\{\int_\Sigma \psi\,d\nu - \int_\Sigma \log\big([\Pi e^\psi]\big)\,d\nu : \psi \in C_b(\Sigma;\mathbb{R})\right\} \\ &\geq \sup\left\{\int_\Sigma \psi\,d\nu - \log\left(\int_\Sigma \big([\Pi e^\psi]\big)\,d\nu\right) : \psi \in C_b(\Sigma;\mathbb{R})\right\} \\ &= I_{\widetilde{\nu\Pi}}(\nu) = \mathbf{H}(\nu|\nu\Pi). \end{aligned}$$

Now apply Exercise 3.2.24.

Finally, suppose that $\mu = \mu\Pi$. Then, by JENSEN's inequality,

$$\begin{aligned} -\int_\Sigma \log\frac{\Pi u}{u}\,d\mu &= \int_\Sigma \log u\,d\mu - \int_\Sigma \log\big(\Pi u\big)\,d\mu \\ &\leq \int_\Sigma \log u\,d\mu - \int_\Sigma \Pi\big(\log u\big)\,d\mu = 0 \end{aligned}$$

for every $u \in C_b(\Sigma;[1,\infty))$. Hence, $J_\Pi(\mu) = 0$. ∎

4.1.48 Exercise.

The condition (**U**) is more than enough to guarantee that there exists precisely one $\mu \in \mathbf{M}_1(\Sigma)$ which is Π-**invariant** (i.e., $\mu = \mu\Pi$). In this exercise, we will give two approaches to the proof of this fact. The first of these approaches makes direct use of the results which we have just proved; the second, and in many ways better, approach is a particularly simple example of the DOEBLIN theory of ergodicity for MARKOV chains [**28**].

(**i**) Let $\mathfrak{I}_\Pi = \{\mu \in \mathbf{M}_1(\Sigma) : \mu = \mu\Pi\}$. Clearly $\mathfrak{I}_\Pi$ is a convex set. Furthermore, by Lemma 4.1.45, $\mathfrak{I}_\Pi = \{\mu \in \mathbf{M}_1(\sigma) : J_\Pi(\mu) = 0\}$; and, by (4.1.44) applied to $\Gamma = \mathbf{M}_1(\Sigma)$, we know that $\inf_{\mathbf{M}_1(\Sigma)} J_\Pi = 0$ and therefore (cf. Lemma 2.1.2) that there is at least one $\mu \in \mathfrak{I}_\Pi$. To show that there is only one such μ, first observe that $\{\mu : J_\Pi(\mu) = 0\} \subset\subset \mathbf{M}_1(\Sigma)$. Now suppose there were more than one element of $\mathfrak{I}_\Pi$, and apply the KREIN–MILLMAN Theorem to deduce that there would then have to exist distinct extreme elements μ_1 and μ_2 of $\mathfrak{I}_\Pi$. Finally, show that this is impossible since, on the one hand, (**U**) says that μ_1 would have to be equivalent to μ_2 while, on the other hand, standard ergodic theoretic considerations (cf. (**iv**) of Exercise 5.2.28 below) guarantee that $\mu_1 \perp \mu_2$.

(ii) We are now going to outline a quite different approach to the same question. Namely, we are going to show that if $(E,\mathcal{F})$ is any measurable space and $\overline{\Pi}$ is a transition probability function on $(E,\mathcal{F})$ with the property that

$$\overline{\Pi}(x,\cdot) \geq \alpha\rho, \quad x \in E, \tag{4.1.49}$$

for some $\alpha \in (0,1]$ and $\rho \in \mathbf{M}_1\big((E,\mathcal{F})\big)$, then there is a unique $\mu \in \mathbf{M}_1\big((E,\mathcal{F})\big)$ such that

$$\big\|\nu\overline{\Pi}^n - \mu\big\|_{\text{var}} \leq 2(1-\alpha)^n, \quad n \in \mathbb{Z}^+ \text{ and } \nu \in \mathbf{M}_1\big((E,\mathcal{F})\big). \tag{4.1.50}$$

Show that, if it exists at all, then such a μ must be the one and only $\overline{\Pi}$-invariant element of $\mathbf{M}_1\big((E,\mathcal{F})\big)$. Also, show that when Π satisfies $(\mathbf{U})$ and $\overline{\Pi} \equiv \frac{1}{N}\sum_{m=1}^{N}\Pi^m$, then $\overline{\Pi}$ satisfies (4.1.49) with $\alpha = \frac{1}{M}$ and $\rho = \Pi^{\ell}(\sigma,\cdot)$ for any $\sigma \in \Sigma$; and conclude that the corresponding $\mu \in \mathbf{M}_1(\Sigma)$ in (4.1.50) is the one and only element of $\mathfrak{I}_{\Pi}$. (**Hint:** check that $\mu\Pi \in \mathfrak{I}_{\overline{\Pi}}$.)

Turning to the proof of (4.1.50), set $\tilde{E} = E \times \{-1,1\}$ and define $\tilde{\Pi}$ to be the transition probability function on $(\tilde{E};\tilde{\mathcal{F}})$, $\tilde{\mathcal{F}} \equiv \mathcal{F} \times \mathcal{B}_{\{-1,1\}}$ determined by

$$\tilde{\Pi}\big((x,\xi),\Gamma \times \{\eta\}\big) = \begin{cases} \alpha\rho(\Gamma) & \text{if } \eta = 1 \\ \overline{\Pi}(x,\Gamma) - \alpha\rho(\Gamma) & \text{if } \eta = -1. \end{cases}$$

Next, let $\big\{\tilde{P}_{(x,\xi)} : (x,\xi) \in \tilde{E}\big\}$ be the Markov chain on $\tilde{\Omega} \equiv \tilde{E}^{\mathbb{N}}$ with transition probability function $\tilde{\Pi}$, and use $X_n(\tilde{\omega})$ and $\Xi_n(\tilde{\omega})$ to denote the projections on E and $\{-1,1\}$, respectively, of the position of $\tilde{\omega} \in \tilde{\Omega}$ at time $n \in \mathbb{N}$. Now check that

$$\tilde{P}_{(x,\xi)}\big(\{\tilde{\omega} : X_n(\tilde{\omega}) \in \Gamma\}\big) = \overline{\Pi}^n(x,\Gamma), \quad n \in \mathbb{Z}^+,\ \Gamma \in \mathcal{F}, \text{ and } (x,\xi) \in \tilde{E}.$$

Finally, define

$$\tau(\tilde{\omega}) = \inf\{n \in \mathbb{Z}^+ : \Xi_n(\tilde{\omega}) = 1\},$$

show that

$$\overline{\Pi}^n(x,\Gamma) = \alpha\sum_{m=1}^{n}(1-\alpha)^{m-1}\rho\overline{\Pi}^{n-m}(\Gamma) + \tilde{P}_{(x,-1)}\big(\{\tilde{\omega} : X_n(\tilde{\omega}) \in \Gamma \text{ and } \tau(\tilde{\omega}) > n\}\big),$$

and conclude that

$$\big\|\nu_1\overline{\Pi}^n - \nu_2\overline{\Pi}^n\big\|_{\text{var}} \leq 2(1-\alpha)^n$$

for all $n \in \mathbb{Z}^+$ and $\nu_1,\ \nu_2 \in \mathbf{M}_1\big((E,\mathcal{F})\big)$. In particular, this means that, for each $\nu \in \mathbf{M}_1\big((E,\mathcal{F})\big)$,

$$\big\|\nu\overline{\Pi}^{m+n} - \nu\overline{\Pi}^n\big\|_{\text{var}} \leq 2(1-\alpha)^n, \quad m,\ n \in \mathbb{Z}^+;$$

and, therefore, not only does $\{\nu\overline{\Pi}^n\}_1^{\infty}$ converge in variation to some $\mu \in \mathbf{M}_1\big((E,\mathcal{F})\big)$, but also the limit is independent of ν and (4.1.50) holds.

4.1.51 Exercise.

Let $\Pi(\sigma, \cdot)$ be a transition probability on Σ.

(**i**) Suppose that $\mu \in \mathbf{M}_1(\Sigma)$ is Π-invariant and define the relative entropy functional $\mathbf{H}(\cdot\,|\mu)$ as in (3.2.14). Show that

$$\mathbf{H}(\nu\Pi|\mu) \le \mathbf{H}(\nu|\mu), \quad \nu \in \mathbf{M}_1(\Sigma).$$

Hint: Apply Lemma 3.2.13 and use JENSEN's inequality.

(**ii**) Assume that (**U**) holds and let μ be the Π-invariant measure produced in Exercise 4.1.48. Note by taking the ν in Lemma 4.1.40 to be μ, one sees that (4.1.41) holds for this μ. From this observation, show that

$$\mathbf{H}(\nu|\mu) \le \ell J_\Pi(\nu) + \log M, \quad \nu \in \mathbf{M}_1(\Sigma). \tag{4.1.52}$$

In particular, conclude that if $J_\Pi(\nu) < \infty$ then $\nu << \mu$ and $\int_\Sigma f \log f\, d\mu \le \ell J_\Pi(\nu) + M$ where $f = \frac{d\nu}{d\mu}$.

4.1.53 Exercise.

Again assume that (**U**) holds. Using Theorem 3.2.21, show that, for every $\Gamma \in \mathcal{B}_{\mathbf{M}_1(\Sigma)}$,

$$\begin{aligned} -\inf_{\Gamma^{\circ\tau}} J_\Pi &\le \varliminf_{n\to\infty} \frac{1}{n} \log\left(\inf_{\sigma\in\Sigma} P_\sigma\big(\{\omega : \mathbf{L}_n(\omega) \in \Gamma\}\big)\right) \\ &\le \varlimsup_{n\to\infty} \frac{1}{n} \log\left(\sup_{\sigma\in\Sigma} P_\sigma\big(\{\omega : \mathbf{L}_n(\omega) \in \Gamma\}\big)\right) \le -\inf_{\overline{\Gamma}^\tau} J_\Pi; \end{aligned} \tag{4.1.54}$$

where the notation is the same as that used in Theorem 3.2.21. In particular, using (4.1.54) and the reasoning which led to Theorem 2.1.10, show that

$$\begin{aligned} \lim_{n\to\infty} \sup_{\sigma\in\Sigma} \Bigg[\frac{1}{n} \log\left(\int_\Omega \exp\Big[n\Phi\big(\mathbf{L}_n(\omega)\big)\Big]\, P_\sigma(d\omega)\right) \\ - \sup\Big\{\Phi(\nu) - J_\Pi(\nu) : \nu \in \mathbf{M}_1(\Sigma)\Big\}\Bigg] = 0 \end{aligned} \tag{4.1.55}$$

for every measurable $\Phi : \mathbf{M}_1(\Sigma) \longrightarrow \mathbb{R}$ which is continuous with respect to the strong topology and satisfies

$$\sup_{n\ge1} \sup_{\sigma\in\Sigma} \left(\int_\Omega \exp\big[\alpha n\Phi\big(\mathbf{L}_n(\omega)\big)\big]\, P_\sigma(d\omega)\right)^{1/n} < \infty$$

for some $\alpha \in (1, \infty)$.

4.1.56 Exercise.

Return to the original setting of this section and assume not only that $(\hat{\mathbf{U}})$ holds but also that E is a separable BANACH space. For $\nu \in \mathbf{M}_1(\hat{\Sigma})$, define $\nu_E \in \mathbf{M}_1(E)$ by $\nu_E(\Gamma) = \nu(\Sigma \times E)$, $\Gamma \in \mathcal{B}_E$. Next, define

$$I(x) = \inf\left\{J_{\hat{\Pi}}(\nu) : \int_E \|y\|\,\nu_E(dy) < \infty \text{ and } m\big(\nu_E\big) = x\right\}, \quad x \in E.$$

(See Theorem 3.3.4 for the definition of $m(\nu_E)$.) Using the technique in Exercise 3.3.12, show that I is a good rate function on E and that (4.1.16) holds with this I in place of $I_{\hat{\Pi}}$. In particular, conclude that $I = \Lambda^*_{\hat{\Pi}} = I_{\hat{\Pi}}$.

4.1.57 Exercise.

In order to provide an example in which $\mathbf{S}_n$ is an additive function other than the empirical distribution, let Π be a transition probability on Σ which satisfies $(\mathbf{U})$, E be a separable BANACH space, ν a probability measure on E, and suppose that $F : \Sigma \times E \longrightarrow E$ is a continuous function satisfying

$$\sup_{\sigma\in\Sigma} \int_E \exp\Big[\alpha\big\|F(\sigma,x)\big\|\Big]\,\nu(dx) < \infty, \quad \alpha \in (0,\infty).$$

Next, define the transition probability function $\hat{\Pi}$ on $\hat{\Sigma} \equiv \Sigma \times E$ so that

$$\hat{\Pi}(\hat{\sigma}, A \times B) = \Pi(\sigma, A)\,\nu\Big(\big\{x : F(\sigma,x) \in B\big\}\Big) \text{ for } A \in \mathcal{B}_\Sigma \text{ and } B \in \mathcal{B}_E.$$

Check that this $\hat{\Pi}$ satisfies $(\hat{\mathbf{U}})$, and note that the corresponding random variables

$$\hat{\omega} \in \hat{\Omega} \longmapsto \overline{\mathbf{S}}_n(\hat{\omega}) \in E, \quad n \in \mathbb{Z}^+,$$

are distributed under $\hat{P}_{(\sigma,x)}$ in the same way as the random variables

$$(\omega, \mathbf{x}) \in \Omega \times E^{\mathbb{Z}^+} \longmapsto \frac{1}{n}\sum_{m=1}^{n} F\big(\Sigma_{m-1}(\omega), x_m\big), \quad n \in \mathbb{Z}^+,$$

are under $P_\sigma \times \nu^{\mathbb{Z}^+}$. Finally, show that

$$I_{\hat{\Pi}}(x) = \sup\Big\{{}_{E^*}\langle\lambda, x\rangle_E - \Lambda_\Pi(V_\lambda) : \lambda \in E^*\Big\}$$

where

$$V_\lambda(\sigma) \equiv \log\left[\int_E \exp\Big[{}_{E^*}\langle\lambda, F(\sigma,x)\rangle_E\Big]\,\nu(dx)\right], \quad \sigma \in \Sigma.$$

4.2 Continuous Time Markov Processes

We now want to carry out the program of Section 4.1 for the case of MARKOV processes having a continuous time-parameter. Again, let Σ be a Polish space and let E, X, ρ, and $\|\cdot\|$ be as in $(\hat{\mathbf{C}})$ at the beginning of Section 3.1. Next, let $(t,\sigma) \in (0,\infty)\times\Sigma \longmapsto \hat{P}(t,\sigma,\cdot) \in \mathbf{M}_1(\Sigma\times E)$ be a measurable map with the property that

$$\hat{P}(s+t,\sigma,\hat{\Gamma}) = \int_{\Sigma\times E} \hat{P}\big(t,\xi,\hat{\Gamma}(s,t,x)\big)\,\hat{P}(s,\sigma,d\xi\times dx)$$

for s, $t \in (0,\infty)$, $\sigma \in \Sigma$, and $\hat{\Gamma} \in \mathcal{B}_{\Sigma\times E}$, where

$$\hat{\Gamma}(s,t,x) \equiv \left\{(\sigma,y) \in \Sigma\times E : \left(\sigma, \frac{s}{s+t}x + \frac{t}{s+t}y\right) \in \hat{\Gamma}\right\}.$$

Using $P(t,\sigma,\Gamma)$ to denote $\hat{P}(t,\sigma,\Gamma\times E)$ for $\Gamma \in \mathcal{B}_\Sigma$, note that $(t,\sigma) \in (0,\infty)\times\Sigma \longmapsto P(t,\sigma,\cdot)$ is a continuous-time transition probability function. That is, it satisfies the **Chapman–Kolmogorov equation**

$$P(s+t,\sigma,\cdot) = \int_\Sigma P(t,\xi,\cdot)P(s,\sigma,d\xi).$$

Throughout, we will be assuming that, in addition, $P(t,\sigma,\cdot)$ is continuous at 0 in the sense that

$$P(t,\sigma,\cdot) \Longrightarrow \delta_\sigma \quad \text{as } t \searrow 0, \quad \sigma \in \Sigma. \tag{4.2.1}$$

Because, in the continuous-time context, there is no canonical choice of the sample space on which to put the associated MARKOV process, we will simply assume that the following items exist:

(i) A measurable space $(\Omega,\mathfrak{M})$ with a non-decreasing family $\{\mathfrak{M}_t : t \geq 0\}$ of sub σ-algebras whose union generates $\mathfrak{M}$.

(ii) A $\{\mathfrak{M}_t : t \geq 0\}$-progressively measurable map

$$(t,\omega) \in (0,\infty)\times\Omega \longmapsto \big(\Sigma_t(\omega), \overline{\mathbf{S}}_t(\omega)\big) \in \Sigma\times E$$

such that, for each $\omega \in \Omega$,

$$\Sigma_0(\omega) \equiv \lim_{t\searrow 0} \Sigma_t(\omega)$$

exists, $\lim_{t\searrow 0} \mathbf{S}_t(\omega) = 0$, and

$$\overline{\mathbf{S}}_t^s(\omega) \equiv \frac{\mathbf{S}_t(\omega) - \mathbf{S}_s(\omega)}{t-s} \in E, \quad 0 \leq s < t < \infty \text{ and } \omega \in \Omega,$$

where $\mathbf{S}_t(\omega) \equiv t\overline{\mathbf{S}}_t(\omega)$ for $t \in (0,\infty)$.

(iii) A measurable family $\{P_\sigma : \sigma \in \Sigma\}$ of probability measures on $(\Omega, \mathfrak{M})$ with the properties that for every $\sigma \in \Sigma$:

$$P_\sigma\big(\{\omega : \Sigma_0(\omega) = \sigma\}\big) = 1$$

and, for every $\hat{\Gamma} \in \mathcal{B}_{\Sigma \times E}$,

$$P_\sigma\Big(\{\omega : (\Sigma_{s+t}(\omega), \overline{\mathbf{S}}^s_{s+t}(\omega)) \in \hat{\Gamma}\} \big| \mathcal{B}_s\Big) = \hat{P}\big(t, \Sigma_s, \hat{\Gamma}\big) \quad (\text{a.s.}, P_\sigma)$$

for all s, $t \in (0, \infty)$.

Finally, for $(t, \sigma) \in (0, \infty) \times \Sigma$, we will use $\mu_{\sigma,t} \in \mathbf{M}_1(E)$ to denote the distribution of $\omega \in \Omega \longmapsto \overline{\mathbf{S}}_t(\omega) \in E$ under P_σ.

4.2.2 Remark.

In the first place, it should be noticed that, even though there is a certain amount of ambiguity about the way in which the family $\{P_\sigma : \sigma \in \Sigma\}$ is realized, the measures $\mu_{\sigma,t}$ are uniquely determined by $\hat{P}(t, \sigma, \cdot)$.

Secondly, as in Section 4.1, our basic example will be the case in which $E = \mathbf{M}_1(\Sigma)$ and $\overline{\mathbf{S}}_t(\omega)$ is **the empirical distribution functional for the position process** $t \in [0, \infty) \longmapsto \Sigma_t(\omega)$ and, as such, is given by

$$\overline{\mathbf{S}}_t(\omega) = \mathbf{L}_t(\omega) \equiv \overline{\lambda}_{[0,t]} \circ \Big(\Sigma_\cdot(\omega)|_{[0,t]}\Big)^{-1}$$

where $\overline{\lambda}_{[0,t]}$ denotes normalized LEBESGUE measure on $[0, t]$. (Note that,

$$\mathbf{L}_t(\omega) = \frac{1}{t} \int_0^t \delta_{\Sigma_s(\omega)}\, ds$$

when the paths

$$t \in [0, \infty) \longmapsto \Sigma_t(\omega) \in \Sigma$$

have any reasonable regularity.) Just as in the case of MARKOV chains, when we are dealing with this situation, there is no need to introduce E as a part of the state space or to have the joint distribution of $\omega \in \Omega \longmapsto \big(\Sigma_t(\omega), \overline{\mathbf{S}}_t(\omega)\big)$ given as part of our data, because it is already determined by $P(t, \sigma, \cdot)$.

For the most part, the argument which we will use in the present situation to get our basic result (Theorem 4.2.16 below) differs very little from that used in the preceding section. Thus, we will avoid repetition and only provide details at those places where new techniques are needed.

4.2.3 Lemma. *For $t \in (0,\infty)$, define*

$$\mathcal{P}_t(\Gamma) = \inf_{\sigma\in\Sigma} \mu_{\sigma,t}(\Gamma), \quad \Gamma \in \mathcal{B}_E.$$

If $\Gamma \in \mathcal{B}_E$ is convex, then $t \in (0,\infty) \longmapsto \mathcal{P}_t(\Gamma)$ is super-multiplicative. In addition, if

$$\text{(4.2.4)} \qquad \lim_{R\to\infty} \sup_{\sigma\in\Sigma} \sup_{t\in[0,T]} P_\sigma\big(\{\omega : \|\mathbf{S}_t(\omega)\| \ge R\}\big) = 0, \quad T \in (0,\infty),$$

then, for every ρ-bounded, convex $\Gamma \in \mathcal{B}_E$, either $\sup_{t>0} \mathcal{P}_t(\Gamma) = 0$ or, for every $\delta > 0$, there exist $0 < a < b < \infty$ such that $\inf_{t\in[a,b]} \mathcal{P}_t\big(\Gamma^{(\delta)}\big) > 0$.

PROOF: The super-multiplicative property is proved in exactly the same way as it was in Lemma 4.1.4. As for the second part, suppose that, for some $s \in [0,\infty)$, $\mathcal{P}_s(\Gamma) > 0$; and, for $t > s$, define $q_t \in [0,\infty)$ and $r_t \in [0,s)$ so that $t = q_t s + r_t$. Since Γ is bounded and

$$\overline{\mathbf{S}}_t - \overline{\mathbf{S}}_t^{r_t} = \frac{r_t}{t}\left(\overline{\mathbf{S}}_{r_t} - \overline{\mathbf{S}}_t^{r_t}\right),$$

one sees that, for sufficiently large $t > s$:

$$\begin{aligned} P_\sigma\big(\{\omega : \overline{\mathbf{S}}_t(\omega) \in \Gamma^{(\delta)}\}\big) &\ge P_\sigma\big(\{\omega : \overline{\mathbf{S}}_t^{r_t}(\omega) \in \Gamma \text{ and } \|\mathbf{S}_{r_t}(\omega)\| < t\delta/2\}\big) \\ &\ge \mathcal{P}_s(\Gamma)^{q_t} P_\sigma\big(\{\omega : \|\mathbf{S}_{r_t}(\omega)\| < t\delta/2\}\big); \end{aligned}$$

from which the desired conclusion is clear. ∎

We next need the following variant of Lemma 3.1.3.

4.2.5 Lemma. *Let $f : (0,\infty) \longrightarrow [0,\infty]$ be a sub-additive function (i.e.,*

$$f(s+t) \le f(s) + f(t)$$

for all $s,\ t \in (0,\infty)$) with the property that $\sup_{t\in[a,b]} f(t) < \infty$ for some $0 < a < b$. Then there is a $T \in (0,\infty)$ such that

$$\lim_{t\to\infty} \frac{f(t)}{t} = \inf_{t\ge T} \frac{f(t)}{t} \in \mathbb{R}.$$

PROOF: Set $M = \sup_{t\in[a,b]} f(t)$, choose $q_0 \in \mathbb{Z}^+$ so that $q_0(b-a) \ge a$, and let $T = q_0 a$. For $t \ge T$ put $q = \big[t/a\big]$, and note that $t/q \in [a,b]$ and therefore that $f(t) \le qM \le \frac{t}{a}M$.

Now let $t_1 \ge T$ be given. For $t \ge 2t_1$, set $q_t = \big[t/t_1\big]$ and $r_t = t - q_t t_1$. Then

$$\frac{f(t)}{t} \le \frac{(q_t - 1)f(t_1) + f(t_1 + r_t)}{t} \le \frac{(q_t-1)f(t_1) + \frac{2t_1}{a}M}{t};$$

and so

$$\overline{\lim_{t\to\infty}} \frac{f(t)}{t} \le \frac{f(t_1)}{t_1}. \quad ∎$$

Assuming that (4.2.4) holds, define

$$\mathcal{L}(p,r) = \begin{cases} -\lim_{t\to\infty} \frac{1}{t} \log \mathcal{P}_t\big(B(p,r)\big) & \text{if } \sup_{t>0} \mathcal{P}_t\big(B(p,r/2)\big) > 0 \\ \infty & \text{otherwise,} \end{cases}$$

and

$$I_{\hat{P}}(p) = \sup_{r>0} \mathcal{L}(p,r) = \lim_{r\searrow 0} \mathcal{L}(p,r) \tag{4.2.6}$$

With the same argument as we used to prove Lemma 4.1.7, one can easily check the following.

4.2.7 Lemma. *Assume that* (4.2.4) *holds and refer to the preceding. Then the map* $I_{\hat{P}} : E \longrightarrow [0,\infty]$ *is lower semi-continuous and convex; and, for open* G *in* E,

$$\varliminf_{t\to\infty} \frac{1}{t} \log \left(\inf_{\sigma\in\Sigma} \mu_{\sigma,t}(G) \right) \geq -\inf_G I_{\hat{P}}.$$

In order to get the complementary upper bound, we must introduce an assumption which will play the same role here as ($\hat{\mathbf{U}}$) played in Section 4.1. Namely, we will assume that

$$(\tilde{\mathbf{U}}) \qquad \begin{cases} \int_{(0,1]} \hat{P}(t,\sigma,\cdot)\,\rho_1(dt) \leq M \int_{(0,1]} \hat{P}(t,\tau,\cdot)\,\rho_2(dt), & \sigma,\tau\in\Sigma, \\ \sup_{\sigma\in\Sigma,\, t\in[0,1]} \int_\Omega \exp\Big[\alpha \big\|\mathbf{S}_t(\omega)\big\|\Big]\, P_\sigma(d\omega) < \infty, & \alpha\in(0,\infty) \end{cases}$$

for some $M \in [1,\infty)$ and ρ_1, $\rho_2 \in \mathbf{M}_1((0,1])$.

4.2.8 Remark.

Although we have stated ($\tilde{\mathbf{U}}$) in such a way that time $t = 1$ appears to have special importance, this is in fact not the case. Indeed, as will be apparent from the development given below, we could deal equally well with the case in which the probability measures ρ_1 and ρ_2 are supported on any bounded interval in $(0,\infty)$. We have chosen the interval $(0,1]$ only for convenience.

As an easy consequence of ($\tilde{\mathbf{U}}$) we can take the following initial step toward the upper bound.

4.2.9 Lemma. *Assume that* ($\tilde{\mathbf{U}}$) *holds.*

(i) *If*

$$H_\alpha \equiv \log\left(\sup_{\sigma\in\Sigma,\, t\in[0,1]} \int_\Omega \exp\bigl[\alpha\|\mathbf{S}_t(\omega)\|\bigr]\, P_\sigma(d\omega)\right), \quad \alpha \in (0,\infty),$$

then

$$\int_\Omega \exp\bigl[\|\alpha \mathbf{S}_t(\omega)\|\bigr]\, P_\sigma(d\omega) \le \exp\bigl[(t+1)H_\alpha\bigr], \tag{4.2.10}$$

for all $\alpha \in (0,\infty)$ *and* $(t,\sigma) \in (0,\infty)\times\Sigma$. *In particular,* (4.2.4) *holds.*

(ii) *For every* $p \in E$

$$\lim_{r\searrow 0}\overline{\lim_{t\to\infty}}\frac{1}{t}\log\left(\sup_{\sigma\in\Sigma}\mu_{\sigma,t}\bigl(B(p,r)\bigr)\right) \le -I_{\hat{P}}(p);$$

and so

$$\overline{\lim_{t\to\infty}}\frac{1}{t}\log\left(\sup_{\sigma\in\Sigma}\mu_{\sigma,t}(K)\right) \le -\inf_K I_{\hat{P}}, \quad K \subset\subset E.$$

PROOF: The proof of (4.2.10) can be accomplished by induction on $[t]$ using the MARKOV property. The details are left to the reader (cf. Lemma 4.1.9).

By the usual HEINE–BOREL argument (cf. **(iv)** of Exercise 2.1.14), we need only prove the first assertion in **(ii)**. To this end, let $\alpha < I_{\hat{P}}(p)$ and choose $\delta > 0$ so that $\mathcal{L}(p,3\delta) > \alpha$. Then, for any $\sigma,\ \tau \in \Sigma$ and $t > 0$,

$$\begin{aligned}
&\mu_{\sigma,t}\bigl(B(p,\delta)\bigr)\\
&\quad\le \int_{(0,1]}\int_\Sigma \mu_{\xi,t}\bigl(B(p,2\delta)\bigr)\,\hat{P}(s,\sigma,d\xi)\rho_1(ds)\\
&\qquad\qquad + \sup_{s\in[0,1]} P_\sigma\bigl(\{\omega : \|\overline{\mathbf{S}}^{s}_{s+t}(\omega) - \overline{\mathbf{S}}_t(\omega)\| \ge \delta\}\bigr)\\
&\quad\le M\int_{(0,1]}\int_\Sigma \mu_{\xi,t}\bigl(B(p,2\delta)\bigr)\,\hat{P}(s,\tau,d\xi)\rho_2(ds)\\
&\qquad\qquad + \sup_{s\in[0,1]} P_\sigma\bigl(\{\omega : \|\overline{\mathbf{S}}^{s}_{s+t}(\omega) - \overline{\mathbf{S}}_t(\omega)\| \ge \delta\}\bigr)\\
&\quad\le M\mu_{\tau,t}\bigl(B(p,3\delta)\bigr)\\
&\qquad\qquad + (1+M)\sup_{\xi\in\Sigma}\sup_{s\in[0,1]} P_\xi\bigl(\{\omega : \|\overline{\mathbf{S}}^{s}_{s+t}(\omega) - \overline{\mathbf{S}}_t(\omega)\| \ge \delta\}\bigr).
\end{aligned}$$

Using (4.2.10) to estimate the second term in the last line of the preceding, we conclude that, for sufficiently large t's and some $C < \infty$,

$$\sup_{\sigma\in\Sigma}\mu_{\sigma,t}\bigl(B(p,\delta)\bigr) \le C\exp[-\alpha t];$$

from which the required estimate is immediate. ∎

Combining Lemma 4.2.7 with Lemma 4.2.9, we now see that, under $(\tilde{\mathbf{U}})$, $\{\mu_{\sigma,t} : t > 0\}$ satisfies the weak large deviation principle, uniformly in $\sigma \in \Sigma$, with rate function $I_{\hat{P}}$. In order to show that $I_{\hat{P}}$ is a good rate function and that the full (uniform) large deviation principle holds under $(\tilde{\mathbf{U}})$, it will be useful to have some additional notation. Set $\tilde{\Omega} = \Omega \times (0,1]^{\mathbb{N}}$; and, for $\sigma \in \Sigma$, define $\tilde{P}_\sigma = P_\sigma \times \rho_1^{\mathbb{N}}$ on $\left(\tilde{\Omega}, \mathcal{B} \times \mathcal{B}_{(0,1]}^{\mathbb{N}}\right)$. Next, for $\tilde{\omega} = (\omega, \mathbf{t}) \in \tilde{\Omega}$, set $\mathbf{S}.(\tilde{\omega}) = \mathbf{S}.(\omega)$, $\overline{\mathbf{S}}.(\tilde{\omega}) = \overline{\mathbf{S}}.(\omega)$; and define

$$\mathcal{T}_n(\tilde{\omega}) = n + \sum_{m=0}^{n} t_m, \quad n \in \mathbb{N},$$

$$\tilde{X}_n(\tilde{\omega}) = \frac{\mathbf{S}_{\mathcal{T}_n(\tilde{\omega})}(\tilde{\omega}) - \mathbf{S}_{\mathcal{T}_{n-1}(\tilde{\omega})}(\tilde{\omega})}{\mathcal{T}_n(\tilde{\omega}) - \mathcal{T}_{n-1}(\tilde{\omega})}, \quad n \in \mathbb{Z}^+,$$

and

$$\tilde{\mathbf{S}}_T(\tilde{\omega}) = \frac{\mathbf{S}_{\mathcal{T}(T,\tilde{\omega})}(\tilde{\omega}) - \mathbf{S}_{\mathcal{T}_0(\tilde{\omega})}(\tilde{\omega})}{\mathcal{T}(T,\tilde{\omega}) - \mathcal{T}_0(\tilde{\omega})} = \sum_{m=1}^{n(T,\tilde{\omega})} \beta_m(T,\tilde{\omega}) \tilde{X}_m(\tilde{\omega}),$$

where

$$n(T,\tilde{\omega}) = \min\{n \in \mathbb{Z}^+ : \mathcal{T}_n(\tilde{\omega}) - \mathcal{T}_0(\tilde{\omega}) \geq T\}, \quad \mathcal{T}(T,\tilde{\omega}) = \mathcal{T}_{n(T,\tilde{\omega})}(\tilde{\omega}),$$

and

$$\beta_m(T,\tilde{\omega}) = \frac{\mathcal{T}_m(\tilde{\omega}) - \mathcal{T}_{m-1}(\tilde{\omega})}{\mathcal{T}(T,\tilde{\omega}) - \mathcal{T}_0(\tilde{\omega})}, \quad 1 \leq m \leq n(T,\tilde{\omega}).$$

Finally, for $(T,\sigma) \in (0,\infty) \times \Sigma$, we denote by $\tilde{\mu}_{\sigma,T} \in \mathbf{M}_1(E)$ the distribution under $\tilde{P}_\sigma$ of $\tilde{\omega} \in \tilde{\Omega} \longmapsto \tilde{\mathbf{S}}_T(\tilde{\omega}) \in E$.

4.2.11 Lemma. *Assume that $(\tilde{\mathbf{U}})$ holds and refer to the preceding. Then, for $\delta \in (0,\infty)$,*

$$\varlimsup_{T\to\infty} \frac{1}{T} \log\left(\sup_{\sigma\in\Sigma} \tilde{P}_\sigma\left(\left\{\tilde{\omega} : \left\|\overline{\mathbf{S}}_T(\tilde{\omega}) - \tilde{\mathbf{S}}_T(\tilde{\omega})\right\| \geq \delta\right\}\right)\right) = -\infty. \tag{4.2.12}$$

Furthermore, given $\nu \in \mathbf{M}_1(\Sigma)$, define $\mathcal{M}_\nu \in \mathbf{M}_1(E)$ by

$$\mathcal{M}_\nu(\Gamma) = \int_{(0,1]^2} \left(\int_{\Sigma^2} \mu_{\xi,1+s}(\Gamma)\, P(t,\sigma,d\xi)\, \nu(d\sigma)\right) (\rho_1 \times \rho_2)(ds \times dt)$$

for $\Gamma \in \mathcal{B}_E$. Then

$$\int_E \exp\left[\alpha\|x\|\right] \mathcal{M}_\nu(dx) < \infty, \quad \alpha \in (0,\infty),$$

and, for every $n \in \mathbb{Z}^+$ and all measurable $F : E^n \longrightarrow [0,\infty)$,

$$\begin{aligned}\int_{\tilde{\Omega}} F\big(\tilde{X}_m(\tilde{\omega}), \tilde{X}_{m+2}(\tilde{\omega}), \dots, \tilde{X}_{m+2(n-1)}(\tilde{\omega})\big)\, \tilde{P}_\sigma(d\tilde{\omega}) \\ \leq M^n \int_{E^n} F(\mathbf{x})\, \mathcal{M}_\nu^n(d\mathbf{x}).\end{aligned} \tag{4.2.13}$$

In particular, when E *is either the space of probability measures on some Polish space or* $E = (X, \|\cdot\|)$ *is a separable* BANACH *space, there exists for each* $L \in [1,\infty)$ *a* $K_L \subset\subset E$ *such that*

$$\overline{\lim_{T\to\infty}} \frac{1}{T} \log\Big(\sup_{\sigma\in\Sigma} \tilde{\mu}_{\sigma,T}\big(K_L^c\big)\Big) \leq -L.$$

Finally, in the case described in Remark *4.2.2, for every measurable function* $V : \Sigma \longrightarrow [0,\infty]$,

$$\begin{aligned}\sup_{\sigma\in\Sigma} \int_{\Omega} \exp\Big[\int_0^T V\big(\Sigma_t(\omega)\big)\, ds\Big]\, P_\sigma(d\omega) \\ \leq e^{\|V\|_B} \left(M^2 \int_\Sigma \exp[16V]\, \mu_\nu(d\sigma)\right)^T,\end{aligned} \tag{4.2.14}$$

where $\mu_\nu \in \mathbf{M}_1(\Sigma)$ *is given by*

$$\mu_\nu(\Gamma) = \int_{\mathbf{M}_1(\Sigma)} \alpha(\Gamma)\, \mathcal{M}_\nu(d\alpha), \quad \Gamma \in \mathcal{B}_\Sigma. \tag{4.2.15}$$

PROOF: To prove (4.2.12), first note that

$$T\big\|\overline{\mathbf{S}}_T(\tilde{\omega}) - \tilde{\mathbf{S}}_T(\tilde{\omega})\big\| \leq \big\|\mathbf{S}_{\mathcal{T}_0(\tilde{\omega})}(\tilde{\omega})\big\| + \big\|\mathbf{S}^T_{\mathcal{T}(T,\tilde{\omega})}(\tilde{\omega})\big\| + 2\big\|\tilde{\mathbf{S}}_T(\tilde{\omega})\big\|.$$

Next, using (4.2.10) and the MARKOV property, observe that, for every $\beta \in (0,\infty)$,

$$\begin{aligned}\int_{\tilde{\Omega}} \exp\Big[\beta\big\|\mathbf{S}_{\mathcal{T}_0(\tilde{\omega})}(\tilde{\omega})\big\|\Big]\, \tilde{P}_\sigma(d\tilde{\omega}) \\ \leq \int_{\tilde{\Omega}} \exp\Big[\big(\mathcal{T}_0(\tilde{\omega}) + 1\big) H_\beta\Big]\, \tilde{P}_\sigma(d\tilde{\omega}) \leq e^{2H_\beta},\end{aligned}$$

$$\begin{aligned}\int_{\tilde{\Omega}} \exp\Big[\beta\big\|\mathbf{S}^T_{\mathcal{T}(T,\tilde{\omega})}(\tilde{\omega})\big\|\Big]\, \tilde{P}_\sigma(d\tilde{\omega}) \\ \leq \int_{\tilde{\Omega}} \exp\Big[\big(\mathcal{T}(T,\tilde{\omega}) - T + 1\big) H_\beta\Big]\, \tilde{P}_\sigma(d\tilde{\omega}) \leq e^{3H_\beta},\end{aligned}$$

and, for $T \in [1, \infty)$,

$$\int_{\tilde{\Omega}} \exp\left[\beta \|\tilde{\mathbf{S}}_T(\tilde{\omega})\|\right] \tilde{P}_\sigma(d\tilde{\omega}) \\ \leq \int_{\tilde{\Omega}} \exp\left[\frac{\mathcal{T}(T,\tilde{\omega}) - \mathcal{T}_0(\tilde{\omega}) + 1}{\mathcal{T}(T,\tilde{\omega}) - \mathcal{T}_0(\tilde{\omega})} H_\beta\right] \tilde{P}_\sigma(d\tilde{\omega}) \leq e^{2H_\beta}.$$

After combining these with the above and again applying HÖLDER's inequality, we arrive at

$$\int_{\tilde{\Omega}} \exp\left[\alpha T \|\overline{\mathbf{S}}_T(\tilde{\omega}) - \tilde{\mathbf{S}}_T(\tilde{\omega})\|\right] \tilde{P}_\sigma(d\tilde{\omega}) \leq \exp\left[5H_{3\alpha} + 2H_{6\alpha}\right]$$

for all $(T, \sigma) \in [1, \infty) \times \Sigma$ and $\alpha \in (0, \infty)$; and clearly (4.2.12) is an easy step from here.

To prove (4.2.13), let $\nu \in \mathbf{M}_1(\Sigma)$ be given and define $\mathcal{M}_\nu \in \mathbf{M}_1(E)$ accordingly. The estimate on the exponential moments of $\mathcal{M}_\nu$ is an easy consequence of (4.2.10). Next, denote by $\tilde{\mu}_\sigma \in \mathbf{M}_1(E)$ the distribution under $\tilde{P}_\sigma$ of $\tilde{\omega} \in \tilde{\Omega} \longmapsto \tilde{X}_1(\tilde{\omega}) \in E$; and note that, by the second part of $(\tilde{\mathbf{U}})$, $\tilde{\mu}_\sigma \leq M\,\mathcal{M}_\nu$ for all $\sigma \in \Sigma$. At the same time, by the MARKOV property, we have that

$$\int_{\tilde{\Omega}} F\big(\tilde{X}_m(\tilde{\omega}), \dots, \tilde{X}_{m+2n}(\tilde{\omega})\big) \tilde{P}_\sigma(d\tilde{\omega}) \\ = \int_{\tilde{\Omega}} \left(\int_E F\big(\tilde{X}_m(\tilde{\omega}), \dots, \tilde{X}_{m+2(n-1)}(\tilde{\omega}), x\big) \tilde{\mu}_{\tilde{\omega}}(dx) \right) \tilde{P}_\sigma(d\tilde{\omega}),$$

where

$$\tilde{\mu}_{\tilde{\omega}} = \tilde{\mu}_{\tilde{\Sigma}_n(\tilde{\omega})} \text{ and } \tilde{\Sigma}_n(\tilde{\omega}) = \Sigma_{1+\mathcal{T}_{m+2(n-1)}(\tilde{\omega})}(\omega)$$

for $\tilde{\omega} = (\omega, \mathbf{t})$. Combining this with the preceding and using induction on n, one quickly arrives at (4.2.13).

In order to use (4.2.13) to check the uniform exponential tightness of $\{\tilde{\mu}_{\sigma,t} : t > 0\}$ when $E = (X, \|\cdot\|)$ is a separable BANACH space, define $\tilde{\omega} \in \tilde{\Omega} \longmapsto \tilde{\mathbf{L}}_T(\tilde{\omega}) \in \mathbf{M}_1(E)$ by (cf. the paragraph preceding the statement of this lemma)

$$\tilde{\mathbf{L}}_T(\tilde{\omega}) = \sum_{m=1}^{n(T,\tilde{\omega})} \beta_m(T, \tilde{\omega}) \delta_{\tilde{X}_m(\tilde{\omega})};$$

and let $\tilde{Q}_{\sigma,T} \in \mathbf{M}_1\big(\mathbf{M}_1(E)\big)$ be the distribution of $\tilde{\mathbf{L}}_T$ under $\tilde{P}_\sigma$. Since (cf. Theorem 3.3.4)

$$\tilde{\mathbf{S}}_T(\tilde{\omega}) = m\big(\tilde{\mathbf{L}}_T(\tilde{\omega})\big),$$

the desired exponential tightness follows immediately from the last part of Lemma 3.3.10 once one notices that, by (4.2.13),

$$\begin{aligned}
&\int_{\mathbf{M}_1(E)} \exp\left[T\int_E V\,d\gamma\right]\tilde{Q}_{\sigma,T}(d\gamma)\\
&\qquad = \int_{\tilde{\Omega}} \exp\left[T\sum_{m=1}^{n(T,\tilde{\omega})}\beta_m(T,\tilde{\omega})V\big(\tilde{X}_m(\tilde{\omega})\big)\right]\tilde{P}_\sigma(d\tilde{\omega})\\
&\qquad \le \left(\int_{\tilde{\Omega}} \exp\left[4\sum_{m=1}^{[T]}V\big(\tilde{X}_{2m}(\tilde{\omega})\big)\right]\tilde{P}_\sigma(d\tilde{\omega})\right)^{1/2}\\
&\qquad\qquad \times\left(\int_{\tilde{\Omega}} \exp\left[4\sum_{m=0}^{[T]}V\big(\tilde{X}_{2m+1}(\tilde{\omega})\big)\right]\tilde{P}_\sigma(d\tilde{\omega})\right)^{1/2}\\
&\qquad \le \left(M\int_E e^{4V}\,d\mathcal{M}_\nu\right)^{T+1} \le \left(M^2\int_E e^{8V}\,d\mathcal{M}_\nu\right)^{T}
\end{aligned}$$

for $T \in [1,\infty)$ and measurable $V : E \longrightarrow [0,\infty]$.

When $E = \mathbf{M}_1(\Sigma')$ for some Polish space Σ', we apply the preceding and JENSEN's inequality to conclude that

$$\begin{aligned}
&\int_{\mathbf{M}_1(\Sigma')} \exp\left[T\int_{\Sigma'} V\,d\nu'\right]\tilde{Q}_{\sigma,T}(d\nu')\\
&\qquad \le \left(M^2\int_{\mathbf{M}_1(\Sigma')} \exp\left[8\int_{\Sigma'} V\,d\nu'\right]\mathcal{M}_\nu(d\nu')\right)^T\\
&\qquad \le \left(M^2\int_{\Sigma'} \exp\big[8V(\sigma')\big]\,\mu'_\nu(d\sigma')\right)^T
\end{aligned}$$

for all $T \in [1,\infty)$ and measurable $V : \Sigma' \longrightarrow [0,\infty]$, where $\mu'_\nu \in \mathbf{M}_1(\Sigma')$ is given by

$$\mu'_\nu = \int_{\mathbf{M}_1(\Sigma')} \alpha'\,\mathcal{M}_\nu(d\alpha').$$

Thus, Lemma 3.2.7 applies and yields the desired tightness.

Finally, to prove (4.2.14) in the situation described in Remark 4.2.2, it suffices to apply the preceding (with $\Sigma' = \Sigma$) and to observe both that the above μ'_ν coincides with the μ_ν in (4.2.15) and that

$$\int_\Sigma V(\sigma)\,\mathbf{L}_T(\omega,d\sigma) \le \frac{\|V\|_B}{T} + 2\int_\Sigma V(\sigma)\,\tilde{\mathbf{S}}_T(\tilde{\omega},d\sigma)$$

for $\tilde{\omega} = (\omega,\mathbf{t})$, $T \in [1,\infty)$, and measurable $V : \Sigma \longrightarrow [0,\infty]$. ∎

We are now in a position to prove the main result of this section.

4.2.16 Theorem. *Assume that* $(\tilde{\mathbf{U}})$ *holds and that* E *is either a separable* BANACH *space or the space of probability measures on a Polish space. Then the* $I_{\hat{P}}$ *in* (4.2.6) *is a good, convex rate function and, for every* $\Gamma \in \mathcal{B}_E$,

$$\begin{aligned} -\inf_{\Gamma^\circ} I_{\hat{P}} &\leq \varliminf_{t\to\infty} \frac{1}{t} \log\left[\inf_{\sigma\in\Sigma} \mu_{\sigma,t}(\Gamma)\right] \\ &\leq \varlimsup_{t\to\infty} \frac{1}{t} \log\left[\sup_{\sigma\in\Sigma} \mu_{\sigma,t}(\Gamma)\right] \leq -\inf_{\overline{\Gamma}} I_{\hat{P}}. \end{aligned}$$

PROOF: We have already proved everything except the goodness of $I_{\hat{P}}$ and the upper bound for closed sets. But, by Lemma 4.2.7 combined with (4.2.12), we see that

$$\varliminf_{t\to\infty} \frac{1}{t} \log\left[\inf_{\sigma\in\Sigma} \tilde{\mu}_{\sigma,t}(G)\right] \geq -\inf_{G} I_{\hat{P}}$$

for all open $G \subseteq E$. In particular, since $\{\tilde{\mu}_{\sigma,t} : t > 0\}$ is uniformly exponentially tight, we now see that, for each $L \in [1,\infty)$ there is a $K_L \subset\subset E$ such that

$$\inf_{K_L^c} I_{\hat{P}} \geq -\varliminf_{t\to\infty} \frac{1}{t} \log\left[\inf_{\sigma\in\Sigma} \tilde{\mu}_{\sigma,t}(K_L^c)\right] \geq L;$$

and, since we already know that $I_{\hat{P}}$ is lower semi-continuous, this completes the proof that $I_{\hat{P}}$ is good.

Turning to the upper bound for closed sets, note that by combining **(ii)** of Lemma 4.2.9 with (4.2.12) one sees that

$$\lim_{r\searrow 0} \varlimsup_{t\to\infty} \frac{1}{t} \log\left[\sup_{\sigma\in\Sigma} \tilde{\mu}_{\sigma,t}\big(B(p,r)\big)\right] \leq -I_{\hat{P}}(p)$$

for every $p \in E$. Thus, again by the uniform exponential tightness of $\{\tilde{\mu}_{\sigma,t} : t > 0\}$, we know that the upper bound holds when $\mu_{\sigma,t}$ is replaced by $\tilde{\mu}_{\sigma,t}$. But, by (4.2.12), this means that

$$\varlimsup_{t\to\infty} \frac{1}{t} \log\left[\sup_{\sigma\in\Sigma} \mu_{\sigma,t}(F)\right] \leq -\inf_{F^{(\delta)}} I_{\hat{P}}$$

for every closed $F \subseteq E$ and $\delta > 0$; and, because (cf. (2.1.3)) $I_{\hat{P}}$ is good, this is all that we need in order to get the upper bound. ∎

Applying the results of Chapter II, we can now take the following preliminary step toward the identification of the rate function $I_{\hat{P}}$.

4.2.17 Corollary. *Let everything be as in the statement of* Theorem *4.2.16. Then, for each continuous* $\Phi : E \longrightarrow \mathbb{R}$ *which satisfies*

$$\sup_{t\geq 1}\left(\sup_{\sigma\in\Sigma}\int_E \exp\big[\alpha t\Phi(x)\big]\,\mu_{\sigma,t}(dx)\right)^{1/t} < \infty$$

for some $\alpha \in (1,\infty)$,

$$\lim_{t\to\infty}\sup_{\sigma\in\Sigma}\left[\frac{1}{t}\log\left(\int_E \exp\big[t\Phi(x)\big]\,\mu_{\sigma,t}(dx)\right) - \sup_{x\in E}\big(\Phi(x) - I_{\hat{P}}(x)\big)\right] = 0.$$

In particular, if

$$\Lambda_{\hat{P}}(\lambda) \equiv \overline{\lim_{t\to\infty}}\,\frac{1}{t}\sup_{\sigma\in\Sigma}\Lambda_{\mu_{\sigma,t}}(t\lambda), \quad \lambda \in X^*, \tag{4.2.18}$$

then $\Lambda_{\hat{P}}(\lambda) \in \mathbb{R}$ *for* $\lambda \in X^*$,

$$\begin{aligned}\Lambda_{\hat{P}}(\lambda) &= \sup\big\{{}_{X^*}\langle\lambda, x\rangle_X - I_{\hat{P}}(x) : x \in E\big\}, \quad \lambda \in X^*,\\ I_{\hat{P}}(x) &= \sup\big\{{}_{X^*}\langle\lambda, x\rangle_X - \Lambda_{\hat{P}}(x) : \lambda \in X^*\big\}, \quad x \in E,\end{aligned} \tag{4.2.19}$$

and

$$\lim_{t\to\infty}\sup_{\sigma\in\Sigma}\left|\frac{1}{t}\Lambda_{\mu_{\sigma,t}}(t\lambda) - \Lambda_{\hat{P}}(\lambda)\right| = 0, \quad \lambda \in X^*. \tag{4.2.20}$$

We are now at the same stage in our development here as we were after proving Corollary 4.1.20 in Section 4.1; and, once again, we want to develop the analogue of the identification made in (4.1.42). Thus, from now on, we will be assuming that we are in the situation described in Remark 4.2.2 and we introduce

$$\Lambda_P(V) \equiv \overline{\lim_{t\to\infty}}\,\frac{1}{t}\log\left(\sup_{\sigma\in\Sigma}\int_\Omega \exp\left[\int_0^t V\big(\Sigma_s(\omega)\big)\,ds\right]P_\sigma(d\omega)\right) \tag{4.2.21}$$

for $V \in B(\Sigma;\mathbb{R})$. By Corollary 4.2.17, we know that, under $(\tilde{\mathbf{U}})$,

$$\begin{aligned}I_P(\nu) &= \Lambda_P^*(\nu)\\ &\equiv \sup\left\{\int_\Sigma V(\sigma)\,\nu(d\sigma) - \Lambda_P(V) : V \in C_b(\Sigma;\mathbb{R})\right\}\end{aligned} \tag{4.2.22}$$

for $\nu \in \mathbf{M}_1(\Sigma)$. Clearly,

$$\big|\Lambda_P(V) - \Lambda_P(W)\big| \leq \|V - W\|_B, \quad V,\, W \in B(\Sigma;\mathbb{R});$$

and, by HÖLDER's inequality, one sees that $V \in B(\Sigma;\mathbb{R}) \longmapsto \Lambda_P(V)$ is convex. Our goal is to find alternative expressions for these functionals; and, just as in the discrete time setting, we point out that the identification itself does not rely hypothesis $(\tilde{\mathbf{U}})$.

What we have to do first is interpret $\Lambda_P(V)$ as the logarithmic spectral radius of an appropriate operator; and this will again involve the FEYNMAN–KAC formula. However, in the present setting, there are a few more technical details which have to be confronted. In the first place, we must do a little elementary perturbation theory for semigroups.

Define

$$[P_t\phi](\sigma) = \int_\Sigma \phi(\xi)\, P(t,\sigma,d\xi), \quad (t,\sigma) \in (0,\infty)\times\Sigma,$$

for $\phi \in B(\Sigma;\mathbb{R})$.

4.2.23 Lemma. *For each $V \in B(\Sigma;\mathbb{R})$ and $\phi \in B(\Sigma;\mathbb{R})$ there is a unique measurable function $u = u_\phi^V : [0,\infty)\times\Sigma \longrightarrow \mathbb{R}$ with the properties that $\sup_{t\in[0,T]} \|u(t,\cdot)\|_B < \infty$ for every $T \in (0,\infty)$ and*

$$u(t,\sigma) = [P_t\phi](\sigma) + \int_{(0,t)} \big[P_{t-s}\big(Vu(s,\cdot)\big)\big](\sigma)\, ds \tag{4.2.24}$$

for $(t,\sigma) \in (0,\infty)\times\Sigma$. In particular, if

$$[P_t^V\phi](\sigma) \equiv u_\phi^V(t,\sigma), \quad (t,\sigma) \in (0,\infty)\times\Sigma,$$

for $\phi \in B(\Sigma;\mathbb{R})$, then $\{P_t^V : t > 0\}$ is a semigroup of bounded operators on $B(\Sigma;\mathbb{R})$; and, in fact, $\|P_t^V\|_{op} \le \exp\big[t\|V^+\|_B\big]$, $t > 0$.

PROOF: The existence as well as the uniqueness of u_ϕ^V is an elementary application of the standard PICARD iteration procedure for solving equations of VOLTERRA type. Furthermore, as a consequence of the uniqueness, one can easily prove the semigroup property for $\{P_t^V : t > 0\}$ by checking that $u(t,\cdot) = u_\phi^V(s+t,\cdot)$ satisfies (4.2.24) with ϕ replaced by $[P_s^V\phi]$. Finally, the asserted bound follows immediately from

$$\|u_\phi^V(t,\cdot)\|_B \le \|\phi\|_B + \|V^+\|_B \int_{(0,t]} \|u_\phi^V(s,\cdot)\|_B\, ds, \quad t > 0. \quad \blacksquare$$

We can now state and prove the **Feynman–Kac formula** in the context of continuous-time processes.

4.2.25 Theorem. *For each $V \in B(\Sigma;\mathbb{R})$ and all $\phi \in B(\Sigma;\mathbb{R})$,*

$$[P_t^V\phi](\sigma) = \int_\Omega \exp\left[\int_{(0,t]} V\big(\Sigma_s(\omega)\big)\, ds\right] \phi\big(\Sigma_t(\omega)\big)\, P_\sigma(d\omega). \tag{4.2.26}$$

In particular, if

$$P^V(t,\sigma,\Gamma) \equiv \int_{\{\omega:\Sigma_t(\omega)\in\Gamma\}} \exp\left[\int_{(0,t]} V(\Sigma_s(\omega))\,ds\right] P_\sigma(d\omega)$$

for $(t,\sigma,\Gamma) \in (0,\infty)\times\Sigma\times\mathcal{B}_\Sigma$, *then* $[P_t^V\phi](\sigma) = \int_\Sigma \phi(\xi)\,P^V(t,\sigma,d\xi)$ *for all* $\phi \in B(\Sigma;\mathbb{R})$ *and* $(t,\sigma)\in(0,\infty)\times\Sigma$.

PROOF: Let $u(t,\sigma)$ denote the right hand side of (4.2.26). Then, by the MARKOV property,

$$\begin{aligned}
&u(t,\sigma) - [P_t\phi](\sigma)\\
&= \int_\Omega \left(\exp\left[\int_{(0,t]} V(\Sigma_s(\omega))\,ds\right] - 1\right)\phi(\Sigma_t(\omega))\,P_\sigma(d\omega)\\
&= \int_{(0,t]}\left(\int_\Omega V(\Sigma_s(\omega))\exp\left[\int_{[s,t]} V(\Sigma_\alpha(\omega))\,d\alpha\right]\phi(\Sigma_t(\omega))\,P_\sigma(d\omega)\right)ds\\
&= \int_{(0,t]}\left[\int_\Omega V(\Sigma_s(\omega))u(t-s,\Sigma_s(\omega))\,P_\sigma(d\omega)\right]ds\\
&= \int_{(0,t]}[P_s(V\,u(t-s,\cdot))](\sigma)\,ds;
\end{aligned}$$

from which it is obvious that $u(t,\cdot)$ satisfies (4.2.24) and is therefore equal to $[P_t^V\phi]$. ∎

Armed with (4.2.26), the interpretation of $\Lambda_P(V)$ as a logarithmic spectral radius is essentially trivial.

4.2.27 Corollary. *For every* $V \in B(\Sigma;\mathbb{R})$, $\Lambda_P(V)$ *is equal to the logarithmic spectral radius*

$$\rho(P_1^V) \equiv \lim_{t\to\infty}\frac{1}{t}\log\left(\|P_t^V\|_{\text{op}}\right) \tag{4.2.28}$$

of the operator P_1^V. *(The limit in* (4.2.28) *exists by sub-additivity.)*

Our problem now is to use (4.2.28) to pass to a more pleasing expression for Λ_P^*; unfortunately, this will involve in a new round of technicalities.

Define B^0 to be the space of $\phi \in B(\Sigma;\mathbb{R})$ with the property that $\phi(\sigma) = \lim_{t\searrow 0}[P_t\phi](\sigma)$ for every $\sigma\in\Sigma$.

4.2.29 Lemma. *If* $\phi\in B(\Sigma;\mathbb{R})$ *and*

$$\lim_{t\searrow 0}[P_t^V\phi](\sigma) = \phi(\sigma), \quad \sigma\in\Sigma, \tag{4.2.30}$$

for some $V \in B(\Sigma;\mathbb{R})$, then $\phi \in B^0$. Conversely, if $\phi \in B^0$, then (4.2.30) holds for every $V \in B(\Sigma;\mathbb{R})$. Further, $C_b(\Sigma;\mathbb{R}) \subseteq B^0$ and B^0 is a closed linear subspace of $B(\Sigma;\mathbb{R})$ which is $\{P_t^V : t > 0\}$-invariant for every $V \in B(\Sigma;\mathbb{R})$. Finally, if $\psi \in C_b(\Sigma;\mathbb{R})$ and $\phi \in B^0$, then $\phi \cdot \psi \in B^0$.

PROOF: The first assertion and its converse are obvious consequences of (4.2.24), and the linearity as well as the closedness need no comment. Moreover, the asserted invariance of B^0 follows immediately from the first part together with the semigroup property; and clearly the inclusion $C_b(\Sigma;\mathbb{R}) \subseteq B^0$ is guaranteed by (4.2.1). In fact, not only do we have the asserted inclusion, but we even have that

$$\lim_{t\searrow 0} \int_\Sigma \left|\psi(\xi) - \psi(\sigma)\right| P(t,\sigma,d\xi) = 0, \quad \sigma \in \Sigma \text{ and } \psi \in C_b(\Sigma;\mathbb{R}).$$

In particular, if $\phi \in B^0$ and $\psi \in C_b(\Sigma;\mathbb{R})$, then

$$\begin{aligned}&\left|\left[P_t(\phi\cdot\psi)\right](\sigma) - \phi\cdot\psi(\sigma)\right|\\ &\qquad\le \|\phi\|_B \int_\Sigma \left|\psi(\xi) - \psi(\sigma)\right| P(t,\sigma,d\xi) + \|\psi\|_B \left|\left[P_t\phi\right](\sigma) - \phi(\sigma)\right| \longrightarrow 0\end{aligned}$$

as $t \searrow 0$ for every $\sigma \in \Sigma$; in other words, $\phi \cdot \psi \in B^0$. ∎

We now define $\mathbf{D}^V$ to be the space of $\phi \in B(\Sigma;\mathbb{R})$ with the property that

$$\left[P_t^V \phi\right](\sigma) = \phi(\sigma) + \int_{(0,t]} \left[P_s^V \psi\right](\sigma)\, ds, \quad (t,\sigma) \in (0,\infty)\times\Sigma,$$

for some $\psi \in B^0$. Note that for $\phi \in \mathbf{D}^V$, the associated $\psi \in B^0$ is uniquely determined by

$$\psi(\sigma) = \left[L^V\phi\right](\sigma) \equiv \lim_{t\searrow 0} \frac{\left[P_t^V\phi\right](\sigma) - \phi(\sigma)}{t}, \qquad \sigma \in \Sigma.$$

When $V \equiv 0$, we will use $\mathbf{D}$ and L, respectively, in place of $\mathbf{D}^0$ and L^0.

4.2.31 Lemma. *For every $V \in B(\Sigma;\mathbb{R})$, $\mathbf{D}^V \subseteq B^0$, $\mathbf{D}^V$ is $\{P_t^V : t > 0\}$-invariant, and $\left[L^V \circ P_t^V\phi\right] = \left[P_t^V \circ L^V\phi\right]$ for $t \in (0,\infty)$ and $\phi \in \mathbf{D}^V$. Moreover, if $\lambda > \rho\left(P_1^V\right)$ and*

$$\left[R_\lambda^V\phi\right](\sigma) \equiv \int_{(0,\infty)} \left(\int_\Sigma e^{-\lambda t}\phi(\xi)\, P^V(t,\sigma,d\xi)\right) dt \tag{4.2.32}$$

for $\phi \in B(\Sigma;\mathbb{R})^+$, *then* R^V_λ *admits a unique extension as a bounded linear operator taking* $B(\Sigma;\mathbb{R})$ *into* B^0; *and, for each* $\phi \in B(\Sigma;\mathbb{R})$,

$$(4.2.33)\qquad \left[P^V_t \circ R^V_\lambda \phi\right](\sigma) = \left[R^V_\lambda \phi\right](\sigma) + \int_{(0,t]} \left[P^V_s\left(\lambda[R^V_\lambda \phi] - \phi\right)\right](\sigma)\, ds$$

for $t \in (0,\infty)$. *In particular, if* $\phi \in B^0$, *then, for every* $\lambda > \rho\big(P^V_1\big)$, $\left[R^V_\lambda \phi\right] \in \mathbf{D}^V$ *and*

$$\left[L^V \circ R^V_\lambda \phi\right] = \lambda\left[R^V_\lambda \phi\right] - \phi.$$

Finally, if $V \in C_b(\Sigma;\mathbb{R})$ *and* $\lambda > \rho\big(P^V_1\big)$, *then, for every* $\phi \in B^0$, $\left[R^V_\lambda \phi\right] \in \mathbf{D} \cap \mathbf{D}^V$ *and*

$$(4.2.34)\qquad \left[L \circ R^V_\lambda \phi\right] = (\lambda - V)\left[R^V_\lambda \phi\right] - \phi.$$

PROOF: The preliminary assertions are all trivial consequences of the definition just given. To see that R^V_λ can be extended as a bounded linear operator on $B(\Sigma;\mathbb{R})$, it suffices to note that R^V_λ is non-negativity preserving and that

$$\left\|\left[R^V_\lambda 1\right]\right\|_B \le \int_{(0,\infty)} e^{-\lambda t}\left\|\left[P^V_\lambda 1\right]\right\|_B\, dt < \infty$$

as long as $\lambda > \rho\big(P^V_1\big)$. Moreover, in proving (4.2.33), we may assume that $\phi \in B(\Sigma;\mathbb{R})^+$, in which case all the steps taken below are easily justified:

$$\begin{aligned}
&\left[P^V_t R^V_\lambda \phi\right](\sigma) - \left[R^V_\lambda \phi\right](\sigma)\\
&\quad = \int_{(0,\infty)} e^{-\lambda s}\Big(\left[P^V_{s+t}\phi\right](\sigma) - \left[P^V_s \phi\right](\sigma)\Big)\, ds\\
&\quad = e^{\lambda t}\int_{[t,\infty)} e^{-\lambda s}\left[P^V_s \phi\right](\sigma)\, ds - \int_{(0,\infty)} e^{-\lambda s}\left[P^V_s \phi\right](\sigma)\, ds\\
&\quad = \left(e^{\lambda t} - 1\right)\int_{(t,\infty)} e^{-\lambda s}\left[P^V_s \phi\right](\sigma)\, ds - \int_{(0,t]} e^{-\lambda t}\left[P^V_s \phi\right](\sigma)\, ds\\
&\quad = \lambda\int_{(0,t]} e^{\lambda\alpha}\left(\int_{(\alpha,\infty)} e^{-\lambda s}\left[P^V_s \phi\right](\sigma)\, ds\right) d\alpha - \int_{(0,t]}\left[P^V_\alpha \phi\right](\sigma)\, d\alpha\\
&\quad = \int_{(0,t]}\Big(\lambda\left[P^V_s \circ R^V_\lambda \phi\right](\sigma) - \left[P^V_s \phi\right](\sigma)\Big)\, ds.
\end{aligned}$$

Clearly this proves (4.2.33) and, therefore, also that $\left[R^V_\lambda \phi\right] \in B^0$ for all $\phi \in B(\Sigma;\mathbb{R})$ and that $\left[R^V_\lambda \phi\right] \in \mathbf{D}^V$ with $\left[L^V \circ R^V_\lambda \phi\right] = \lambda\left[R^V_\lambda \phi\right] - \phi$ when $\phi \in B^0$.

Finally, suppose that $V \in C_b(\Sigma;\mathbb{R})$ and that $\lambda > \rho(P_1^V) \vee 0$. It is then easy to check from (4.2.24) and (4.2.32) that

$$[R_\lambda^V \phi] = [R_\lambda^0 \phi] + [R_\lambda^0(V[R_\lambda^V \phi])], \quad \phi \in B(\Sigma;\mathbb{R}).$$

Hence, in this case, if $\phi \in B^0$, then not only is $[R_\lambda^V \phi] \in \mathbf{D}^V$ but also (cf. the last part of Lemma 4.2.29) $[R_\lambda^V \phi] \in \mathbf{D}$. Thus, for such ϕ's, we also have that

$$\lambda[R_\lambda^V \phi] - [L \circ R_\lambda^V \phi] = \phi - V[R_\lambda^V \phi];$$

from which (4.2.34) is immediate. To handle the case when $\lambda \in (\rho(P_1^V), 0]$, simply observe that $\rho(P_1^{V+\alpha}) = \rho(P_1^V) + \alpha$ and that $R_{\lambda+\alpha}^{V+\alpha} = R_\lambda^V$ for every $\alpha \in \mathbb{R}$. ∎

We are now ready to return to the problem of finding alternate descriptions of Λ_P^*.

4.2.35 Lemma. *For $\nu \in \mathbf{M}_1(\Sigma)$ define*

$$\overline{J}_P(\nu) = \sup\left\{-\int_\Sigma \frac{Lu}{u}\,d\nu : u \in \mathbf{D} \cap B(\Sigma;[1,\infty))\right\}. \tag{4.2.36}$$

Then

$$\begin{aligned}\Lambda_P^*(\nu) &\le \overline{J}_P(\nu) \le \overline{\Lambda}_P^{\,*}(\nu)\\ &\equiv \sup\left\{\int_\Sigma V\,d\nu - \Lambda_P(V) : V \in B(\Sigma;\mathbb{R})\right\}, \quad \nu \in \mathbf{M}_1(\Sigma).\end{aligned} \tag{4.2.37}$$

Moreover, if $\{P_t : t > 0\}$ is FELLER *continuous (i.e., $C_b(\Sigma;\mathbb{R})$ is $\{P_t : t > 0\}$-invariant) and $\mathbf{D}_c \equiv \{u \in \mathbf{D} \cap C_b(\Sigma;\mathbb{R}) : Lu \in C_b(\Sigma;\mathbb{R})\}$, then*

$$\Lambda_P^*(\nu) = J_P(\nu) \equiv \sup\left\{-\int_\Sigma \frac{Lu}{u}\,d\nu : u \in \mathbf{D}_c \cap C_b(\Sigma;[1,\infty))\right\}. \tag{4.2.38}$$

PROOF: Let $u \in \mathbf{D} \cap B(\Sigma;[1,\infty))$ and set $V_u = -\frac{Lu}{u}$. It is then a trivial matter to check that the function $w : [0,\infty) \times \Sigma \longrightarrow \mathbb{R}$ given by $w(t,\cdot) = u$ satisfies

$$w(t,\sigma) = [P_t u](\sigma) + \int_{(0,t]} [P_{t-s}(V_u \cdot w(s,\cdot))](\sigma)\,ds, \quad t > 0.$$

Hence, $u = [P_t^{V_u} u]$, $t > 0$. But this means that $\Lambda_P(V_u) = \rho(P_1^{V_u}) = 0$; and so, for every $\nu \in \mathbf{M}_1(\Sigma)$,

$$-\int_\Sigma \frac{Lu}{u}\,d\nu \le \overline{\Lambda}_P^{\,*}(\nu).$$

Clearly, this proves the second part of (4.2.37). Moreover, if $u \in \mathbf{D}_c \cap C_b(\Sigma; [1,\infty))$ and therefore $V_u \in C_b(\Sigma;\mathbb{R})$, then the same argument shows that $J_P \leq \Lambda_P^*$.

To complete the proof, let $V \in C_b(\Sigma;\mathbb{R})$ and $\lambda > \Lambda_P(V)$ be given. Set $u = [R_\lambda^V 1]$ and observe that, by Corollary 4.2.27 and the last part of Lemma 4.2.31, $u \in \mathbf{D}$ and that $\lambda u - (Lu + Vu) = 1$. In addition, by the FEYNMAN–KAC formula, one sees that $u \geq \epsilon$ for some $\epsilon > 0$. Hence,

$$\int_\Sigma V\,d\nu - \lambda \leq -\int_\Sigma \frac{Lu}{u}\,d\nu \leq \overline{J}_P(\nu);$$

from which the first part of (4.2.37) follows after one lets $\lambda \searrow \Lambda_P(V)$ and then takes the supremum over $V \in C_b(\Sigma;\mathbb{R})$. Finally, in the FELLER continuous case, one can easily check that $[R_\lambda^V 1] \in \mathbf{D}_c$; and so the preceding shows that $\Lambda_P^*(\nu) \leq J_P(\nu)$. ∎

4.2.39 Theorem. *For $h > 0$ let Π_h be the transition probability function given by $\Pi_h(\sigma,\cdot) = P(h,\sigma,\cdot)$. Then,*

$$(4.2.40)\qquad \begin{aligned} &J_{\Pi_h}(\nu) \leq h\overline{\Lambda_P}^*(\nu) \text{ for } h > 0 \\ &\text{and } \overline{J}_P(\nu) \leq \varliminf_{h\searrow 0} \frac{1}{h} J_{\Pi_h}(\nu) \text{ for } \nu \in \mathbf{M}_1(\Sigma); \end{aligned}$$

and so, $\mu \in \mathbf{M}_1(\Sigma)$ is $\{P_t : t > 0\}$-invariant if $\overline{\Lambda_P}^*(\mu) = 0$, and $\overline{J}_P(\mu) = 0$ if μ is $\{P_t : t > 0\}$-invariant. (See (4.1.38) for the definition of J_{Π_h}.) In particular, if $\Lambda_P^* = \overline{\Lambda_P}^*$, then

$$(4.2.41)\quad \overline{J}_P(\nu) = \overline{\Lambda_P}^*(\nu),\ J_{\Pi_h}(\nu) \leq h\overline{J}_P(\nu),\ \text{and } \overline{J}_P(\nu) = \lim_{h\searrow 0}\frac{1}{h}J_{\Pi_h}(\nu)$$

for all $\nu \in \mathbf{M}_1(\Sigma)$; and so $\mu \in \mathbf{M}_1(\Sigma)$ is $\{P_t : t > 0\}$-invariant if and only if $\overline{J}_P(\mu) = 0$. Finally, when $\{P_t : t > 0\}$ is FELLER continuous, then

$$(4.2.42)\quad J_P(\nu) = \Lambda_P^*(\nu),\ J_{\Pi_h}(\nu) \leq hJ_P(\nu),\ \text{and } J_P(\nu) = \lim_{h\searrow 0}\frac{1}{h}J_{\Pi_h}(\nu)$$

for $\nu \in \mathbf{M}_1(\Sigma)$; and so, in this case, $\mu \in \mathbf{M}_1(\Sigma)$ is $\{P_t : t > 0\}$-invariant if and only if $J_P(\mu) = 0$.

PROOF: To prove the first part of (4.2.40), first use (4.1.39) to see that

$$J_{\Pi_h}(\nu) \leq \sup\left\{\int_\Sigma V\,d\nu - \Lambda_{\Pi_h}(V) : V \in B(\Sigma;\mathbb{R})\right\}.$$

Thus, the inequality will be established once we note that, by JENSEN's inequality and Lemma 4.1.33,

$$\begin{aligned}
&\int_\Omega \exp\left[\int_0^{nh} V\big(\Sigma_s(\omega)\big)\, ds\right] P_\sigma(d\omega) \\
&\quad = \int_\Omega \exp\left[\sum_{k=0}^{n-1}\int_0^h V\big(\Sigma_{s+kh}(\omega)\big)\, ds\right] P_\sigma(d\omega) \\
&\quad \le \frac{1}{h}\int_0^h \left(\int_\Omega \exp\left[\sum_{k=0}^{n-1} hV\big(\Sigma_{s+kh}(\omega)\big)\, ds\right] P_\sigma(d\omega)\right) ds \\
&\quad \le \left\|\big(\Pi_h\big)^n_{hV}\right\|_{op}.
\end{aligned}$$

To prove the second part of (4.2.40), let $u \in \mathbf{D} \cap B\big(\Sigma; [1, \infty)\big)$ be given and note that, since $1 - x \le -\log x$ for $x \in (0, \infty)$,

$$\int_\Sigma \frac{u - P_h u}{u}\, d\nu \le -\int_\Sigma \log\frac{\Pi_h u}{u}\, d\nu \le J_{\Pi_h}(\nu),$$

and therefore that

$$-\int_\Sigma \frac{Lu}{u}\, d\nu = \lim_{h\searrow 0}\frac{1}{h}\int_\Sigma \frac{u - P_h u}{u}\, d\nu \le \varliminf_{h\searrow 0}\frac{1}{h} J_{\Pi_h}(\nu).$$

Clearly, this completes the proof of (4.2.40). Moreover, if $\overline{\Lambda_P}^*(\mu) = 0$, then, by (4.2.40) and Lemma 4.1.45, one sees that $\mu P_h = \mu$ for all $h > 0$. On the other hand, if μ is $\{P_t : t > 0\}$-invariant, then, by Lemma 4.1.45, $J_{\Pi_h}(\mu) = 0$ for all $h > 0$, and therefore $\overline{J}_P(\mu) = 0$ by (4.2.40).

Next, suppose that $\Lambda_P^* = \overline{\Lambda_P}^*$. Then (4.2.41) follows immediately from (4.2.37) and (4.2.40).

Finally, suppose that $\{P_t : t > 0\}$ is FELLER continuous. Then, by Lemma 4.1.36, $J_{\Pi_h} = \Lambda_{\Pi_h}^*$. At the same time, by the same argument as the one which led to the first part of (4.2.40), $\Lambda_{\Pi_h}^*(\nu) \le h\Lambda_P^*(\nu)$; and, by (4.2.38), $\Lambda_P^* = J_P$. Clearly this proves that $J_{\Pi_h} \le hJ_P$. Hence, by the last part of (4.2.40), we now see that $J_P(\nu) \le \overline{J}_P(\nu) \le \varliminf_{h\searrow 0}\frac{1}{h}J_{\Pi_h}(\nu) \le J_P(\nu)$. ∎

We have now proved the following version of a result which was originally derived by DONSKER and VARADHAN [**30**].

4.2.43 Theorem. (DONSKER & VARADHAN) *Assume that* ($\tilde{\mathbf{U}}$) *holds and define* $\overline{J}_P$ *as in* (4.2.36). *Then* $\overline{J}_P$ *is a good rate function,* $\overline{\Lambda_P}^* = \overline{J}_P = \Lambda_P^*$

and, for every $\Gamma \in \mathcal{B}_{\mathbf{M}_1(\Sigma)}$,

$$
(4.2.44)\qquad
\begin{aligned}
-\inf_{\Gamma^\circ} \overline{J}_P &\le \varliminf_{t\to\infty} \frac{1}{t}\log\Big(\inf_{\sigma\in\Sigma} P_\sigma\big(\{\omega : \mathbf{L}_t(\omega)\in\Gamma\}\big)\Big)\\
&\le \varlimsup_{t\to\infty} \frac{1}{t}\log\Big(\sup_{\sigma\in\Sigma} P_\sigma\big(\{\omega : \mathbf{L}_t(\omega)\in\Gamma\}\big)\Big) \le -\inf_{\overline{\Gamma}} \overline{J}_P.
\end{aligned}
$$

In particular, if, in addition, $\{P_t : t>0\}$ *is* FELLER *continuous and* J_P *is defined as in* (4.2.38), *then* $J_P = \overline{J}_P$; *and so* (4.2.44) *holds with* J_P *in place of* $\overline{J}_P$.

PROOF: In view of Theorem 4.2.16, (4.2.19), and the second part of Theorem 4.2.39, all that we have to do is show that $\Lambda_P^* = \overline{\Lambda}_P^{\,*}$. But, with the aid of (4.2.14), this follows by the same argument as we used to prove Lemma 4.1.40. ∎

A truth which is familiar to MARKOV process devotees is that life often becomes simpler when one deals with symmetric transition probability functions. Thus, it should come as no surprise that the preceding theory of large deviations takes a more pleasing form when applied to such processes. In particular, we will close this section by showing that the rate function can often be expressed in terms of the DIRICHLET form associated with the symmetric process. We begin by recalling a few of the basic facts about symmetric MARKOV semigroups.

The σ-finite measure m on $(\Sigma, \mathcal{B}_\Sigma)$ is said to be **reversing** for the transition probability function $P(t,\sigma,\cdot)$ if the measures m_t, $t\in(0,\infty)$, defined on $\big(\Sigma^2, \mathcal{B}_\Sigma^2\big)$ by

$$
(4.2.45)\qquad m_t(d\sigma_1\times d\sigma_2) = P(t,\sigma_1,d\sigma_2)m(d\sigma_1)
$$

are symmetric (i.e., $m_t(\Gamma_1\times\Gamma_2) = m_t(\Gamma_2\times\Gamma_1)$ for all Γ_1, $\Gamma_2\in\mathcal{B}_\Sigma$). Clearly, m being reversing for $P(t,\sigma,\cdot)$ is equivalent to the statement that the semigroup $\{P_t : t>0\}$ is m-**symmetric** in the sense that, for each $t\in(0,\infty)$,

$$
\int_\Sigma \phi P_t\psi\, dm = \int_\Sigma \psi P_t\phi\, dm, \qquad \phi,\ \psi\in B\big(\Sigma;[0,\infty)\big).
$$

In particular, by taking $\psi = 1$ in the preceding, we see that for any $t\in(0,\infty)$ and $\phi\in B\big(\Sigma;[0,\infty)\big)$,

$$
\int_\Sigma P_t\phi\, dm = \int_\Sigma \phi\, dm.
$$

In other words, m is $\{P_t : t > 0\}$-invariant; and therefore, by JENSEN's inequality,

$$\|P_t\phi\|^2_{L^2(m)} \le \|P_t(\phi^2)\|_{L^1(m)} = \|\phi\|^2_{L^2(m)}$$

for all $t \in (0,\infty)$ and $\phi \in B(\Sigma;\mathbb{R})$. After combining this with the fact that $[P_t\phi](\sigma) \longrightarrow \phi(\sigma)$, $\sigma \in \Sigma$, for $\phi \in C_b(\Sigma;\mathbb{R})$, one can easily show that $\{P_t : t > 0\}$ determines a unique strongly continuous semigroup $\{\overline{P}_t : t > 0\}$ of self-adjoint contractions on $L^2(m)$ such that $\overline{P}_t\phi = P_t\phi$ for $\phi \in B(\Sigma;\mathbb{R}) \cap L^2(m)$.

Use $\overline{L}$ to denote the generator of the semigroup $\{\overline{P}_t : t > 0\}$ and note that $\overline{L}$ is a non-positive self-adjoint operator on $L^2(m)$. (The self-adjointness of $\overline{L}$ follows from that of the $\overline{P}_t$'s and the non-positivity is a consequence of their contractive property.) Moreover, by either STONE's Theorem or the HILLE–YOSHIDA Theorem, one knows that

$$\overline{P}_t = \int_{[0,\infty)} e^{-\lambda t}\, dE_\lambda, \quad t \in (0,\infty), \tag{4.2.46}$$

where $\{E_\lambda : \lambda \in [0,\infty)\}$ is the spectral resolution of the identity for $-\overline{L}$. Finally, define the **Dirichlet form** $\mathcal{E}$ to be the quadratic mapping given by

$$\mathcal{E}(\phi,\phi) = \int_{[0,\infty)} \lambda\, d\big(E_\lambda\phi,\phi\big)_{L^2(m)}, \quad \phi \in L^2(m); \tag{4.2.47}$$

let $\mathbf{D}(\mathcal{E}) = \{\phi \in L^2(m) : \mathcal{E}(\phi,\phi) < \infty\}$; and note that, by (4.2.46),

$$\frac{1}{t}\big(\phi - \overline{P}_t\phi,\phi\big)_{L^2(m)} \nearrow \mathcal{E}(\phi,\phi) \text{ as } t \searrow 0, \quad \phi \in L^2(m). \tag{4.2.48}$$

What we want to show is that, under reasonable assumptions, the function $J_{\mathcal{E}} : \mathbf{M}_1(\Sigma) \longrightarrow [0,\infty]$ given by

$$J_{\mathcal{E}}(\mu) \equiv \begin{cases} \mathcal{E}\big(f^{1/2}, f^{1/2}\big) & \text{if } \mu << m \text{ and } f = \frac{d\mu}{dm} \\ \infty & \text{otherwise.} \end{cases} \tag{4.2.49}$$

governs the large deviations of $\{\mathbf{L}_t : t > 0\}$.

4.2.50 Lemma. *For $V \in B(\Sigma;\mathbb{R})$ define $\{P^V_t : t > 0\}$ as in* Lemma 4.2.23. *Assuming that* (4.2.45) *holds,*

$$\big\|P^V_t\phi\big\|_{L^2(m)} \le e^{t\|V\|_B}\|\phi\|_{L^2(m)}, \quad t \in (0,\infty) \text{ and } \phi \in B(\Sigma;\mathbb{R});$$

and so, for each $t \in (0,\infty)$, there is a unique continuous extension $\overline{P}_t^V$ to $L^2(m)$ of P_t^V on $B(\Sigma;\mathbb{R}) \cap L^1(m)$. Moreover, $\{\overline{P}_t^V : t > 0\}$ is a strongly continuous semigroup of bounded, self-adjoint operators on $L^2(m)$; and

$$\tag{4.2.51} \begin{aligned} \Lambda_{\mathcal{E}}(V) &\equiv \sup\left\{\int_\Sigma V\,d\mu - J_{\mathcal{E}}(\mu) : \mu \in \mathbf{M}_1(\Sigma)\right\} \\ &= \lim_{t\to\infty} \frac{1}{t}\log\left(\|\overline{P}_t^V\|_{L^2(m)\to L^2(m)}\right) \leq \Lambda_P(V), \end{aligned}$$

where we have used $\|\cdot\|_{L^2(m)\to L^2(m)}$ to denote the norm for operators on $L^2(m)$ into itself.

PROOF: By (4.2.26), it is obvious that $\left|[P_t^V\phi](\sigma)\right| \leq e^{t\|V\|_B}[P_t|\phi|](\sigma)$, $\sigma \in \Sigma$. Hence, the first assertion follows immediately from the fact that P_t itself acts contractively on $L^2(m)$; and so there is no problem about proving the existence and uniqueness of the extensions $\overline{P}_t^V$. In addition, it is clear that $\{\overline{P}_t^V : t > 0\}$ forms a semigroup and that this semigroup is strongly continuous on $L^2(m)$. In order to show that the $\overline{P}_t^V$'s are self-adjoint, first observe that

$$\tag{4.2.52} [P_t^V\phi](\sigma) = [P_t\phi](\sigma) + \int_{(0,t)} \left[P_s^V\left(V[P_{t-s}\phi]\right)\right](\sigma)\,ds,$$

for $(t,\sigma) \in (0,\infty)\times\Sigma$ and $\phi \in B(\Sigma;\mathbb{R})$. Indeed, using the expression in (4.2.26) for $[P_t^V\phi](\sigma)$, one sees that

$$\begin{aligned} &[P_t^V\phi](\sigma) - [P_t\phi](\sigma) \\ &= \int_\Omega \left(\exp\left[\int_{[0,t]} V\left(\Sigma_s(\omega)\right)ds\right] - 1\right)\phi\left(\Sigma_t(\omega)\right)P_\sigma(d\omega) \\ &= \int_{(0,t)}\left[\int_\Omega V\left(\Sigma_s(\omega)\right)\exp\left[\int_{[0,s]} V\left(\Sigma_\alpha(\omega)\right)d\alpha\right]\phi\left(\Sigma_t(\omega)\right)P_\sigma(d\omega)\right]ds \\ &= \int_{(0,t)}\left[\int_\Omega V\left(\Sigma_s(\omega)\right)\exp\left[\int_{[0,s]} V\left(\Sigma_\alpha(\omega)\right)d\alpha\right]\right. \\ &\qquad\qquad \left.\times\left[P_{t-s}\phi\right]\left(\Sigma_s(\omega)\right)P_\sigma(d\omega)\right]ds \\ &= \int_{(0,t)} \left[P_s^V\left(V[P_{t-s}\phi]\right)\right](\sigma)\,ds. \end{aligned}$$

Now let $\overline{Q}_t^V$ denote the adjoint of $\overline{P}_t^V$. Then, for ϕ, $\psi \in B(\Sigma;\mathbb{R}) \cap L^1(m)$, one sees from (4.2.52) and (4.2.24), respectively, that

$$(\overline{Q}_t^V\phi,\psi)_{L^2(m)} = (\overline{P}_t\phi,\psi)_{L^2(m)} + \int_{(0,t)}\left(\overline{P}_{t-s}(V\overline{Q}_s^V\phi),\psi\right)_{L^2(m)}ds$$

and

$$\left(\overline{P}_t^V \phi, \psi\right)_{L^2(m)} = \left(\overline{P}_t \phi, \psi\right)_{L^2(m)} + \int_{(0,t)} \left(\overline{P}_{t-s}\left(V \overline{P}_s^V \phi\right), \psi\right)_{L^2(m)} ds,$$

where we have used the self-adjointness of $\overline{P}_t$ to get the first of these expressions. Starting from the above, it is an easy step to

$$\left\|\overline{Q}_t^V \phi - \overline{P}_t^V \phi\right\|_{L^2(m)} \le \|V\|_B \int_{(0,t)} \left\|\overline{Q}_s^V \phi - \overline{P}_s^V \phi\right\|_{L^2(m)} ds, \quad t \in (0,\infty);$$

and thence to $\overline{Q}_t^V = \overline{P}_t^V$.

Having established that $\{\overline{P}_t^V : t > 0\}$ is a strongly continuous semi-group of bounded, self-adjoint operators on $L^2(m)$, we can now say that

$$\overline{P}_t^V = \int_{[\lambda_V,\infty)} e^{-\lambda t} \, dE_\lambda^V, \quad t \in (0,\infty), \tag{4.2.53}$$

where $\{E_\lambda^V : \lambda \in [\lambda_V, \infty)\}$ is the spectral resolution of the identity for $-\overline{L}^V$ and $\overline{L}^V$ is the generator of $\{\overline{P}_t^V : t > 0\}$. In particular, $-\lambda_V = \lim_{t\to\infty} \frac{1}{t} \log\left(\|\overline{P}_t^V\|_{L^2(m)\to L^2(m)}\right)$. Thus, we will be done once we show that $-\lambda_V \le \Lambda_P(V)$ and that $-\lambda_V = \Lambda_{\mathcal{E}}(V)$. To see the first of these, let $\lambda > \lambda_V$ be given. Then there is a $\psi \in L^2(m)$ such that $\psi = E_\lambda^V \psi \neq 0$. Thus, we can find a $\phi \in B(\Sigma; \mathbb{R}) \cap L^1(m)$ such that $E_\lambda^V \phi \neq 0$. But this means, on the one hand, that

$$\lim_{t\to\infty} \frac{1}{t} \log\left(\left(\overline{P}_t^V \phi, \phi\right)_{L^2(m)}\right) \ge -\lambda,$$

and, on the other hand, that

$$\begin{aligned} &\lim_{t\to\infty} \frac{1}{t} \log\left(\left(\overline{P}_t^V \phi, \phi\right)_{L^2(m)}\right) \\ &\le \lim_{t\to\infty} \frac{1}{t} \log\left(\|\phi\|_{L^1(m)} \|\phi\|_B \|P_t^V\|_{\mathrm{op}}\right) = \Lambda_P(V). \end{aligned}$$

In other words, $-\lambda_V \le \Lambda_P(V)$. To prove that $-\lambda_V = \Lambda_{\mathcal{E}}(V)$, first note that

$$\frac{1}{t}\left(\phi - \overline{P}_t^V \phi, \phi\right)_{L^2(m)} \nearrow \int_{[\lambda_V,\infty)} \lambda \, d\left(E_\lambda^V \phi, \phi\right)_{L^2(m)} \text{ as } t \searrow 0;$$

and therefore that

$$-\lambda_V = \sup\left\{\lim_{t\to 0} \frac{1}{t}\left(\overline{P}_t^V \phi - \phi, \phi\right)_{L^2(m)} : \|\phi\|_{L^2(m)} = 1\right\}.$$

Next, using (4.2.24) and (4.2.48), check that

$$\lim_{t\to 0}\frac{1}{t}\big(\overline{P}_t^V\phi-\phi,\phi\big)_{L^2(m)}=\int_\Sigma V\phi^2\,dm-\mathcal{E}(\phi,\phi).$$

By combining these, we see that

$$-\lambda_V=\sup\left\{\int_\Sigma V\phi^2\,dm-\mathcal{E}(\phi,\phi):\ \|\phi\|_{L^2(m)}=1\right\}.$$

Thus, all that remains is to check that the preceding supremum is unchanged if we restrict ourselves to non-negative ϕ's. But, for $\phi\in L^1(m)\cap B(\Sigma;\mathbb{R})$,

$$\begin{aligned}\big(\phi-\overline{P}_t\phi,\phi\big)_{L^2(m)}&=\int_{\Sigma^2}\big(\phi(\sigma)-\phi(\tau)\big)\phi(\sigma)\,m_t(d\sigma\times d\tau)\\&=\frac{1}{2}\int_{\Sigma^2}\big(\phi(\sigma)-\phi(\tau)\big)^2\,m_t(d\sigma\times d\tau);\end{aligned}$$

and so, by (4.2.48) and an easy limit argument, we see that

$$\frac{1}{2t}\int_{\Sigma^2}\big(\phi(\sigma)-\phi(\tau)\big)^2\,m_t(d\sigma\times d\tau)\nearrow\mathcal{E}(\phi,\phi)\tag{4.2.54}$$

as $t\searrow 0$ for every $\phi\in L^2(m)$. In particular, we now know that

$$\mathcal{E}\big(|\phi|,|\phi|\big)\le\mathcal{E}(\phi,\phi).\quad\blacksquare$$

4.2.55 Lemma. *Assume that m is a reversing measure for $P(t,\sigma,\cdot)$ and define $\mathcal{E}$ and $J_\mathcal{E}$ accordingly. Then (cf. (4.2.37) and* Theorem *4.2.39 for the notation)*

$$\overline{J}_P(\mu)\le\overline{\Lambda}_P{}^*(\mu)\le J_\mathcal{E}(\mu)\quad\textit{for all }\mu\in\mathbf{M}_1(\Sigma)\tag{4.2.56}$$

and

$$J_\mathcal{E}(\mu)=\lim_{h\searrow 0}\frac{1}{h}J_{\Pi_h}(\mu)\quad\textit{for }\mu\in\mathbf{M}_1(\Sigma)\textit{ satisfying }\mu<<m.\tag{4.2.57}$$

PROOF: Obviously,

$$J_\mathcal{E}(\mu)\ge\sup\left\{\int_\Sigma V\,d\mu-\Lambda_\mathcal{E}(V):\ V\in B(\Sigma;\mathbb{R})\right\},$$

and so (4.2.56) follows immediately from (4.2.37) and (4.2.51). (See Exercise 4.2.63 below for more information about the relationship between $J_{\mathcal{E}}$ and $\Lambda_{\mathcal{E}}$.)

To prove (4.2.57), let $\mu \in \mathbf{M}_1(\Sigma)$ with $\mu << m$ be given and set $f = \frac{d\mu}{dm}$. Noting that $\log(1-x) \le -x$, $x \in (-\infty, 1]$, and that $\frac{u - P_h u}{u} \le 1$ for any $u \in B\big(\Sigma; (0,\infty)\big)$, we see (cf. (4.1.39)) that

$$\int_\Sigma \frac{f^{1/2}\wedge n - P_h\big(f^{1/2}\wedge n\big)}{f^{1/2}\wedge n + \epsilon}\, f\, dm \\ \le -\int_\Sigma \log\left(\frac{P_h\big(f^{1/2}\wedge n + \epsilon\big)}{f^{1/2}\wedge n + \epsilon}\right) d\mu \le J_{\Pi_h}(\mu)$$

for all $n \in \mathbb{Z}^+$ and $\epsilon > 0$. At the same time, for all $n \in \mathbb{Z}^+$ and $\epsilon > 0$,

$$0 \le \frac{f^{1/2}\wedge n}{f^{1/2}\wedge n + \epsilon} f \le f \in L^1(m)$$

and

$$0 \le \frac{P_h\big(f^{1/2}\wedge n\big)}{f^{1/2}\wedge n + \epsilon} f \le \Big(\big(P_h f^{1/2}\big)\wedge n\Big)\left(f^{1/2} \vee \frac{f}{n}\right) \\ \le \Big(f^{1/2}\big(P_h f^{1/2}\big)\Big) \vee f \in L^1(m).$$

Thus, by LEBESGUE's Dominated Convergence Theorem, we have that

$$\big(f^{1/2} - P_h f^{1/2}, f^{1/2}\big)_{L^2(m)} \le J_{\Pi_h}(\mu).$$

Clearly the desired result follows from this together with (4.2.40) and (4.2.48). ∎

By combining the preceding with our earlier results, we arrive at the following version of a result which, once again, is due originally to DONSKER and VARADHAN **[30]**.

4.2.58 Theorem. *Assume that m is $P(t,\sigma,\cdot)$- reversing. If $\mu << m$ whenever $\overline{\Lambda}_P^{\,*}(\mu) < \infty$, then $\overline{\Lambda}_P^{\,*} = J_{\mathcal{E}}$; and, when $P(t,\sigma,\cdot)$ is FELLER continuous, $J_P = J_{\mathcal{E}}$ if $\mu << m$ whenever $J_P(\mu) < \infty$. Finally, under the condition $(\tilde{\mathbf{U}})$, $\overline{J}_P = J_{\mathcal{E}}$ and so $\overline{J}_P$ can be replaced by $J_{\mathcal{E}}$ throughout (4.2.44).*

PROOF: The only assertion that has not been covered already is the last one. However, this one would be obvious if we knew that $(\tilde{\mathbf{U}})$ implied that $\nu << m$ whenever $\overline{J}_P(\nu) < \infty$; and this latter fact is established in Exercise 4.2.59 below. ∎

4.2.59 Exercise.

Assume that $(\tilde{\mathbf{U}})$ holds.

(i) Proceeding as in Exercise 4.1.48, show that there is one and only one $\mu \in \mathbf{M}_1(\Sigma)$ which is $\{P_t : t > 0\}$-invariant (i.e., $\mu = \mu P_t$, $t > 0$) and observe that when $\nu = \mu$ the measure μ_ν in (4.2.15) coincides with μ itself. Use this observation together with (3.2.20) to check that

$$\mathbf{H}(\nu|\mu) \leq 16\big(\overline{J}_P(\nu) + \log(2M)\big), \quad \nu \in \mathbf{M}_1(\Sigma). \tag{4.2.60}$$

(ii) Conclude from **(i)** that: $\mu = m$ if m is any $\big\{P_t : t > 0\big\}$-reversing σ-finite measure, $\nu << \mu$ whenever $\overline{J}_P(\nu) < \infty$, and $\{\nu : \overline{J}_P(\nu) \leq L\}$ is compact with respect to the strong topology on $\mathbf{M}_1(\Sigma)$ for each $L \geq 0$.

4.2.61 Exercise.

Again assume that $(\tilde{\mathbf{U}})$ holds. By **(ii)** of the preceding exercise, we know that $\{\overline{J}_P \leq L\}$ is compact in the strong topology for each $L \geq 0$. The purpose of the present exercise is to show that (4.2.44) continues to hold when Γ° and $\overline{\Gamma}$ are replaced by $\Gamma^{\circ\tau}$ and $\overline{\Gamma}^\tau$, respectively. (See the last part of Section 3.2 for the notation here.) This can be done in two quite easy steps.

(i) Use the results in Lemma 4.2.11 together with Theorem 3.2.21 to show that

$$\varliminf_{t\to\infty} \frac{1}{t} \log\left(\inf_{\sigma\in\Sigma} \mu_{\sigma,t}(\Gamma)\right) \geq -\inf_{\Gamma^{\circ\tau}} \overline{J}_P$$

and

$$\varlimsup_{t\to\infty} \frac{1}{t} \log\left(\sup_{\sigma\in\Sigma} \mu_{\sigma,t}(\Gamma)\right) \leq -\inf_{\overline{\Gamma}^\tau} \overline{J}_P$$

for every $\Gamma \in \mathcal{B}_{\mathbf{M}_1(\Sigma)}$. (Recall that, in the course of proving Theorem 4.2.16, we showed that $\{\mu_{\sigma,t} : t > 0\}$ satisfies the uniform full large deviation principle with rate $I_{\hat{P}}$.)

(ii) Starting with **(i)** and using (4.2.12), complete the program. In particular, show that

$$\begin{aligned}\lim_{t\to\infty} \sup_{\sigma\in\Sigma} \Bigg[\frac{1}{t} \log\bigg(\int_\Omega \exp\big[t\Phi(\omega)\big]\, P_\sigma(d\omega)\bigg) \\ - \sup\Big\{\Phi(\nu) - \overline{J}_P(\nu) : \nu \in \mathbf{M}_1(\Sigma)\Big\}\Bigg] = 0\end{aligned} \tag{4.2.62}$$

for every measurable $\Phi : \mathbf{M}_1(\Sigma) \longrightarrow \mathbb{R}$ which is continuous with respect to the strong topology and satisfies

$$\sup_{t\geq 1} \sup_{\sigma\in\Sigma} \left(\int_\Omega \exp\big[\alpha t \Phi\big(\mathbf{L}_t(\omega)\big)\big] \, P_\sigma(d\omega) \right)^{1/t} < \infty$$

for some $\alpha \in (1, \infty)$.

4.2.63 Exercise.

Assume that m is reversing for $P(t, \sigma, \cdot)$ and define the DIRICHLET form $\mathcal{E}$ and the associated function $J_\mathcal{E} : \mathbf{M}_1(\Sigma) \longrightarrow [0, \infty]$ accordingly. In Theorem 4.2.58, we saw that $J_\mathcal{E} = \overline{\Lambda}_P{}^*$ under the condition that $\mu << m$ whenever $\overline{\Lambda}_P{}^*(\mu) < \infty$. In this exercise, we will show that in general

$$J_\mathcal{E}(\mu) = \sup\left\{ \int_\Sigma V \, d\mu - \Lambda_\mathcal{E}(V) : V \in B(\Sigma; \mathbb{R}) \right\}. \tag{4.2.64}$$

Notice that although (4.2.64) proves that $\mu \in \mathbf{M}_1(\Sigma) \longmapsto J_\mathcal{E}(\mu)$ is lower semi-continuous with respect to the strong topology on $\mathbf{M}_1(\Sigma)$, it does **not** show that it is lower semi-continuous with respect to the weak topology!

(**i**) Show that the map $\phi \in L^2(m) \longmapsto \mathcal{E}(\phi, \phi)$ is lower semi-continuous.

(**ii**) Using (4.2.54) and the preceding, show that the map $f \in L^1(m)^+ \longmapsto \mathcal{E}\big(f^{1/2}, f^{1/2}\big)$ is lower semi-continuous and convex. (**Hint:** Because of (4.2.54), the convexity assertion comes down to an application of the triangle inequality for $\mathbb{R}^2$.) Conclude from this that $J_\mathcal{E}$ is convex and that

$$\mathcal{E}\big(f^{1/2}, f^{1/2}\big) = \sup\left\{ \int_\Sigma V f \, dm - \Lambda_\mathcal{E}(V) : V \in B(\Sigma; \mathbb{R}) \right\}$$

for $f \in L^1(m)^+$ with $\|f\|_{L^1(m)} = 1$. (**Hint:** Use Theorem 2.2.15 with $X = L^1(m)$.)

(**iii**) In view of (**ii**), we see that (4.2.64) holds when $\mu << m$. Thus, to complete the proof of (4.2.64), all that one has to do is check that the right side of (4.2.64) is infinite when μ is not absolutely continuous with respect to m. To this end, note that $\Lambda_\mathcal{E}(V) \leq 0$ if $V \in B(\Sigma; \mathbb{R})$ vanishes m-almost everywhere and then use this fact to complete the derivation of (4.2.64).

4.2.65 Exercise.

Let $P(t, \sigma, \cdot)$ be a transition probability function on Σ and assume that $P(t, \sigma, \cdot) \Longrightarrow \delta_\sigma$ as $t \searrow 0$ for each $\sigma \in \Sigma$. Next, let $\{P_t : t > 0\}$ be the semigroup on $B(\Sigma;\mathbb{R})$ determined by $P(t,\sigma,\cdot)$ and suppose that $\mu \in \mathbf{M}_1(\Sigma)$ is $\{P_t : t > 0\}$-invariant. Using (**i**) in Exercise 4.1.51, show that

$$\mathbf{H}(\nu P_t|\mu) \nearrow \mathbf{H}(\nu|\mu), \quad \nu \in \mathbf{M}_1(\Sigma), \quad \text{as } t \searrow 0.$$

4.2.66 Exercise.

Let $(s,\sigma,t) \in \mathbb{R}\times\Sigma\times(0,\infty) \longmapsto P(s,\sigma;t,\cdot) \in \mathbf{M}_1(\Sigma)$ be a time inhomogeneous transition probability function in the sense that it is measurable,

$$P(s,\sigma;t,\cdot) \Longrightarrow \delta_\sigma \quad \text{as } t \searrow 0 \quad \text{for all } (s,\sigma) \in \mathbb{R}\times\Sigma,$$

and it satisfies the CHAPMAN–KOLMOGOROV equation:

$$P(s,\sigma;t+t',\cdot) = \int_\Sigma P(s+t,\tau;t',\cdot)P(s,\sigma;t,d\tau) \tag{4.2.67}$$

for all $(s,\sigma) \in \mathbb{R}\times\Sigma$ and all $t,\ t' \in (0,\infty)$. In addition, we will assume that $(s,\sigma,t) \in \mathbb{R}\times\Sigma\times(0,\infty) \longmapsto P(s,\sigma;t,\cdot)$ is **periodic** in the sense that

$$P(s+1,\sigma;t,\cdot) = P(s,\sigma;t,\cdot), \quad (s,\sigma) \in \mathbb{R}\times\Sigma \text{ and } t \in [0,\infty);$$

and, finally, we impose the condition that there exist $M \in (0,\infty)$ and $T \in (0,\infty)$ for which

$$P(s,\sigma;t,\cdot) \le MP(s,\tau;t,\cdot), \quad t \in [T,T+1] \tag{4.2.68}$$

for all $(s,\sigma,\tau) \in \mathbb{R}\times\Sigma^2$.

(**i**) Define $]r[= r-[r]$ for $r \in \mathbb{R}$; and, using (4.2.67), (4.2.68), and periodicity, show that

$$\begin{aligned}\frac{1}{M}&P(s,\sigma;T+]t-T-\delta[,\,\cdot)\\ &\le P(s',\sigma';t,\,\cdot) \le MP(s,\sigma;T+]t-T-\delta[,\,\cdot)\end{aligned} \tag{4.2.69}$$

for all $t \in [T,T+1]$ and $(s,\sigma),\ (s',\sigma') \in \mathbb{R}\times\Sigma$ with $\delta \equiv s'-s \in [0,1]$.

(**ii**) Set $\tilde\Sigma = [0,1)\times\Sigma$, where we think of $[0,1)$ as the compact metric space in which the distance between points ξ and η is given by

$$|\xi-\eta| \wedge |\xi-\eta-1| \wedge |\xi-\eta+1|.$$

Next, define

$$(t,\tilde\sigma) \in (0,\infty)\times\tilde\Sigma \longmapsto \tilde P(t,\tilde\sigma,\,\cdot) \in \mathbf{M}_1(\tilde\Sigma)$$

so that

$$\tilde{P}(t,\tilde{\sigma},\tilde{\Gamma}) = \int_{\Sigma} \chi_{\tilde{\Gamma}}(]\xi + t[,\tau)\, P(\xi,\sigma;t,d\tau), \quad t \in (0,\infty),$$

for $\tilde{\sigma} = (\xi,\sigma) \in [0,1)\times\Sigma$ and $\tilde{\Gamma} \in \mathcal{B}_{\tilde{\Sigma}}$. Using periodicity, check that $(t,\tilde{\sigma}) \in (0,\infty) \times \tilde{\Sigma} \longmapsto \tilde{P}(t,\tilde{\sigma},\,\cdot\,)$ is a (time homogeneous) transition probability function and show, from (4.2.69), that

$$\int_T^{T+1} \tilde{P}(t,\tilde{\sigma},\,\cdot\,)\,dt \le M \int_T^{T+1} \tilde{P}(t,\tilde{\tau},\,\cdot\,)\,dt$$

for all $\tilde{\sigma},\ \tilde{\tau} \in \tilde{\Sigma}$. In particular, if $\{\tilde{P}_{\tilde{\sigma}} : \tilde{\sigma} \in \tilde{\Sigma}\}$ on $(\tilde{\Omega},\tilde{\mathcal{B}})$ is a MARKOV family corresponding to $\tilde{P}(t,\tilde{\sigma},\cdot)$ and if

$$(t,\tilde{\omega}) \in (0,\infty) \times \tilde{\Sigma} \longmapsto \mathbf{L}_t(\tilde{\omega}) \in \mathbf{M}_1(\tilde{\Sigma})$$

is defined accordingly, conclude that the large deviations of

$$\{\mathbf{L}_t(\tilde{\omega}) : t > 0\} \text{ under } \{\tilde{P}_{\tilde{\sigma}} : \tilde{\sigma} \in \tilde{\Sigma}\}$$

are uniformly governed by the good rate function

$$\overline{J}_{\tilde{P}}(\tilde{\nu}) = \sup\left\{ -\int_{\tilde{\Sigma}} \frac{\tilde{L}\tilde{u}}{\tilde{u}}\, d\tilde{\nu} : \tilde{u} \in \tilde{\mathbf{D}} \cap B(\tilde{\Sigma};[1,\infty)) \right\},$$

where the generator $\tilde{L}$ on $\tilde{\mathbf{D}}$ is defined for $\tilde{P}(t,\tilde{\sigma},\cdot)$ as in the discussion preceding Lemma 4.2.31.

(**iii**) Given $\tilde{\nu} \in \mathbf{M}_1(\tilde{\Sigma})$, let $\nu_1 \in \mathbf{M}_1([0,1))$ denote its marginal distribution on $[0,1)$ (i.e., $\nu_1(A) = \tilde{\nu}(A\times\Sigma)$ for $A \in \mathcal{B}_{[0,1)}$). Referring to part (**ii**), show that $\overline{J}_{\tilde{P}}(\tilde{\nu}) = \infty$ unless ν_1 is equal to LEBESGUE's measure on $[0,1)$. (For more information on this subject, the reader might want to consult [**27**].)

4.2.70 Exercise.

Let everything be as it was at the beginning of this section, assume that condition $(\tilde{\mathbf{U}})$ holds, and let $I = I_{\hat{P}}$ be the corresponding good rate function which appears in Theorem 4.2.12. In addition, assume that $t \in [0,\infty) \longmapsto \mathbf{S}_t(\omega) \in E$ is continuous for every $\omega \in \Omega$, and consider the process

$$t \in [0,\infty) \longmapsto \mathfrak{S}_t(\omega) \in \mathcal{C} \equiv C([0,1];E), \quad \omega \in \Omega$$

given by

$$\mathfrak{S}_t(\omega)(s) = \mathbf{S}_{st}(\omega), \quad s \in [0,1].$$

The purpose of this exercise is to investigate the large deviations as $t \longrightarrow \infty$ of

$$\overline{\mathfrak{S}}_t(\omega) \equiv \frac{\mathfrak{S}_t(\omega)}{t}$$

under the measures P_σ.

(**i**) For $n \in \mathbb{N}$, define

$$\Delta_n : \mathcal{C} \longrightarrow E^{2^n} \text{ by } \left(\Delta_n \psi\right)_\ell = 2^n \Big(\psi(\ell/2^n) - \psi((\ell-1)/2^n)\Big)$$

for $1 \le \ell \le 2^n$, and define

$$I_n : E^{2^n} \longrightarrow [0,\infty] \text{ by } I_n(\mathbf{x}) = \frac{1}{2^n}\sum_{\ell=1}^{2^n} I(x_\ell), \quad \mathbf{x} \in E^{2^n}.$$

Given $\Gamma = \Gamma_1 \times \cdots \times \Gamma_{2^n}$, where $\Gamma_\ell \in \mathcal{B}_E$ for each $1 \le \ell \le 2^n$, show that

$$\begin{aligned}\prod_{\ell=1}^{2^n} &\left[\inf_{\tau\in\Sigma} P_\tau\Big(\big\{\omega : \overline{\mathbf{S}}_{t/2^n}(\omega) \in \Gamma_\ell\big\}\Big)\right] \\ &\le P_\sigma\Big(\big\{\omega : \Delta_n \overline{\mathfrak{S}}_t \in \Gamma\big\}\Big) \\ &\le \prod_{\ell=1}^{2^n}\left[\sup_{\tau\in\Sigma} P_\tau\Big(\big\{\omega : \overline{\mathbf{S}}_{t/2^n}(\omega) \in \Gamma_\ell\big\}\Big)\right].\end{aligned}$$

Conclude from this both that the E^{2^n}-valued family $\big\{\Delta_n \overline{\mathfrak{S}}_t : t > 0\big\}$ is exponentially tight and second that I_n is a good rate function which governs the large deviations of this family uniformly under $\big\{P_\sigma : \sigma \in \Sigma\big\}$.

(**ii**) Define $\pi_n : E^{2^n} \longrightarrow \mathcal{C}$ so that

$$\pi_n(\mathbf{x})\big(\ell/2^n\big) = \begin{cases} 0 & \text{if } \ell = 0 \\ 2^{-n}\sum_{k=0}^{\ell} x_k & \text{if } 1 \le \ell \le 2^n \end{cases}$$

and $t \in \big[(\ell-1)/2^n, \ell/2^n\big] \longmapsto \pi_n(\mathbf{x})(t)$ is linear for each $1 \le \ell \le 2^n$. Next, set $\Pi_n = \pi_n \circ \Delta_n$, $\mathcal{C}_n = \pi_n\big(E^{2^n}\big)$, and define

$$\mathcal{I}_n : \mathcal{C} \longrightarrow [0,\infty] \text{ by } \mathcal{I}_n(\psi) = I_n\big(\Delta_n\psi\big).$$

Check that $\mathcal{I}_n$ is a convex rate function and that

$$\begin{aligned}-\inf_{\Gamma^\circ\cap\mathcal{C}_n} \mathcal{I}_n &\le \varliminf_{t\to\infty} \frac{1}{t}\log\left[\inf_{\sigma\in\Sigma} P_\sigma\Big(\big\{\omega : \Pi_n\overline{\mathfrak{S}}_t(\omega) \in \Gamma\big\}\Big)\right] \\ &\le \varlimsup_{t\to\infty}\frac{1}{t}\log\left[\sup_{\sigma\in\Sigma} P_\sigma\Big(\big\{\omega : \Pi_n\overline{\mathfrak{S}}_t(\omega) \in \Gamma\big\}\Big)\right] \le -\inf_{\overline{\Gamma}\cap\mathcal{C}_n} \mathcal{I}_n\end{aligned}$$

for every $\Gamma \in \mathcal{B}_{\mathcal{C}}$. (We give $\mathcal{C}$ the topology of uniform convergence.)

(iii) Show that $\mathcal{I}_n \leq \mathcal{I}_{n+1}$, $n \in \mathbb{N}$, and define

$$\mathcal{I} : \mathcal{C} \longrightarrow [0, \infty] \text{ by } \mathcal{I} = \lim_{n\to\infty} \mathcal{I}_n = \sup_{n\in\mathbb{N}} \mathcal{I}_n.$$

Note that $\mathcal{I}$ is a convex rate function, and show that

$$\varliminf_{t\to\infty} \frac{1}{t} \log \left[\inf_{\sigma\in\Sigma} P_\sigma\Big(\big\{ \omega : \overline{\mathfrak{S}}_t(\omega) \in G \big\} \Big) \right] \geq - \inf_G \mathcal{I}$$

for every open $G \subseteq \mathcal{C}$ and that

$$\varlimsup_{t\to\infty} \frac{1}{t} \log \left[\sup_{\sigma\in\Sigma} P_\sigma\Big(\big\{ \omega : \overline{\mathfrak{S}}_t(\omega) \in K \big\} \Big) \right] \leq - \inf_K \mathcal{I}$$

for every $K \subset\subset \mathcal{C}$. Conclude, in particular, that if $\big\{\overline{\mathfrak{S}}_t : t > 0\big\}$ is exponentially tight uniformly under $\big\{P_\sigma : \sigma \in \Sigma\big\}$, then $\mathcal{I}$ is good and governs the large deviations of $\big\{\overline{\mathfrak{S}}_t : t > 0\big\}$ uniformly under $\big\{P_\sigma : \sigma \in \Sigma\big\}$.

Hint: In proving the upper bound, let $\epsilon > 0$ be given and choose

$$\psi_1, \ldots, \psi_M \in K \text{ and } \delta_1, \ldots, \delta_M \in (0, \infty)$$

so that

$$K \subseteq \bigcup_{\ell=1}^{M} B(\psi_\ell, \delta_\ell) \text{ and } \mathcal{I}(\psi_\ell) \leq \inf_{B(\psi_\ell, 3\delta_\ell)} \mathcal{I} + \epsilon, \quad 1 \leq \ell \leq M.$$

Next, set $\delta = \delta_1 \wedge \cdots \wedge \delta_M$ and choose $n \in \mathbb{N}$ so that

$$\big\|\Pi_n \psi - \psi\big\|_{\mathcal{C}} < \delta \text{ for every } \psi \in K.$$

Conclude that

$$\begin{aligned} &\varlimsup_{t\to\infty} \frac{1}{t} \log \left[\sup_{\sigma\in\Sigma} P_\sigma\Big(\big\{ \omega : \Pi_n \overline{\mathfrak{S}}_t(\omega) \in K \big\} \Big) \right] \\ &\qquad \leq \max_{1\leq\ell\leq M} \sup_{K_\ell \cap \mathcal{C}_n} -\mathcal{I}_n \leq \max_{1\leq\ell\leq M} -\mathcal{I}(\psi_\ell) + \epsilon \leq -\inf_K \mathcal{I} + \epsilon, \end{aligned}$$

where we have used K_ℓ to denote $K \cap B(\psi_\ell, 2\delta_\ell)$.

4.3 The Wiener Sausage

In the preceding two sections, we discussed the large deviations of the empirical distribution measured in the weak topology on $\mathbf{M}_1(\Sigma)$. Although, as we pointed out in Exercise 4.1.53 and Exercise 4.2.61, the conditions $(\mathbf{U})$ and $(\tilde{\mathbf{U}})$ allow one to transform weak topology results into ones about the strong topology, thus far we have said nothing about the **variation-norm** (or **uniform**) topology. The reason for this should be obvious; namely, the empirical distributions are likely to be mutually singular to one another and therefore behave very badly in the variation norm. (This is particularly clear in the case of MARKOV chains, when the empirical distributions are purely atomic.) Thus, in some sense, it is not even reasonable to hope for a satisfactory theory with respect the variation norm unless one first carries out a "mollification procedure" to make the empirical distributions more "friendly" to one another. The purpose of the present section is to show how this can be done.

Throughout this section we will be working in the setting of Remark 4.2.2 in Section 4.2 and will be assuming (without always mentioning it) that $(\tilde{\mathbf{U}})$ holds. In particular, all the notation is the same as it was in Section 4.2.

4.3.1 Lemma. *For each $V \in B(\Sigma;\mathbb{R})$,*

$$\int_\Omega \exp\left[\int_0^t V(\Sigma_u(\omega))\,du\right] P_\sigma(d\omega) \le M\exp\Big[4\|V\|_B + t\Lambda_P(V)\Big] \tag{4.3.2}$$

for $(t,\sigma) \in (0,\infty)\times\Sigma$.

PROOF: Given $V \in B(\Sigma;\mathbb{R})$, set

$$\Phi(s,t;\sigma) = \int_\Omega \exp\left[\int_s^{s+t} V\big(\Sigma_u(\omega)\,du\big)\right] P_\sigma(d\omega)$$

for $s,\ t \in (0,\infty)$ and $\sigma \in \Sigma$. Note that, on the one hand,

$$\exp\big[-2\|V\|_B\big] \le \frac{\Phi(0,t;\sigma)}{\Phi(s,t;\sigma)} \le \exp\big[2\|V\|_B\big]$$

for $(s,t,\sigma) \in [0,1]\times(0,\infty)\times\Sigma$; while, on the other hand, the MARKOV property leads to

$$\Phi(s,t;\sigma) = \int_\Sigma \Phi(0,t;\xi)\,P(s,\sigma,d\xi), \quad (s,t,\sigma) \in [0,\infty)^2\times\Sigma.$$

Hence, by ($\tilde{\mathbf{U}}$), for all $(t,\sigma,\tau) \in (0,\infty) \times \Sigma^2$,

$$\Phi(0,t;\sigma) \leq \exp\big[2\|V\|_B\big] \int_{(0,1]\times\Sigma} \Phi(0,t;\xi)\,\rho_1(ds)P(s,\sigma,d\xi)$$

$$\leq M \exp\big[2\|V\|_B\big] \int_{(0,1]\times\Sigma} \Phi(0,t;\xi)\,\rho_2(ds)P(s,\tau,d\xi)$$

$$\leq M \exp\big[4\|V\|_B\big]\Phi(0,t;\tau);$$

and so, if

$$\Phi(t) \equiv \inf_{\sigma\in\Sigma} \Phi(0,t;\sigma), \quad t \in (0,\infty),$$

then

$$\Phi(0,t;\sigma) \leq M \exp\big[4\|V\|_B\big]\Phi(t), \quad (t,\sigma) \in (0,\infty) \times \Sigma.$$

Therefore, all that we have to do is show that $\Phi(t) \leq \exp\big[t\Lambda_P(V)\big]$ for $t \in (0,\infty)$. But, by (4.2.20), we know that

$$\lim_{t\to\infty} \frac{1}{t} \log\big(\Phi(t)\big) = \Lambda_P(V).$$

In addition, by the MARKOV property,

$$\Phi(0,s+t;\sigma) = \int_\Omega \Phi\big(0,t;\Sigma_s(\omega)\big) \exp\left[\int_0^s V\big(\Sigma_u(\omega)\big)\,du\right] P_\sigma(d\omega)$$

$$\geq \Phi(0,s;\sigma)\Phi(t)$$

for $(s,t) \in (0,\infty)^2$; and so $t \in (0,\infty) \longmapsto \Phi(t) \in \mathbb{R}$ is super-multiplicative. Thus, by Lemma 4.2.5,

$$\sup_{t\in(0,\infty)} \frac{\log\big(\Phi(t)\big)}{t} = \lim_{t\to\infty} \frac{\log\big(\Phi(t)\big)}{t} = \Lambda_P(V). \quad \blacksquare$$

Before stating our next result, we need to introduce a little more structure. Namely, for each $m \in \mathbb{Z}^+$ let β_m be a probability measure on $(0,1/m]$ and define

$$\mathcal{K}_m = \int_{(0,1/m]} P_t\,\beta_m(dt), \quad m \in \mathbb{Z}^+. \tag{4.3.3}$$

When $\mathcal{K}_m$ acts on a function $\phi \in B(\Sigma;\mathbb{R})$, we will write $\big[\mathcal{K}_m\phi\big]$; when it acts on a $\nu \in \mathbf{M}_1(\Sigma)$, we write $\nu\mathcal{K}_m$. In addition, for each $t \in (0,\infty)$ we will suppose that we are given a FELLER continuous transition probability $S_t(\sigma,\cdot)$ on Σ; and we will use $\big[S_t\phi\big]$, $\phi \in B(\Sigma;\mathbb{R})$, and νS_t, $\nu \in \mathbf{M}_1(\Sigma)$, to denote

$$\int_\Sigma \phi(\tau)\,S_t(\cdot,d\tau) \in B(\Sigma;\mathbb{R}) \quad \text{and} \quad \int_\Sigma S_t(\tau,\cdot)\nu(d\tau) \in \mathbf{M}_1(\Sigma),$$

respectively. (Recall that FELLER continuity is simply the statement that $\big[S_t\phi\big] \in C_b(\Sigma;\mathbb{R})$ whenever $\phi \in C_b(\Sigma;\mathbb{R})$.) Our development from here on will rely on our making the following hypotheses about these quantities.

(i) There exists a separable, closed subspace E of the BANACH space $\big(\mathbf{M}(\Sigma), \|\cdot\|_{\text{var}}\big)$ such that

$$\nu \in \mathbf{M}_1(\Sigma) \longmapsto \nu\mathcal{K}_m \in E_1 \equiv E \cap \mathbf{M}_1(\Sigma)$$

is a continuous mapping into E_1 (with the $\|\cdot\|_{\text{var}}$-topology) for each $m \in \mathbb{Z}^+$.

(ii) For every $\nu \in \mathbf{M}_1(\Sigma)$, $\nu S_t \in E_1$ for each $t \in (0,\infty)$ and $\nu S_t \Longrightarrow \nu$ as $t \longrightarrow \infty$; and for $\nu \in E_1$, $\|\nu - \nu S_t\|_{\text{var}} \longrightarrow 0$ as $t \longrightarrow \infty$.

(iii) For each $\alpha > 0$ and $K \subset\subset \Sigma$, there is a function $t \in (0,\infty) \longmapsto n(t;\alpha,K) \in \mathbb{Z}^+$ satisfying

$$\sup_{t\in[1,\infty)} \frac{n(t;\alpha,K)}{t} < \infty$$

such that $\big\{S_t(\sigma,\cdot) : \sigma \in K\big\}$ can be covered by $n(t;\alpha,K)$ open $\|\cdot\|_{\text{var}}$-balls of radius α.

Throughout, the topology on E and E_1 will be the one determined by $\|\cdot\|_{\text{var}}$.

The following lemma summarizes a few elementary consequences of the above assumptions.

4.3.4 Lemma. *For each $L \in (0,\infty)$, $\big\{\nu \in \mathbf{M}_1(\Sigma) : \overline{J}_P(\nu) \le L\big\} \subset\subset E_1$ and*

$$\lim_{t\to\infty} \sup\Big\{\big\|\nu - \nu S_t\big\|_{\text{var}} : \overline{J}_P(\nu) \le L\Big\} = 0.$$

Moreover, for each $t \in (0,\infty)$, $m \in \mathbb{Z}^+$, $\ell \in \mathbb{Z}^+$, and $K \subset\subset \Sigma$ there exists an

$$N(t;\ell,K) = (2\ell)^{n(t;1/\ell,K)}$$

element subset $\mathcal{W}(t,m,\ell,K)$ of $\mathcal{V} \equiv \{V \in C_b(\Sigma;\mathbb{R}) : \|V\|_B \le 1\}$ such that

$$\Big\{B\Big(\big[S_t \circ \big(\mathcal{K}_m - I\big)W\big], 10/\ell\Big) : W \in \mathcal{W}(t,m,\ell,K)\Big\}$$

covers

$$\Big\{\big[S_t \circ \big(\mathcal{K}_m - I\big)V\big] : V \in \mathcal{V}\Big\}.$$

(The balls here are taken with respect to the uniform norm $\|\cdot\|_B$ on $C_b(\Sigma;\mathbb{R})$.)

PROOF: To see that the level sets of $\overline{J}_P$ are compact in E_1, recall (cf. (4.1.46) and (4.2.41)) that

$$\|\nu - \mathcal{K}_m\nu\|_{\text{var}}^2 \le \frac{2\overline{J}_P(\nu)}{m}. \tag{4.3.5}$$

In particular, since $\nu\mathcal{K}_m \in E_1$ for every $m \in \mathbb{Z}^+$, this means that $\nu \in E_1$ if $\overline{J}_P(\nu) < \infty$. In fact, since, as the continuous image of a compact set,

$$\{\nu\mathcal{K}_m : \overline{J}_P(\nu) \le L\} \subset\subset E_1$$

for each $m \in \mathbb{Z}^+$ and $L \in (0,\infty)$, it also means that $\{\nu : \overline{J}_P(\nu) \le L\}$ is totally bounded in E_1 and is therefore compact there.

To prove the uniform convergence in E of νS_t to ν for ν's in $\overline{J}_P$-level sets, simply note that since $\|\nu S_t - \nu\|_{\text{var}} \longrightarrow 0$ for each $\nu \in E_1$ this convergence must take place uniformly fast over compact sets.

To check the last assertion, let $t \in (0,\infty)$, $m \in \mathbb{Z}^+$, $\ell \in \mathbb{Z}^+$, and $K \subset\subset \Sigma$ be given. Then, by the third property listed above, there exist $n = n(t; 1/\ell, K)$ points $\sigma_1, \dots, \sigma_n \in K$ with the property that the open sets

$$U_k \equiv \{\sigma \in \Sigma : \|S_t(\sigma,\cdot) - S_t(\sigma_k,\cdot)\|_{\text{var}} < 2/\ell\}, \quad 1 \le k \le n,$$

cover K. Now let $\{\eta_k\}_{k=1}^n \subseteq C_b(\Sigma;[0,1])$ be chosen so that $\{\sigma : \eta_k(\sigma) > 0\}$ is a relatively compact subset of U_k and $\sum_{k=1}^n \eta_k \equiv 1$ on K; and define ψ_α for

$$\alpha = (\alpha_1, \dots, \alpha_n) \in \mathfrak{A} \equiv \{-2\ell+1, \dots, 2\ell\}^n$$

by

$$\psi_\alpha = \sum_{k=1}^n \frac{\alpha_k}{\ell}\eta_k.$$

It is then an easy matter to check that for each $V \in \mathcal{V}$ there is an $\alpha \in \mathfrak{A}$ with the property that

$$\sup_{\sigma\in K}\left|[S_t \circ (\mathcal{K}_m - I)V](\sigma) - \psi_\alpha(\sigma)\right| < 5/\ell.$$

Finally, given $\alpha \in \mathfrak{A}$, choose $W_\alpha \in \mathcal{V}$ so that

$$\sup_{\sigma\in K}\left|[S_t \circ (\mathcal{K}_m - I)W_\alpha](\sigma) - \psi_\alpha(\sigma)\right| < 5/\ell$$

if such an element of $\mathcal{V}$ exists, and set $W_\alpha \equiv 0$ otherwise. ∎

4.3.6 Lemma. *For every $\delta \in (0,\infty)$,*

$$\lim_{\delta\searrow 0}\overline{\lim_{t\to\infty}}\frac{1}{t}\log\left[\sup_{\sigma\in\Sigma}\mu_{\sigma,t}\Big(\{\nu : \|\nu S_t - \nu S_t\mathcal{K}_m\|_{\text{var}} \ge \delta\}\Big)\right] = -\infty.$$

PROOF: Let $L \in (0,\infty)$ be given. Using the estimates obtained in Lemma 4.2.11, choose $K \subset\subset \Sigma$ so that

$$\varlimsup_{t\to\infty} \frac{1}{t} \log\left(\sup_{\sigma\in\Sigma} \mu_{\sigma,t}\big(\{\nu : \nu(K^c) \geq \delta/16\}\big)\right) \leq -L.$$

Next, let ℓ be the smallest element of $\mathbb{Z}^+$ which dominates $80/\delta$, and (referring to Lemma 4.3.4) note that

$$\begin{aligned}\|\nu S_t - \nu S_t \mathcal{K}_m\|_{\text{var}} &= \sup\left\{\int_\Sigma \big[S_t \circ (\mathcal{K}_m - I)V\big]\, d\nu : V \in \mathcal{V}\right\}\\ &\leq \frac{\delta}{8} + \sup\left\{\int_K \big[S_t \circ (\mathcal{K}_\delta - I)V\big]\, d\nu : V \in \mathcal{V}\right\}\\ &\leq \frac{\delta}{4} + \max\left\{\int_K \big[S_t \circ (\mathcal{K}_\delta - I)W\big]\, d\nu : W \in \mathcal{W}(t,m,\ell,K)\right\}\\ &\leq \frac{3\delta}{8} + \max\left\{\int_\Sigma \big[S_t \circ (\mathcal{K}_\delta - I)W\big]\, d\nu : W \in \mathcal{W}(t,m,\ell,K)\right\}\end{aligned}$$

for any $\nu \in \mathbf{M}_1(\Sigma)$ satisfying $\nu(K^c) \leq \delta/16$. Thus,

$$\begin{aligned}&\varlimsup_{t\to\infty} \frac{1}{t} \log\left[\sup_{\sigma\in\Sigma} \mu_{\sigma,t}\Big(\big\{\nu : \big\|\nu S_t - \nu S_t \mathcal{K}_m\big\|_{\text{var}} \geq \delta\big\}\Big)\right]\\ &\qquad \leq (-L) \vee (A + F(m)),\end{aligned}$$

where $A = \sup_{t\in[1,\infty)} \frac{\log\big(N(t;\ell,K)\big)}{t}$ and $F(m)$ is defined to be

$$\varlimsup_{t\to\infty} \frac{1}{t} \log\left[\sup_{\sigma\in\Sigma}\ \sup_{V\in\mathcal{V}} \mu_{\sigma,t}\left(\left\{\nu : \int_\Sigma \big[S_t \circ (\mathcal{K}_m - I)V\big]\, d\nu \geq \frac{5\delta}{8}\right\}\right)\right].$$

Since, by Lemma 4.3.1, $F(m) \leq -L + G(m)$, where

$$G(m) \equiv \varlimsup_{t\to\infty} \sup\left\{\Lambda_P\Big(R\big[S_t \circ (\mathcal{K}_m - I)V\big]\Big) : V \in \mathcal{V}\right\}$$

and $R = \frac{8L}{5\delta}$, all that remains is to check that $\lim_{m\to\infty} G(m) = 0$.

To this end, recall that

$$\Lambda_P(V) = \sup\left\{\int_\Sigma V\, d\nu - \overline{J}_P(\nu) : \nu \in \mathbf{M}_1(\Sigma)\right\}$$

for any $V \in C_b(\Sigma;\mathbb{R})$. Hence, for any $V \in \mathcal{V}$,

$$\begin{aligned}&\Lambda_P\Big(R\big[S_t \circ (\mathcal{K}_m - I)V\big]\Big)\\ &\quad \leq 0 \vee \left(\sup\left\{R\int_\Sigma \big[S_t \circ (\mathcal{K}_m - I)V\big]\, d\nu : \overline{J}_P(\nu) \leq 2R\right\}\right)\\ &\quad \leq R \sup\left\{\big\|\nu S_t \mathcal{K}_m - \nu S_t\big\|_{\text{var}} : \overline{J}_P(\nu) \leq 2R\right\}\\ &\quad \leq R \sup\left\{2\|\nu S_t - \nu\|_{\text{var}} + \|\nu - \nu\mathcal{K}_m\|_{\text{var}} : \overline{J}_P(\nu) \leq 2R\right\},\end{aligned}$$

where we have used

$$\begin{aligned}&\|\nu S_t\mathcal{K}_m - S_t\nu\|_{\text{var}}\\ &\quad\le \|(\nu S_t - \nu)\mathcal{K}_m\|_{\text{var}} + \|\nu\mathcal{K}_m - \nu\|_{\text{var}} + \|\nu - S_t\nu\|_{\text{var}}\\ &\quad\le 2\|\nu S_t - \nu\|_{\text{var}} + \|\nu\mathcal{K}_m - \nu\|_{\text{var}}\end{aligned}$$

in the derivation of the last line. Thus, the desired result follows immediately from Lemma 4.3.4 and (4.3.5). ∎

With these preliminaries, we can now state and prove the following version of the main theorem in [**31**].

4.3.7 Theorem. *Assume that* ($\tilde{\mathbf{U}}$) *holds and refer to the preceding. Then* $\overline{J}_P\big|_{E_1}$ *is a good rate function on* $(E_1, \|\cdot\|_{\text{var}})$; *and, for every* $\Gamma \in \mathcal{B}_{E_1}$,

$$(4.3.8)\qquad \begin{aligned}-\inf_{\Gamma^{\circ\,\text{var}}} \overline{J}_P &\le \underline{\lim}_{t\to\infty} \frac{1}{t}\log\left(\inf_{\sigma\in\Sigma} \mu_{\sigma,t}\big(\{\nu : \nu S_t \in \Gamma\}\big)\right)\\ &\le \overline{\lim}_{t\to\infty} \frac{1}{t}\log\left(\sup_{\sigma\in\Sigma} \mu_{\sigma,t}\big(\{\nu : \nu S_t \in \Gamma\}\big)\right) \le -\inf_{\overline{\Gamma}^{\text{var}}} \overline{J}_P,\end{aligned}$$

where $\Gamma^{\circ\,\text{var}}$ *and* $\overline{\Gamma}^{\text{var}}$ *are, respectively, the* $\|\cdot\|_{\text{var}}$*-interior and closure of the set* Γ.

PROOF: Define $f : \mathbf{M}_1(\Sigma) \longrightarrow E_1$ by

$$f(\nu) = \begin{cases}\nu & \text{if } \nu\in E_1\\ \nu S_1 & \text{if } \nu \in \mathbf{M}_1(\Sigma)\setminus E_1;\end{cases}$$

and, for $m \in \mathbb{Z}^+$, let f_m denote the map $\nu \in \mathbf{M}_1(\Sigma) \longmapsto \nu\mathcal{K}_m \in E_1$. Then, by (4.3.5), $\|f_n(\nu) - f(\nu)\|_{\text{var}} \longrightarrow 0$ uniformly on each level set of $\overline{J}_P$; and so, by the first part of Lemma 2.1.4, $\overline{J}_P|_{E_1}$ is a good rate function on E_1.

Next, observe that, for every $\Gamma \in \mathcal{B}_{\mathbf{M}_1(\Sigma)}$ and $\delta \in (0,\infty)$, ($\tilde{\mathbf{U}}$) can be used to easily prove that

$$\inf_{\sigma\in\Sigma} \mu_{\sigma,t}\Big(\big\{\nu : \|\nu S_t - \Gamma\|_{\text{var}} \le \delta\big\}\Big) \ge \frac{1}{M}\sup_{\sigma\in\Sigma} \mu_{\sigma,t}\Big(\{\nu : \nu S_t \in \Gamma\}\Big)$$

for all sufficiently large $t \in (0,\infty)$; and therefore, (4.3.8) reduces to proving that

$$(4.3.9)\qquad \begin{aligned}-\inf_{\Gamma^{\circ\,\text{var}}} \overline{J}_P &\le \underline{\lim}_{t\to\infty} \frac{1}{t}\log\Big(\mu_{\sigma_0,t}\big(\{\nu : \nu S_t \in \Gamma\}\big)\Big)\\ &\le \overline{\lim}_{t\to\infty} \frac{1}{t}\log\Big(\mu_{\sigma_0,t}\big(\{\nu : \nu S_t \in \Gamma\}\big)\Big) \le -\inf_{\overline{\Gamma}^{\text{var}}} \overline{J}_P\end{aligned}$$

for all $\Gamma \in \mathcal{B}_{\mathbf{M}_1(\Sigma)}$ and some $\sigma_0 \in \Sigma$. Indeed, suppose that (4.3.9) holds. Then for any open subset G of E_1 and $\nu \in G$ we would have that

$$\begin{aligned}-\overline{J}_P(\nu) &\le \lim_{\delta\searrow 0} \varliminf_{t\to\infty} \frac{1}{t}\log\Big(\mu_{\sigma_0,t}\big(\{\nu : \|\nu S_t - \mu\|_{\text{var}} < \delta\}\big)\Big) \\ &\le \varliminf_{t\to\infty} \frac{1}{t}\log\Big(\inf_{\sigma\in\Sigma}\mu_{\sigma,t}\big(\{\nu : \nu S_t \in G\}\big)\Big).\end{aligned}$$

At the same time, for any closed subset F of E_1, we would have that

$$\begin{aligned}&\varlimsup_{t\to\infty} \frac{1}{t}\log\Big(\sup_{\sigma\in\Sigma}\mu_{\sigma,t}\big(\{\nu : \nu S_t \in F\}\big)\Big) \\ &\quad\le \lim_{\delta\searrow 0}\varlimsup_{t\to\infty} \frac{1}{t}\log\Big(\mu_{\sigma_0,t}\big(\{\nu : \|\nu S_t - F\|_{\text{var}} \le \delta\}\big)\Big) \\ &\qquad\le -\lim_{\delta\searrow 0}\inf\Big\{\overline{J}_P(\nu) : \|\nu - F\|_{\text{var}} < \delta\Big\} = -\inf_F \overline{J}_P,\end{aligned}$$

where we have used Lemma 2.1.2 to get the final equality.

In order to prove (4.3.9), we will again apply Lemma 2.1.4. Namely, for $t \in (0,\infty)$, denote by Q_t the distribution of $\nu \in \mathbf{M}_1(\Sigma) \longmapsto \nu S_t \in E_1$ under $\mu_{\sigma_0,t}$; and observe that, since $\nu S_t \Longrightarrow \nu$ as $t \longrightarrow \infty$ uniformly for ν's in compact subsets of $\mathbf{M}_1(\Sigma)$, the estimates in Lemma 4.2.11 are sufficient to justify applying part **(ii)** of Exercise 2.1.20 and thereby conclude that $\overline{J}_P$ governs the large deviations of $\{Q_t : t > 0\}$ as a family of probability measures on $\mathbf{M}_1(\Sigma)$. In addition, if f and $\{f_m\}_1^\infty$ are the functions defined above, then (4.3.5) and Lemma 4.3.6 tell us that all the hypotheses of Lemma 2.1.4 are met by these functions and the family $\{Q_t : t > 0\}$. Hence, as a consequence of Lemma 2.1.4, we now see that $\overline{J}_P|_{E_1}$ governs the large deviations of $\{Q_t : t > 0\}$ as a family of probability measures on E_1; and this is just another way of saying that (4.3.9) holds. ∎

The principle reason for DONSKER and VARADHAN's interest in Theorem 4.3.7 is that they wanted to apply it to the following rather strange computation. Namely, let $N \in \mathbb{Z}^+$ be given and, as in Section 1.3, denote by $\mathcal{W}$ WIENER's measure on Θ. Given $\epsilon > 0$, $t \in (0,\infty)$, and $\theta \in \Theta$, define

$$\mathfrak{S}_t^{(\epsilon)}(\theta) = \Big\{x \in \mathbb{R}^N : \big|x - \theta(s)\big| < \epsilon \text{ for some } s \in [0,t]\Big\}$$

to be the **ϵ-sausage** around $\theta|_{[0,t]}$. Using $|\Gamma|$ to denote the LEBESGUE measure of $\Gamma \in \mathcal{B}_{\mathbb{R}^N}$, note that $\theta \in \Theta \longmapsto \big|\mathfrak{S}_t^{(\epsilon)}(\theta)\big|$ is measurable and set

$$\mathcal{A}^{(\epsilon)}(t;\gamma) = \int_\Theta \exp\Big[-\gamma\big|\mathfrak{S}_t^{(\epsilon)}(\theta)\big|\Big]\,\mathcal{W}(d\theta), \quad t \in (0,\infty)$$

for fixed $\gamma \in (0,\infty)$. In order to verify a conjecture made by some physicists, what DONSKER and VARADHAN wanted to do is compute the asymptotic behavior of $\mathcal{A}^{(\epsilon)}(t;\gamma)$ as $t \longrightarrow \infty$; and we will devote the rest of this section to showing what they did.

The first step is to rewrite $\mathcal{A}^{(\epsilon)}(t;\gamma)$ in such a way that it becomes clearer what one should expect. To this end, observe that, by BROWNIAN scaling (cf. (**iv**) of Theorem 1.3.2), for each $\alpha \in (0,\infty)$:

$$\theta \in \Theta \longmapsto \left|\mathfrak{S}_t^{(\epsilon)}(\theta)\right| \text{ and } \theta \in \Theta \longmapsto \left|\mathfrak{S}_{t/\alpha}^{(\epsilon)}(\alpha^{1/2}\theta)\right|$$

have the same distribution under $\mathcal{W}$. Thus, since

$$\left|\mathfrak{S}_{t/\alpha}^{(\epsilon)}(\alpha^{1/2}\theta)\right| = \alpha^{N/2}\left|\mathfrak{S}_{t/\alpha}^{(\epsilon/\alpha^{1/2})}(\theta)\right|,$$

we see, upon taking $\alpha = t^{2/N}$, that

$$\mathcal{A}^{(\epsilon)}\left(t^{1+2/N};\gamma\right) = \overline{\mathcal{A}}^{(\epsilon)}(t;\gamma) \equiv \mathcal{A}^{(\epsilon(t))}(t;t\gamma), \tag{4.3.10}$$

where $\epsilon(t) \equiv \epsilon/t^{1/N}$. Looking at the form of $\overline{\mathcal{A}}^{(\epsilon)}(t;\gamma)$, one is led to guess that

$$\lim_{t\to\infty} \frac{1}{t} \log\left(\overline{\mathcal{A}}^{(\epsilon)}(t;\gamma)\right)$$

and therefore, by (4.3.10),

$$\lim_{t\to\infty} \frac{1}{t^{N/(N+2)}} \log\left(\mathcal{A}^{(\epsilon)}(t;\gamma)\right)$$

might be the appropriate limit to compute.

Further evidence that the preceding is a step in the right direction is provided by the following relatively simple computation.

4.3.11 Lemma. *Let G be a bounded, non-empty, open subset of $\mathbb{R}^N$ and set*

$$\lambda(G) = \frac{1}{2}\inf\left\{\int_G \left|\nabla\phi(x)\right|^2 dx : \phi \in C_c^\infty(G;\mathbb{R}) \text{ with } \|\phi\|_{L^2(G)} = 1\right\}.$$

(The space $C_c^\infty(G;\mathbb{R})$ consists of those $\phi \in C^\infty(\mathbb{R}^N;\mathbb{R})$ with compact support in G.) Then

$$\varliminf_{t\to\infty} \frac{1}{t} \log\left(\overline{\mathcal{A}}^{(\epsilon)}(t;\gamma)\right) \geq -\left(\gamma|G| + \lambda(G)\right). \tag{4.3.12}$$

(*See* Remark *4.3.33 below.*)

PROOF: For $x \in \mathbb{R}^N$ and $\theta \in \Theta$, let θ_x be the path $t \in [0,\infty) \longmapsto x+\theta(t) \in \mathbb{R}^N$ and define $\mathfrak{S}_t^{(\epsilon(t))}(\theta_x) \subseteq \mathbb{R}^N$ accordingly. It is then clear, by the translation invariance of LEBESGUE's measure, that

$$\overline{\mathcal{A}}^{(\epsilon)}(t;\gamma) = \int_\Theta \exp\Big[-t\gamma\big|\mathfrak{S}_t^{(\epsilon(t))}(\theta_x)\big|\Big]\,\mathcal{W}(d\theta)$$

for all $x \in \mathbb{R}^N$. Next, define

$$\zeta(x,\theta) = \inf\big\{t \geq 0 : \theta_x(t) \notin G\big\}.$$

Then, since

$$\mathfrak{S}_t^{(\epsilon(t))}(\theta_x) \subseteq G^{(\epsilon(t))} \text{ if } \zeta(x,\theta) > t,$$

we see that

$$\begin{aligned}\big|G\big|\,\overline{\mathcal{A}}^{(\epsilon(t))}(t;\gamma) &= \int_G \left(\int_\Theta \exp\Big[-t\gamma\big|\mathfrak{S}_t^{(\epsilon(t))}(\theta_x)\big|\Big]\,\mathcal{W}(d\theta)\right)\,dx \\ &\geq \exp\Big[-t\gamma\big|G^{(\epsilon(t))}\big|\Big]\int_G u_G(t,x)\,dx,\end{aligned}$$

where $u_G(t,x) = \mathcal{W}\big(\{\theta : \zeta(x,\theta) > t\}\big)$. Thus, all that we have to do is show that

$$\varliminf_{t\to\infty} \frac{1}{t}\log\left(\int_G u_G(t,x)\,dx\right) \geq -\lambda(G). \tag{4.3.13}$$

The proof of (4.3.13) depends on an elementary fact about the relation between WIENER's measure and the FRIEDRICHS' extension $\overline{L}$ of $\frac{1}{2}\Delta$ on $C_c^\infty(G;\mathbb{R})$. (We use Δ here to denote the standard LACLACE operator on $\mathbb{R}^N$.) Namely, if Q_t is the operator on $B(G;\mathbb{R})$ defined by

$$\big[Q_t\phi\big](x) = \int_{\{\theta:\zeta(x,\theta)>t\}} \phi\big(\theta_x(t)\big)\,\mathcal{W}(d\theta), \quad x \in G \text{ and } \phi \in B(G;\mathbb{R}),$$

then $\{Q_t : t > 0\}$ is a sub-MARKOVian semigroup on $B(G;\mathbb{R})$ which is weakly continuous on $C_b(G;\mathbb{R})$ and satisfies

$$\Big(\big[Q_t\phi\big],\psi\Big)_{L^2(G)} = \Big(\phi,\big[Q_t\psi\big]\Big)_{L^2(G)}$$

for all ϕ, $\psi \in B(G;\mathbb{R})$. In particular, each Q_t admits a unique extension as a self-adjoint contraction $\overline{Q}_t$ on $L^2(G)$, and $\{\overline{Q}_t : t > 0\}$ becomes a strongly continuous semigroup of self-adjoint contractions whose generator coincides with $\overline{L}$. That is, $Q_t = e^{t\overline{L}}$, $t \in [0,\infty)$. (For more information on such matters, the reader might want to consult [**50**] or [**51**].)

With the preceding in hand, we now see that

$$\int_G u_G(t,x)\,dx = \left(1, \left[e^{t\overline{L}}1\right]\right)_{L^2(G)};$$

and so (4.3.13) comes down to checking that

$$\varliminf_{t\to\infty} \frac{1}{t}\log\left(1, \left[e^{t\overline{L}}1\right]\right)_{L^2(G)} \geq -\lambda(G). \tag{4.3.14}$$

But, because $\overline{L}$ is the FRIEDRICHS' extension of $\frac{1}{2}\Delta$ on $C_c^\infty(G;\mathbb{R})$,

$$\begin{aligned}
&\inf\left\{\left(\phi, \left[-\overline{L}\phi\right]\right)_{L^2(G)} : \phi \in \mathrm{Dom}(\overline{L}) \text{ and } \|\phi\|_{L^2(G)} = 1\right\} \\
&\quad = \inf\left\{\left(\phi, \left[-\frac{1}{2}\Delta\phi\right]\right)_{L^2(G)} : \phi \in C_c^\infty(G;\mathbb{R}) \text{ and } \|\phi\|_{L^2(G)} = 1\right\} \\
&\quad = \lambda(G);
\end{aligned}$$

and therefore

$$e^{-t\lambda(G)} = \sup\left\{\left(\phi, \left[Q_t\phi\right]\right)_{L^2(G)} : \phi \in B(G;\mathbb{R}) \text{ and } \|\phi\|_{L^2(G)} = 1\right\}.$$

At the same time, if $\phi \in B(G;\mathbb{R})$ and $\|\phi\|_{L^2(G)} = 1$, then

$$\begin{aligned}
e^{-(t+2)\lambda(G)} &\leq \left(\phi, \left[Q_{t+2}\phi\right]\right)_{L^2(G)} = \left(\left[Q_1\phi\right], \left[Q_t \circ Q_1\phi\right]\right)_{L^2(G)} \\
&\leq \left(1, \left[Q_t 1\right]\right)_{L^2(G)} \left\|Q_1\phi\right\|^2_{L^\infty(G)} \leq K\left(1, \left[Q_t 1\right]\right)_{L^2(G)},
\end{aligned}$$

where

$$K \equiv \sup_{x\in G} \frac{1}{(2\pi)^N}\int_G \exp\left[-|y-x|^2\right] dy \in (0,\infty)$$

and we have used

$$\begin{aligned}
\left|\left[Q_1\phi\right](x)\right| &= \left|\int_{\{\theta:\zeta(x,\theta)>1\}} \phi\big(\theta_x(1)\big)\,\mathcal{W}(d\theta)\right| \\
&\leq \int_{\{\theta:\theta_x(1)\in G\}} \left|\phi\big(\theta_x(1)\big)\right| \mathcal{W}(d\theta) \\
&= \int_G \left|\phi(y)\right| \gamma_1(y-x)\,dy \leq K^{1/2}.
\end{aligned}$$

After combining these we see that

$$\left(1, \left[Q_t 1\right]\right)_{L^2(G)} \geq \frac{1}{K} e^{-(t+2)\lambda(G)}, \quad t \in (0,\infty),$$

and obviously (4.3.14) is an immediate consequence of this. ∎

Considering how crude the idea behind (4.3.12) appears to be, one may be surprised that, after making the optimal choice of G, the right hand side of (4.3.12) turns out to be the limit which we are seeking. The intuitive explanation for this is that a WIENER path θ either takes an excursion which carries it far away from the origin, with the result that $\left|\mathfrak{S}_t^{(\epsilon(t))}(\theta)\right|$ becomes very large as $t \longrightarrow \infty$, or θ remains in some fixed bounded open G, in which case its "sausage" eventually fills up the whole of G. Although this intuitive picture is appealing, it does not lend itself easily to a rigorous proof. Instead, our derivation of the upper bound will rely on an application of Theorem 4.3.7 and will not make any direct reference to the preceding intuition.

In order to arrive at a situation to which that theorem is applicable, we will need to make some preliminary preparations. Let $R \in (0,\infty)$ be chosen and fixed, and set

$$\Sigma(R) = \left\{x \in \mathbb{R}^N : 0 \le x_j < R \text{ for } 1 \le j \le N\right\}.$$

Next, introduce on $\Sigma(R)$ the metric

$$D_R(x,y) \equiv \min\left\{|x + R\mathbf{k} - y| : \mathbf{k} \in \mathbb{Z}^N\right\}, \quad x,\, y \in \Sigma(R);$$

and observe that $\left(\Sigma(R), D_R\right)$ becomes a compact metric space for which the corresponding BOREL field $\mathcal{B}_{\Sigma(R)}$ coincides with the field $\mathcal{B}_{\mathbb{R}^N}\left[\Sigma(R)\right]$ of $\mathcal{B}_{\mathbb{R}^N}$-measurable subsets of $\Sigma(R)$. Also, define $F_R : \mathbb{R}^N \longrightarrow \Sigma(R)$ by

$$F_R(x) = \left(x_1 - R\left[\frac{x_1}{R}\right], \dots, x_N - R\left[\frac{x_N}{R}\right]\right)$$

($[\xi] \equiv \max\{n \in \mathbb{Z} : n \le \xi\}$ for $\xi \in \mathbb{R}$), and note that F_R is a continuous surjection which is locally isometric.

Using λ_R to denote the restriction to $\mathcal{B}_{\Sigma(R)}$ of LEBESGUE's measure, define $(t,x) \in (0,\infty) \times \Sigma(R) \longmapsto P_R(t,x,\cdot) \in \mathbf{M}_1\left(\Sigma(R)\right)$ by

$$P_R(t,x,dy) = \sum_{\mathbf{k} \in \mathbb{Z}^N} \gamma_t(x + R\mathbf{k} - y)\, \lambda_R(dy).$$

It is then an easy matter to check that $P_R(t,x,\cdot)$ is a λ_R-symmetric, FELLER-continuous transition probability function on $\Sigma(R)$ and that

$$P_R(t,x,\cdot) \Longrightarrow \delta_x \quad \text{as} \quad t \searrow 0 \quad \text{for each } x \in \Sigma(R).$$

We will use $\mathcal{E}_R$ to denote the corresponding DIRICHLET form (cf. (4.2.47) and (4.2.48)) and will denote the associated function $J_{\mathcal{E}_R} : \mathbf{M}_1\left(\Sigma(R)\right) \longrightarrow$

$[0,\infty]$ (cf. (4.2.49)) by I_R. Finally, choose and fix an even function $\rho \in C_c^\infty\big(B_{\mathbb{R}^N}(0,1);[0,\infty)\big)$ having total integral one, and define

$$(t,x) \in (0,\infty) \times \Sigma(R) \longmapsto S_{t,R}(x,\cdot) \in \mathbf{M}_1\big(\Sigma(R)\big)$$

by

$$S_{t,R}(x,dy) = \frac{1}{\epsilon(t)^N}\left(\sum_{\mathbf{k}\in\mathbb{Z}^N} \rho\left(\frac{x+R\mathbf{k}-y}{\epsilon(t)}\right)\right)\lambda_R(dy).$$

(Recall that $\epsilon(t) \equiv \frac{\epsilon}{t^{1/N}}$.)

4.3.15 Lemma. *Set*

$$E(R) = \{\mu \in \mathbf{M}\big(\Sigma(R)\big) : \mu << \lambda_R\},$$

give $E(R)$ the topology induced by the variation norm, and set $E_1(R) = E(R) \cap \mathbf{M}_1\big(\Sigma(R)\big)$. Then $I_R : E_1(R) \longrightarrow [0,\infty]$ is a good rate function. Moreover, if $(t,\theta) \in (0,\infty) \times \Theta \longmapsto \mathbf{L}_{t,R}(\theta) \in \mathbf{M}_1\big(\Sigma(R)\big)$ is defined by

$$\mathbf{L}_{t,R}(\theta) = \frac{1}{t}\int_0^t \delta_{F_R(\theta(s))}\,ds,$$

then, for every measurable subset Γ of $E(R)$,

$$\begin{aligned} -\inf_{\Gamma^{\circ\,\mathrm{var}}} I_R &\le \varliminf_{t\to\infty} \frac{1}{t}\log\Big(\mathcal{W}\big(\{\theta : \mathbf{L}_{t,R}(\theta)S_{t,R} \in \Gamma\}\big)\Big) \\ &\le \varlimsup_{t\to\infty} \frac{1}{t}\log\Big(\mathcal{W}\big(\{\theta : \mathbf{L}_{t,R}(\theta)S_{t,R} \in \Gamma\}\big)\Big) \le -\inf_{\overline{\Gamma}^{\mathrm{var}}} I_R. \end{aligned} \tag{4.3.16}$$

PROOF: Set $\Omega(R) = C\big([0,\infty);\Sigma(R)\big)$ and turn $\Omega(R)$ into a Polish space by giving it the topology of uniform convergence on finite intervals. Next, for $x \in \Sigma(R)$, let $P_{x,R} \in \mathbf{M}_1\big(\Omega(R)\big)$ be the distribution of $\theta \in \Theta \longmapsto F_R \circ \theta_x \in \Omega(R)$ under $\mathcal{W}$. Using $\Sigma_{t,R}(\omega) \in \Sigma(R)$ to denote the position of $\omega \in \Omega(R)$ at time $t \in [0,\infty)$ and setting

$$\mathbf{L}_{t,R}(\omega) = \frac{1}{t}\int_0^t \delta_{\Sigma_{s,R}(\omega)}\,ds, \quad (t,\omega) \in (0,\infty) \times \Omega(R),$$

we see that (4.3.16) is equivalent to

$$\begin{aligned} -\inf_{\Gamma^{\circ\,\mathrm{var}}} I_R &\le \varliminf_{t\to\infty} \frac{1}{t}\log\Big(P_{0,R}\big(\{\theta : \mathbf{L}_{t,R}(\theta)S_{t,R} \in \Gamma\}\big)\Big) \\ &\le \varlimsup_{t\to\infty} \frac{1}{t}\log\Big(P_{0,R}\big(\{\omega : \mathbf{L}_{t,R}(\omega)S_{t,R} \in \Gamma\}\big)\Big) \le -\inf_{\overline{\Gamma}^{\mathrm{var}}} I_R; \end{aligned}$$

and in order to prove this, it suffices to check that Theorem 4.3.7 applies to $\omega \in \Omega(R) \longmapsto \mathbf{L}_{t,R}(\omega)S_{t,R} \in \mathbf{M}_1(\Omega(R))$ under the family $\{P_{x,R} : x \in \Sigma(R)\}$.

We begin by showing that $\{P_{x,R} : x \in \Sigma(R)\}$ is a time-homogeneous MARKOV family with transition probability function $P_R(t, x, \cdot)$. To this end, let $\mathcal{B}_{t,R}$ be the σ-algebra over $\Omega(R)$ generated by the maps $\omega \in \Omega(R) \longmapsto \Sigma_{s,R}(\omega) \in \Sigma(R)$ for $s \in [0,t]$. Next, given $0 \le s < t < \infty$, $A \in \mathcal{B}_{s,R}$, and $\Gamma \in \mathcal{B}_{\Sigma(R)}$, note that

$$A_{x,R} \equiv \{\theta : F_R \circ \theta_x \in A\} \in \mathcal{B}_s$$

and therefore, by **(ii)** of Theorem 1.3.2,

$$\begin{aligned}
P_{x,R}&\Big(\{\omega \in A : \Sigma_{t,R}(\omega) \in \Gamma\}\Big) \\
&= \sum_{\mathbf{k}\in\mathbb{Z}^N} \mathcal{W}\Big(\{\theta \in A_{x,R} : \theta(t) \in -R\mathbf{k} + \Gamma\}\Big) \\
&= \sum_{\mathbf{k}\in\mathbb{Z}^N} \int_{A_{x,R}} \left(\int_{-R\mathbf{k}+\Gamma} \gamma_{t-s}\big(y - \theta(s)\big)\, dy\right) \mathcal{W}(d\theta) \\
&= \int_{A_{x,R}} P_R\big(t-s, F_R(\theta(s)), \Gamma\big)\, \mathcal{W}(d\theta) \\
&= \int_A P_R\big(t-s, \Sigma_{s,R}(\omega), \Gamma\big)\, P_{x,R}(d\omega).
\end{aligned}$$

That is, $\{P_{x,R} : x \in \Sigma(R)\}$ is indeed a MARKOV family with transition probability function $P_R(t, x, \cdot)$.

We next note that

$$P_R(1, x, \cdot) \le M_R P_R(1, y, \cdot), \quad x,\, y \in \Sigma(R),$$

for some $M_R \in [1,\infty)$, and therefore that $P_R(t, x, \cdot)$ satisfies $(\tilde{\mathbf{U}})$. Further, it is clear that $x \in \Sigma(R) \longmapsto P_R(1/m, x, \cdot) \in E_1(R)$ is continuous for each $m \in \mathbb{Z}^+$. Thus, all that remains is to check that $(t,x) \in (0,\infty) \times \Sigma(R) \longmapsto S_{t,R}(x,\cdot) \in \mathbf{M}_1(\Sigma(R))$ satisfies the conditions **(ii)** and **(iii)** stated prior to Lemma 4.3.4. But it is easy to see that

$$\big\|S_{t,R}(x,\cdot) - S_{t,R}(y,\cdot)\big\|_{\text{var}} \le \frac{K_R}{\epsilon(t)} D_R(x,y), \quad x,\, y \in \Sigma(R),$$

for some $K_R \in (0,\infty)$. Hence, **(ii)** certainly holds. In addition, since there is a $C_R \in (0,\infty)$ such that, for every $r \in (0,\infty)$, $\Sigma(R)$ can be covered by fewer than $[C_R/r^N]$ D_R-balls of radius r, one can easily use this same estimate to check that (**iii**) is also satisfied. ∎

In order to apply Lemma 4.3.15 to our problem, we make a sequence of simple observations. In the first place, it is clear that $F_R(\Gamma) \in \mathcal{B}_{\Sigma(R)}$ and that $|F_R(\Gamma)| \leq |\Gamma|$ for every $\Gamma \in \mathcal{B}_{\mathbb{R}^N}$. Secondly, if

$$\Gamma^{(\delta,R)} \equiv \{x \in \Sigma(R) : D_R(x,\Gamma) < \delta\} \quad \text{for } \delta \in (0,\infty) \quad \text{and} \quad \Gamma \subseteq \Sigma(R),$$

then for every $\Gamma \subseteq \mathbb{R}^N$,

$$\left(F_R(\Gamma)\right)^{(\delta,R)} \subseteq F_R\left(\Gamma^{(\delta)}\right).$$

In particular, these remarks lead to

$$\left|\left(\left\{F_R(\theta(s)) : s \in [0,t]\right\}\right)^{(\epsilon(t),R)}\right| \leq \left|\mathfrak{S}_t^{\epsilon(t)}(\theta)\right| ;$$

and therefore, since

$$\operatorname{supp}\left(\mathbf{L}_{t,R}(\theta)S_{t,R}\right) \subseteq \left(\left\{F_R(\theta(s)) : s \in [0,t]\right\}\right)^{(\epsilon(t),R)},$$

we now have that

$$\left|\operatorname{supp}\left(\mathbf{L}_{t,R}(\theta)S_{t,R}\right)\right| \leq \left|\mathfrak{S}_t^{(\epsilon(t))}(\theta)\right|, \quad (t,\theta) \in (0,\infty) \times \Theta. \tag{4.3.17}$$

4.3.18 Lemma. *For every $R \in (0,\infty)$,*

$$\varlimsup_{t\to\infty} \frac{1}{t} \log\left(\overline{\mathcal{A}}^{(\epsilon)}(t;\gamma)\right) \leq -\inf\left\{\gamma\left|\{f > 0\}\right| + \mathcal{E}_R\left(f^{1/2}, f^{1/2}\right)\right\}, \tag{4.3.19}$$

where the infimum is taken over the set

$$B_R^+ \equiv \left\{f \in L^1(\lambda_R)^+ : \|f\|_{L^1(\lambda_R)} = 1\right\}.$$

PROOF: Define $\Phi_R : E_1(R) \longrightarrow [0, R^N]$ by

$$\Phi_R(\mu) = \gamma\left|\left\{\frac{d\mu}{d\lambda_R} > 0\right\}\right|.$$

Since, for $\mu \in E_1(R)$,

$$\gamma \int_{\Sigma(R)} \frac{f}{f+\delta}\, d\lambda_R \nearrow \Phi_R(\mu) \quad \text{as } \delta \searrow 0,$$

where $f = \frac{d\mu}{d\lambda_R}$, it is an easy matter to see that Φ_R is lower semi-continuous on $E_1(R)$. Furthermore, by (4.3.17)

$$\Phi_R\big(\mathbf{L}_{t,R}(\theta)S_{t,R}\big) \le \gamma \left|\mathfrak{S}_t^{(\epsilon(t))}(\theta)\right|, \quad (t,\theta) \in (0,\infty)\times\Theta.$$

Thus, (4.3.19) follows from Lemma 4.3.15 together with Lemma 2.1.8. ∎

Let $\mathfrak{G}_{\mathrm{b}}$ denote the collection of all non-empty, bounded, open subsets of $\mathbb{R}^N$. Then, by combining (4.3.12) with (4.3.19), we arrive at

$$\begin{aligned}&\inf\big\{|G| + \lambda(G) : G \in \mathfrak{G}_{\mathrm{b}}\big\}\\ &\ge \sup_{R\in(0,\infty)} \inf\Big\{\gamma\big|\{f>0\}\big| + \mathcal{E}_R\big(f^{1/2}, f^{1/2}\big) : f \in B_R^+\Big\}.\end{aligned}$$

Thus we will know that our limit exists as soon as we show that

$$\begin{aligned}&\inf\big\{|G| + \lambda(G) : G \in \mathfrak{G}_{\mathrm{b}}\big\}\\ &\qquad\le \sup_{R\in(0,\infty)} \inf\Big\{\gamma\big|\{f>0\}\big| + \mathcal{E}_R\big(f^{1/2}, f^{1/2}\big) : f \in B_R^+\Big\}.\end{aligned} \tag{4.3.21}$$

The proof of (4.3.21) requires some work. In particular, we must find a more tractable expression for the right hand side of (4.3.21), and this is most easily done by introducing some SOBOLEV-space terminology. Thus, define the SOBOLEV space $H^1(\mathbb{R}^N)$ to be the completion of $C_{\mathrm{c}}^\infty(\mathbb{R}^N;\mathbb{R})$ with respect to the HILBERT norm

$$\|\phi\|_{H^1(\mathbb{R}^N)} \equiv \Big(\|\phi\|^2_{L^2(\mathbb{R}^N)} + \|\,|\nabla\phi|\,\|^2_{L^2(\mathbb{R}^N)}\Big)^{1/2}. \tag{4.3.22}$$

(Throughout this section we will use the classical notation $\nabla\phi$ to denote the EUCLIDean gradient of the function ϕ.) It is then a familiar and elementary fact that $\phi \in L^2(\mathbb{R}^N)$ is an element of $H^1(\mathbb{R}^N)$ if and only if there is a (necessarily unique) $\nabla\phi \in \big(L^2(\lambda_R)\big)^N$ with the property that

$$\int_{\mathbb{R}^N} \nabla\phi\cdot\Psi\,dx = -\int_{\mathbb{R}^N}\phi\,(\nabla\cdot\Psi)\,dx, \quad \Psi \in \big(C_{\mathrm{c}}^\infty(\mathbb{R}^N;\mathbb{R})\big)^N,$$

where

$$\nabla\cdot\Psi \equiv \sum_{j=1}^N \frac{\partial\Psi_j}{\partial x_j}$$

is the EUCLIDean divergence of Ψ. Moreover, if $\phi \in H^1(\mathbb{R}^N)$, then (4.3.22) continues to hold, and therefore $\phi \in H^1(\mathbb{R}^N) \longmapsto \nabla\phi \in \big(L^2(\mathbb{R}^N)\big)^N$ is a continuous map.

By analogy with the preceding, we next introduce the SOBOLEV spaces $H^1(\Sigma(R))$ for $R \in (0,\infty)$. To this end, define $C^\infty(\Sigma(R);\mathbb{R})$ to be the space of $\phi \in C(\Sigma(R);\mathbb{R})$ with the property that $\phi \circ F_R \in C^\infty(\mathbb{R}^N;\mathbb{R})$; and, for $\phi \in C^\infty(\Sigma(R);\mathbb{R})$, define $\nabla_R\phi$ to be the restriction of $\nabla(\phi \circ F_R)$ to $\Sigma(R)$. Then, $H^1(\Sigma(R))$ is the HILBERT space obtained by completing $C^\infty(\Sigma(R);\mathbb{R})$ with respect to the HILBERT norm

$$\|\phi\|_{H^1(\Sigma(R))} \equiv \left(\|\phi\|^2_{L^2(\lambda_R)} + \|\,|\nabla_R\phi|\,\|^2_{L^2(\lambda_R)}\right)^{1/2}; \tag{4.3.23}$$

and, just as before, $\phi \in L^2(\lambda_R)$ is an element of $H^1(\Sigma(R))$ if and only if there is a (unique) $\nabla_R\phi \in \left(L^2(\lambda_R)\right)^N$ such that

$$\int_{\Sigma(R)} \nabla_R\phi \cdot \Psi \, d\lambda_R = -\int_{\Sigma(R)} \phi\left(\nabla_R \cdot \Psi\right) d\lambda_R, \quad \Psi \in \left(C^\infty(\Sigma(R);\mathbb{R})\right)^N,$$

where $\nabla_R \cdot \Psi$ is the restriction to $\Sigma(R)$ of $\nabla \cdot (\Psi \circ F_R)$. In addition, $\|\phi\|_{H^1(\Sigma(R))}$ continues to be given by (4.3.23) for all $\phi \in H^1(\Sigma(R))$.

4.3.24 Lemma. *Let $R \in (0,\infty)$. If $\phi \in L^2(\lambda_R)$ and $\mathcal{E}_R(\phi,\phi) < \infty$, then $\phi \in H^1(\Sigma(R))$ and*

$$\mathcal{E}_R(\phi,\phi) = \frac{1}{2}\int_{\Sigma(R)} \left|\nabla_R\phi\right|^2 d\lambda_R. \tag{4.3.25}$$

Furthermore, if $R > 4$ and

$$\Gamma(R) \equiv \left\{x \in \mathbb{R}^N : R^{1/2} \le x_j \le R - R^{1/2} \text{ for } 1 \le j \le N\right\},$$

then, for each $\psi \in H^1(\Sigma(R))^+$, there is a $\phi \in H^1(\Sigma(R))^+$ such that

$$\begin{aligned}\gamma\left|\{\phi > 0\}\right| &+ \frac{1}{2}\int_{\Sigma(R)} \left|\nabla_R\phi\right|^2 d\lambda_R \\ &= \gamma\left|\{\psi > 0\}\right| + \frac{1}{2}\int_{\Sigma(R)} \left|\nabla_R\psi\right|^2 d\lambda_R\end{aligned} \tag{4.3.26}$$

and

$$\int_{\Gamma(R)^c} \phi^2 \, d\lambda_R \le \frac{\left|\Sigma(R) \setminus \Gamma(R)\right|}{R^N}\|\phi\|^2_{L^2(\lambda_R)}. \tag{4.3.27}$$

PROOF: We begin the proof of the first assertion by checking that (4.3.25) holds for $\phi \in C^\infty(\Sigma(R);\mathbb{R})$. To this end, we use (4.2.54) to see that $2\mathcal{E}_R(\phi,\phi)$ is equal to

$$\lim_{t\searrow 0} \frac{1}{t}\int_{\Sigma(R)} \left(\int_{\mathbb{R}^N} \left(\phi \circ F_R(x+y) - \phi \circ F_R(x)\right)^2 \gamma_t(y)\, dy\right) dx;$$

and we then apply TAYLOR's theorem to get (4.3.25). Next, let $\phi \in L^2(\lambda_R)$ with $\mathcal{E}_R(\phi,\phi) < \infty$ be given, and set

$$\begin{aligned}\phi_t(x) &= \int_{\Sigma(R)} \phi(y)\, P_R(t,x,dy)\\ &= \int_{\mathbb{R}^N} \phi\circ F_R(x-y)\,\gamma_t(y)\,dy, \quad (t,x)\in(0,\infty)\times\Sigma(R).\end{aligned}$$

Clearly $\phi_t \in C^\infty\big(\Sigma(R);\mathbb{R}\big)$ for each $t\in(0,\infty)$. Moreover, $\phi_t \longrightarrow \phi$ in $L^2(\lambda_R)$ and, by (4.2.46) and (4.2.47),

$$\mathcal{E}_R\big(\phi-\phi_t,\phi-\phi_t\big) \longrightarrow 0$$

as $t \longrightarrow 0$. Thus, by (4.3.25) for elements of $C^\infty\big(\Sigma(R);\mathbb{R}\big)$, we know that ϕ_t converges in $H^1\big(\Sigma(R)\big)$ as $t\longrightarrow 0$; and, because $\phi_t \longrightarrow \phi$ in $L^2(\lambda_R)$, ϕ_t must be converging to ϕ in $H^1\big(\Sigma(R)\big)$. In particular, $\phi\in H^1\big(\Sigma(R)\big)$; and, since, again by (4.2.46), $\mathcal{E}_R\big(\phi_t,\phi_t\big) \longrightarrow \mathcal{E}_R(\phi,\phi)$ as $t\longrightarrow 0$, this completes the proof of (4.3.25).

In order to prove the second assertion, we introduce on $L^2(\lambda_R)$ the translation operators $\tau_{a,R}$, $a\in\mathbb{R}^N$, defined by

$$\big[\tau_{a,R}\phi\big](x) = \phi\circ F_R(x-a), \quad \phi\in L^2(\lambda_R).$$

Since, as is easily checked for all $\mathcal{B}_{\Sigma(R)}$-measurable ϕ 's, $a\in\mathbb{R}^N$, and $t\in\mathbb{R}$,

$$\lambda_R\big(\{x\in\Sigma(R):\ [\tau_{a,R}\phi] > t\}\big) = \lambda_R\big(\{x\in\Sigma(R):\ \phi > t\}\big),$$

one sees that each $\tau_{a,R}$ induces an isometry on both $L^2(\lambda_R)$ and $H^1\big(\Sigma(R)\big)$. Moreover, both

$$(a,\phi)\in\mathbb{R}^N\times L^2(\lambda_R) \longmapsto \big[\tau_{a,R}\phi\big]\in L^2(\lambda_R)$$

and

$$(a,\phi)\in\mathbb{R}^N\times H^1\big(\Sigma(R)\big) \longmapsto \big[\tau_{a,R}\phi\big]\in H^1\big(\Sigma(R)\big)$$

are continuous mappings. Now let $\psi\in H^1\big(\Sigma(R)\big)^+$ be given. Then (4.3.26) holds when $\phi = \big[\tau_{a,R}\psi\big]$ for any $a\in\mathbb{R}^N$. In addition,

$$\begin{aligned}\frac{1}{R^N}&\int_{\Sigma(R)}\left(\int_{\Gamma(R)^c}\big[\tau_{a,R}\psi\big](x)^2\,\lambda_R(dx)\right)\lambda_R(da)\\ &= \frac{1}{R^N}\int_{\Gamma(R)^c}\left(\int_{\Sigma(R)}\big[\tau_{x,R}\psi\big](a)^2\,\lambda_R(da)\right)\lambda_R(dx)\\ &= \frac{\big|\Sigma(R)\setminus\Gamma(R)\big|}{R^N}\|\psi\|^2_{L^2(\lambda_R)};\end{aligned}$$

and therefore, there must exist an $a\in\mathbb{R}^N$ for which (4.3.27) holds with $\phi = \big[\tau_{a,R}\psi\big]$. ∎

4.3.28 Lemma. *The right hand side of* (4.3.21) *dominates*

$$\inf\left\{\gamma\left|\{\phi>0\}\right|+\frac{1}{2}\int_{\mathbb{R}^N}\left|\nabla\phi\right|^2dx : \phi\in H^1(\mathbb{R}^N)^+ \text{ with } \|\phi\|_{L^2(\mathbb{R}^N)}=1\right\}.$$

PROOF: Clearly there is an $R_0 \geq 16$ and a $C_1 \in (0,\infty)$ such that

$$\frac{\left|\Sigma(R)\setminus\Gamma(R)\right|}{R^N}\leq\frac{C_1}{R^{1/2}}\leq\frac{1}{2},\quad R\geq R_0.$$

For $R \geq R_0$, set

$$\Gamma'(R)=\left\{x\in\mathbb{R}^N : \frac{R^{1/2}}{2}\leq x_j\leq R-\frac{R^{1/2}}{2}\text{ for }1\leq j\leq N\right\},$$

and define

$$\eta_R(x)=\left(\frac{2}{R^{1/2}}\right)^N\int_{\Gamma'(R)}\rho\left(\frac{2(x-y)}{R^{1/2}}\right)dy,\quad x\in\mathbb{R}^N,$$

where $\rho\in C_c^\infty\big(B_{\mathbb{R}^N}(0,1);[0,\infty)\big)$ has total integral one. Obviously: $\eta_R=1$ on $\Gamma(R)$; $\eta_R\in C_c^\infty\big(\Sigma^\circ(R);[0,1]\big)$, where

$$\Sigma^\circ(R)=\left\{x\in\mathbb{R}^N : 0<x_j<R\text{ for }1\leq j\leq N\right\};$$

and there exists a $C_2\in(0,\infty)$ for which

$$\left|\nabla\eta_R(x)\right|\leq\frac{C_2}{R^{1/2}},\quad R\in[R_0,\infty)\text{ and }x\in\mathbb{R}^N.$$

Hence, for any $\phi\in H^1\big(\Sigma(R)\big)$, not only is $\eta_R\phi$ an element of $H^1(\mathbb{R}^N)$ but also

$$\begin{aligned}\left|\nabla(\eta_R\phi)\right|^2&=\left|\phi\nabla\eta_R+\eta_R\nabla_R\phi\right|^2\\&\leq\left(1+\frac{1}{R^{1/2}}\right)\left|\nabla_R\phi\right|^2+C_2^2\left(\frac{1}{R}+\frac{1}{R^{1/2}}\right)|\phi|^2.\end{aligned}$$

With these preliminaries, we can now complete the proof as follows. By Lemma 4.3.24, we know that the right hand side of (4.3.21) dominates

$$\varlimsup_{R\to\infty}\inf\left\{\gamma\left|\{\phi>0\}\right|+\frac{1}{2}\int_{\Sigma(R)}\left|\nabla_R\phi\right|^2d\lambda_R : \phi\in H^1\big(\Sigma(R)\big)^+\right.$$
$$\left.\text{with }\|\phi\|_{L^2(\lambda_R)}=1\text{ and }\int_{\Gamma(R)^c}\phi^2\,d\lambda_R\leq\frac{C_1}{R^{1/2}}\right\}.$$

Now let $R > R_0$ and $\phi \in H^1(\Sigma(R))^+$ with $\|\phi\|_{L^2(\lambda_R)} = 1$ and

$$\int_{\Sigma(R)\setminus\Gamma(R)} \phi^2 \, d\lambda_R \leq \frac{|\Sigma(R) \setminus \Gamma(R)|}{R^N}$$

be given; and set $\psi = \eta_R\phi/\|\eta_R\phi\|_{L^2(\lambda_R)}$. Then $\psi \in H^1(\mathbb{R}^N)^+$, $\|\psi\|_{L^2(\lambda_R)} = 1$, and

$$\begin{aligned}
&\gamma\big|\{\psi > 0\}\big| + \frac{1}{2}\int_{\mathbb{R}^N} \left|\nabla\psi\right|^2 dx \\
&\quad \leq \gamma\big|\{\phi > 0\}\big| + \frac{R^{1/2}}{2(R^{1/2} - C_1)}\left[2C_2^2\left(\frac{1}{R} + \frac{1}{R^{1/2}}\right)\right. \\
&\qquad\qquad \left. + \left(1 + \frac{1}{R^{1/2}}\right)\int_{\Sigma(R)} \left|\nabla_R\phi\right|^2 d\lambda_R\right] \\
&\quad \leq \left(1 + \frac{C_3}{R^{1/2}}\right)\left[\gamma\big|\{\phi > 0\}\big| + \frac{1}{2}\int_{\Sigma(R)} \left|\nabla_R\phi\right|^2 d\lambda_R\right] + \frac{C_4}{R^{1/2}}
\end{aligned}$$

for some C_3, $C_4 \in (0, \infty)$; and clearly the desired result follows from this. ∎

At this point what we know is that

$$\begin{aligned}
&-\inf\big\{\gamma|G| + \lambda(G) : G \in \mathfrak{G}_b\big\} \\
&\qquad \leq \varliminf_{t\to\infty} \frac{1}{t}\log\left(\overline{\mathcal{A}}^{(\epsilon)}(t;\gamma)\right) \leq \varlimsup_{t\to\infty} \frac{1}{t}\log\left(\overline{\mathcal{A}}^{(\epsilon)}(t;\gamma)\right) \\
&\qquad \leq -\inf\left\{\gamma\big|\{\phi > 0\}\big| + \frac{1}{2}\int_{\mathbb{R}^N} \left|\nabla\phi\right|^2 dx\right. \\
&\qquad\qquad\qquad \left. : \phi \in H^1(\mathbb{R}^N)^+ \text{ with } \|\phi\|_{L^2(\mathbb{R}^N)} = 1\right\}.
\end{aligned} \tag{4.3.29}$$

Although (4.3.29) appears to be still some distance from our goal, it, in conjunction with a beautiful result from classical potential theory, turns out to be all that we need. To be precise, for measurable $\phi : \mathbb{R}^N \longrightarrow [0, \infty)$ define the **decreasing rearrangement** of ϕ to be the non-negative measurable function $\tilde{\phi}$ on $\mathbb{R}^N$ with the property that

$$\{\tilde{\phi} > t\} = B_{\mathbb{R}^N}\left(0, \left(\frac{\big|\{\phi > t\}\big|}{\Omega_N}\right)^{1/N}\right), \quad t \in [0, \infty),$$

where $\Omega_N \equiv \left|B_{\mathbb{R}^N}(0,1)\right|$. Obviously, $\left|\{\tilde{\phi} > t\}\right| = \left|\{\phi > t\}\right|$ for every $t \in [0,\infty)$, and therefore $\phi \in L^2(\mathbb{R}^N) \longmapsto \tilde{\phi} \in L^2(\mathbb{R}^N)$ is an isometry. The beautiful result alluded to states that $\tilde{\phi} \in H^1(\mathbb{R}^N)$ and

$$\int_{\mathbb{R}^N} \left|\nabla\tilde{\phi}\right|^2 dx \le \int_{\mathbb{R}^N} \left|\nabla\phi\right|^2 dx \tag{4.3.30}$$

if $\phi \in H^1(\mathbb{R}^N)$. For an elegant proof of this statement, see [**74**].

4.3.31 Theorem. (DONSKER & VARADHAN) *Set*

$$\kappa_N(\gamma) = \frac{N+2}{N}\left(\frac{2\lambda_N}{N}\right)^{N/(N+2)} \gamma^{2/(N+2)}, \quad \gamma \in (0,\infty),$$

where

$$\lambda_N = \inf\left\{\frac{1}{2}\int_{\mathbb{R}^N} \left|\nabla\phi\right|^2 dx \right.$$
$$\left. : \phi \in C_{\mathrm{c}}^\infty\left(B_{\mathbb{R}^N}\left(0, \left(1/\Omega_N\right)^{1/N}\right)\right) \text{ with } \|\phi\|_{L^2(\mathbb{R}^N)} = 1\right\}.$$

Then, for every $\epsilon \in (0,\infty)$,

$$\lim_{t\to\infty} \frac{1}{t^{N/(N+2)}} \log\left(\int_\Theta \exp\left[-\gamma\left|\mathfrak{S}_t^{(\epsilon)}(\theta)\right|\right] \mathcal{W}(d\theta)\right) = -\kappa_N(\gamma).$$

PROOF: In view of (4.3.10) and (4.3.29), all that we have to do is check that

$$\begin{aligned} &\inf\{\gamma|G| + \lambda(G) : G \in \mathfrak{G}_{\mathrm{b}}\} \le \kappa_N(\gamma) \\ &\le \inf\left\{\gamma\left|\{\phi > 0\}\right| + \frac{1}{2}\int_{\mathbb{R}^N} \left|\nabla\phi\right|^2 dx \right. \\ &\qquad\qquad \left. : \phi \in H^1(\mathbb{R}^N) \text{ with } \|\phi\|_{L^2(\mathbb{R}^N)} = 1\right\}. \end{aligned} \tag{4.3.32}$$

To this end, note that, by an obvious scaling argument,

$$\lambda(B_A) = \frac{\lambda_N}{A^{2/N}}, \quad A \in (0,\infty),$$

where B_A denotes the open ball in $\mathbb{R}^N$ around the origin with volume A. Hence,

$$\begin{aligned} &\inf\{\gamma|G| + \lambda(G) : G \in \mathfrak{G}_{\mathrm{b}}\} \le \inf\{\gamma|B_A| + \lambda(B_A) : A \in (0,\infty)\} \\ &= \inf\left\{\gamma A + \frac{\lambda_N}{A^{2/N}}\right\} = \kappa_N(\gamma); \end{aligned}$$

which is the left hand side of (4.3.32).

To prove the right hand side of (4.3.32), suppose that $\phi \in H^1(\mathbb{R}^N)^+$ with $\|\phi\|_{L^2(\mathbb{R}^N)} = 1$ and $A \equiv \big|\{\phi > 0\}\big| < \infty$ is given. Then, by the result cited above,

$$\gamma\big|\{\phi > 0\}\big| + \frac{1}{2}\int_{\mathbb{R}^N} \left|\nabla\phi\right|^2 dx \geq \gamma A + \frac{1}{2}\int_{\mathbb{R}^N} \left|\nabla\tilde{\phi}\right|^2 dx,$$

where $\tilde{\phi}$ is the decreasing rearrangement of ϕ. At the same time, by an elementary mollification procedure, one can easily check that

$$\frac{1}{2}\int_{\mathbb{R}^N} \left|\nabla\tilde{\phi}\right|^2 dx \geq \lambda\big(B_{A+\delta}\big)$$

for every $\delta \in (0,\infty)$. Thus, after letting $\delta \searrow 0$, we conclude that

$$\gamma\big|\{\phi > 0\}\big| + \frac{1}{2}\int_{\mathbb{R}^N} \left|\nabla\phi\right|^2 dx \geq \gamma A + \lambda\big(B_A\big). \quad \blacksquare$$

4.3.33 Remark.

The reader who is uncomfortable with the sort of DIRICHLET-form technology used in the proof of Lemma 4.3.11 should note that the proof of Theorem 4.3.31 only required our knowing (4.3.12) when G is a ball around the origin, in which case (4.3.12) can be easily derived from familiar, classical facts about the eigenvalues and eigenfunctions for $\frac{1}{2}\Delta$ with boundary condition 0.

4.4 Process Level Large Deviations

In the preceding three sections, we discussed the large deviation theory for the empirical distribution of the position of a MARKOV process. In this section, we will develop the same theory for the empirical distribution of the whole process.

We begin in the setting of MARKOV chains. Thus, let Π be a transition probability function on a Polish space Σ and denote by $\{P_\sigma : \sigma \in \Sigma\}$ the associated MARKOV family of probability measures on $\Omega = \Sigma^{\mathbb{N}}$. For $n \in \mathbb{N}$, define $\theta_n : \Omega \longrightarrow \Omega$ so that $\Sigma_m(\theta_n\omega) = \Sigma_{m+n}(\omega)$ (recall that $\Sigma_n(\omega)$ is the position of $\omega \in \Omega$ at time $n \in \mathbb{N}$); and, given $n \in \mathbb{Z}^+$, define

$$\omega \in \Omega \longmapsto \mathbf{R}_n(\omega) \equiv \frac{1}{n}\sum_{k=1}^{n} \delta_{\theta_k\omega} \in \mathbf{M}_1(\Omega). \tag{4.4.1}$$

Once again, under the conditions introduced in Section 4.1, ergodic considerations predict that $\mathbf{R}_n(\omega) \Longrightarrow P_\mu$ almost surely, where $P_\mu \equiv \int_\Sigma P_\sigma\,\mu(d\sigma)$ and $\mu \in \mathbf{M}_1(\Sigma)$ is the Π-invariant discussed in Exercise 4.1.48. Our goal is to describe the large deviation theory for the families

$$\{P_\sigma \circ (\mathbf{R}_n)^{-1} : n \geq 1\}, \quad \sigma \in \Sigma.$$

Note that $\mathbf{L}_n(\omega) = \mathbf{R}_n(\omega) \circ \Sigma_0^{-1}$ and therefore that the result which we are now pursuing is "higher" than the earlier one.

We will begin by considering the more modest task of dealing with a study of the analogous problem for the finite dimensional marginals of the $\mathbf{R}_n(\omega)$'s. Namely, for $1 \leq k < \ell < \infty$, define

$$\omega \longmapsto \Sigma_{[k,\ell)}(\omega) \equiv \big(\Sigma_k(\omega), \dots, \Sigma_{\ell-1}(\omega)\big) \in \Sigma^{\ell-k};$$

and, for $d \geq 2$, consider the map

$$\omega \in \Omega \longmapsto \mathbf{L}_n^{(d)}(\omega) = \frac{1}{n}\sum_{k=1}^{n} \delta_{\Sigma_{[k,k+d)}}(\omega) \in \mathbf{M}_1(\Sigma^d); \tag{4.4.2}$$

and let $\mu_{\sigma,n}^{(d)} \in \mathbf{M}_1\big(\mathbf{M}_1(\Sigma^d)\big)$ denote the distribution of $\omega \longmapsto \mathbf{L}_n^{(d)}(\omega)$ under P_σ.

We will now develop the large deviation theory for the families $\{\mu_{\sigma,n}^{(d)} : n \geq 1\}$ when Π satisfies (**U**). To this end, define the transition probability function $\Pi^{(d)}$ on Σ^d by

$$\Pi^{(d)}\big(\sigma^{(d)}, \Gamma\big) = \int_\Sigma \chi_\Gamma\big(\sigma_2^{(d)}, \dots, \sigma_d^{(d)}, \tau\big)\Pi\big(\sigma_d^{(d)}, d\tau\big) \tag{4.4.3}$$

for $\sigma^{(d)} \in \Sigma^d$ and $\Gamma \in \mathcal{B}_{\Sigma^d}$; and let $\{P^{(d)}_{\sigma^{(d)}} : \sigma^{(d)} \in \Sigma^d\}$ be the associated MARKOV family on $\Omega^{(d)} \equiv (\Sigma^d)^{\mathbb{N}}$. Noting that

$$\left(\Pi^{(d)}\right)^{d+\ell-1}\left(\sigma^{(d)}, d\tau^{(d)}\right) = \Pi^\ell\left(\sigma^{(d)}_d, d\tau^{(d)}_1\right)\Pi\left(\tau^{(d)}_1, d\tau^{(d)}_2\right)\cdots\Pi\left(\tau^{(d)}_{d-1}, d\tau^{(d)}_d\right)$$

for $\ell \in \mathbb{Z}^+$, one sees that $(\mathbf{U})$ implies that

$$(\mathbf{U^{(d)}}) \qquad \left(\Pi^{(d)}\right)^{d+\ell-1}\left(\sigma^{(d)}, \cdot\right) \leq \frac{M}{N}\sum_{m=1}^{d+N-1}\left(\Pi^{(d)}\right)^m\left(\tau^{(d)}, \cdot\right)$$

for $\sigma^{(d)}$, $\tau^{(d)} \in \Sigma^d$.

Thus, when Π satisfies $(\mathbf{U})$, Theorem 4.1.43 applies to the empirical distribution of the position of the MARKOV chain $\{P^{(d)}_{\sigma^{(d)}} : \sigma^{(d)} \in \Sigma^d\}$ and tells us that

$$(4.4.4) \qquad J^{(d)}_\Pi(\nu) \equiv J_{\Pi^{(d)}}(\nu), \quad \nu \in \mathbf{M}_1(\Sigma^d)$$

is a good rate function and that

$$-\inf_{\Gamma^\circ} J^{(d)}_\Pi \leq \varliminf_{n\to\infty} \frac{1}{n}\log\left(\inf_{\sigma^{(d)}\in\Sigma^d} P^{(d)}_{\sigma^{(d)}}\left(\{\omega^{(d)} : \mathbf{L}_n(\omega^{(d)}) \in \Gamma\}\right)\right)$$
$$\leq \varlimsup_{n\to\infty} \frac{1}{n}\log\left(\sup_{\sigma^{(d)}\in\Sigma^d} P^{(d)}_{\sigma^{(d)}}\left(\{\omega^{(d)} : \mathbf{L}_n(\omega^{(d)}) \in \Gamma\}\right)\right) \leq -\inf_{\overline{\Gamma}} J^{(d)}_\Pi$$

for every $\Gamma \in \mathcal{B}_{\mathbf{M}_1(\Sigma^d)}$; where

$$\omega^{(d)} \in \Omega^{(d)} \longmapsto \mathbf{L}_n\left(\omega^{(d)}\right) = \frac{1}{n}\sum_{k=1}^n \delta_{\Sigma^{(d)}_k(\omega^{(d)})}$$

and $\Sigma^{(d)}_n\left(\omega^{(d)}\right)$ is the position of $\omega^{(d)}$ at time $n \in \mathbb{N}$. Since, by the MARKOV property, it is an easy matter to check that for any $n \in \mathbb{Z}^+$, $\sigma \in \Sigma$, and $\sigma^{(d)} \in \Sigma^d$ with $\sigma^{(d)}_d = \sigma$:

$$\mu^{(d)}_{\sigma,n}(\Gamma) = \int_{\Sigma^d} P^{(d)}_{\xi^{(d)}}\left(\{\omega^{(d)} : \mathbf{L}_n(\omega^{(d)}) \in \Gamma\}\right)\Pi^{(d)}\left(\sigma^{(d)}, d\xi^{(d)}\right), \quad \Gamma \in \mathcal{B}_E,$$

and therefore that

$$\inf_{\xi^{(d)}\in\Sigma^d} P_{\xi^{(d)}}\left(\{\omega^{(d)} : \mathbf{L}_n(\omega^{(d)}) \in \Gamma\}\right)$$
$$\leq \mu^{(d)}_{\sigma,n}(\Gamma) \leq \sup_{\xi^{(d)}\in\Sigma^d} P_{\xi^{(d)}}\left(\{\omega^{(d)} : \mathbf{L}_n(\omega^{(d)}) \in \Gamma\}\right)$$

for all $n \in \mathbb{Z}^+$ and $\sigma \in \Sigma$, we have now proved the following uniform large deviation result.

4.4.5 Lemma. *Assume that* (**U**) *holds. Then the function* $J_\Pi^{(d)}$ *is a good rate function on* $\mathbf{M}_1(\Sigma^d)$ *and*

$$\begin{aligned} -\inf_{\Gamma^\circ} J_\Pi^{(d)} &\leq \varliminf_{n\to\infty} \frac{1}{n} \log\left(\inf_{\sigma\in\Sigma} \mu_{\sigma,n}^{(d)}(\Gamma)\right) \\ &\leq \varlimsup_{n\to\infty} \log\left(\sup_{\sigma\in\Sigma} \mu_{\sigma,n}^{(d)}(\Gamma)\right) \leq -\inf_{\overline{\Gamma}} J_\Pi^{(d)} \end{aligned} \tag{4.4.6}$$

for all $\Gamma \in \mathcal{B}_{\mathbf{M}_1(\Sigma^d)}$.

We next want to give an alternative expression for $J_\Pi^{(d)}$. In order to develop this other expression, it will be necessary to recall a basic property of probability measures on a Polish space. Namely, given a Polish space E, a countably generated sub-σ-algebra $\mathcal{F}$ of $\mathcal{B}_E$, and a $P \in \mathbf{M}_1(E)$, there is a map $x \in E \longmapsto P^{\mathcal{F}}(x,\cdot) \in \mathbf{M}_1(\Sigma)$ with the properties that

(1) $x \in E \longmapsto P^{\mathcal{F}}(x,B)$ is $\mathcal{F}$-measurable for every $B \in \mathcal{B}_E$;
(2) $P^{\mathcal{F}}(x,A) = \chi_A(x)$, $x \in E$, for each $A \in \mathcal{F}$;
(3) $P(A\cap B) = \int_A P^{\mathcal{F}}(x,B)\,P(dx)$ for all $A \in \mathcal{F}$ and $B \in \mathcal{B}_E$.

The map $x \in E \longmapsto P^{\mathcal{F}}(x,\cdot)$ is called a **regular conditional probability distribution of** P **given** $\mathcal{F}$ (abbreviated by r.c.p.d. of P given $\mathcal{F}$). The existence of a regular probability distribution is a well-known but non-trivial fact (cf. Theorem 1.1.8 in [**104**]) about the measure theory of Polish spaces. On the other hand, it is easy to see that any two r.c.p.d.'s of P given $\mathcal{F}$ can differ only on a $\mathcal{F}$-measurable, P-null set.

4.4.7 Lemma. *Let* E *be a Polish space and* $\mathcal{F}$ *a countably generated sub-*σ*-algebra of* $\mathcal{B}_E$*. Given* P, $Q \in \mathbf{M}_1(E)$, *let* $x \in E \longmapsto P^{\mathcal{F}}(x,\cdot)$ *and* $x \in E \longmapsto Q^{\mathcal{F}}(x,\cdot)$ *be, respectively, r.c.p.d.'s of* P *and* Q *given* $\mathcal{F}$*. Then* $x \in E \longmapsto \mathbf{H}\big(Q^{\mathcal{F}}(x,\cdot)\big|P^{\mathcal{F}}(x,\cdot)\big)$ *is* $\mathcal{F}$*-measurable; and*

$$\mathbf{H}(Q|P) = \mathbf{H}\big(Q|_{\mathcal{F}}\big|P|_{\mathcal{F}}\big) + \int_E \mathbf{H}\big(Q^{\mathcal{F}}(x,\cdot)\big|P^{\mathcal{F}}(x,\cdot)\big)\,Q(dx), \tag{4.4.8}$$

where $P|_{\mathcal{F}}$ *and* $Q|_{\mathcal{F}}$ *are the restrictions of* P *and* Q *to* $\mathcal{F}$.

PROOF: First note that since, by Lemma 3.2.13,

$$\mathbf{H}(\nu|\mu) = \sup\left\{\int_E \phi\,d\nu - \log\left(\int_E e^\phi\,d\mu\right) : \phi \in C_b(E;\mathbb{R})\right\},$$

we have that

$$(\nu,\mu) \in \big(\mathbf{M}_1(E)\big)^2 \longmapsto \mathbf{H}(\nu|\mu)$$

is a lower semi-continuous function; and therefore the $\mathcal{F}$-measurability of

$$x \in E \longmapsto \mathbf{H}\big(Q^{\mathcal{F}}(x,\cdot)\big|P^{\mathcal{F}}(x,\cdot)\big)$$

is established.

Second, observe that if either side of (4.4.8) is finite, then $Q << P$. Indeed, $Q << P$ by definition if the left hand side is finite. On the other hand, if the right hand side is finite, then $Q|_{\mathcal{F}} << P|_{\mathcal{F}}$ and there is a Q-null set $A \in \mathcal{F}$ such that $Q^{\mathcal{F}}(x,\cdot) << P^{\mathcal{F}}(x,\cdot)$ for $x \notin A$. Thus, if $\Gamma \in \mathcal{B}_E$ is a P-null set, then there is a P-null set $B \in \mathcal{F}$ such that $Q^{\mathcal{F}}(x,\Gamma) = 0$ for all $x \notin B$. Since $Q(B) = 0$, we conclude that $Q(\Gamma) = 0$.

In view of the preceding, it remains only to handle P and Q for which $Q << P$. But if $Q << P$, then we may and will assume that $Q^{\mathcal{F}}(x,\cdot) << P^{\mathcal{F}}(x,\cdot)$ for all $x \in E$. Hence, because $\mathcal{F}$ is countably generated, one can use the Martingale Convergence Theorem to construct an $\mathcal{F} \times \mathcal{B}_E$-measurable $f : E^2 \longrightarrow [0,\infty)$ with the property that

$$Q^{\mathcal{F}}(x,\Gamma) = \int_\Gamma f(x,y)\, P^{\mathcal{F}}(x,dy), \quad x \in E \text{ and } \Gamma \in \mathcal{B}_E.$$

Noting that $P^{\mathcal{F}}\big(x,[x]_{\mathcal{F}}\big) = 1$ where $[x]_{\mathcal{F}} = \bigcap\{A \in \mathcal{F} : x \in A\}$ is the $\mathcal{F}$-atom containing x, one sees that, for each $x \in E$, $f(x,y) = f(y,y)$ (a.e., $P^{\mathcal{F}}(x,\cdot)$); and therefore

$$Q^{\mathcal{F}}(x,\Gamma) = \int_\Gamma h(y)\, P^{\mathcal{F}}(x,dy), \quad x \in E \text{ and } \Gamma \in \mathcal{B}_E,$$

where $h(y) = f(y,y)$, $y \in E$. Hence, if $g : E \longrightarrow [0,\infty)$ is an $\mathcal{F}$-measurable function such that $Q(A) = \int_A g(y)\, P(dy)$, $A \in \mathcal{F}$, then $g \cdot h = \frac{dQ}{dP}$. In particular,

$$\begin{aligned}\mathbf{H}(Q|P) &= \int_E \log\big(g(x)\big)\, Q(dx) + \int_E \log\big(h(y)\big)\, Q(dy) \\ &= \mathbf{H}\big(Q|_{\mathcal{F}}\big|P|_{\mathcal{F}}\big) + \int_E \left(\int_E \log\big(h(y)\big)\, Q^{\mathcal{F}}(x,dy)\right) Q(dx) \\ &= \mathbf{H}\big(Q|_{\mathcal{F}}\big|P|_{\mathcal{F}}\big) + \int_E \mathbf{H}\big(Q^{\mathcal{F}}(x,\cdot)\big|P^{\mathcal{F}}(x,\cdot)\big)\, Q(dx). \quad \blacksquare\end{aligned}$$

With the preceding result in hand, we will be ready to give another expression for $J_\Pi^{(d)}$ as soon as we have introduced a couple of notions. In the first place, we will say that $\nu \in \mathbf{M}_1\big(\Sigma^d\big)$ is **shift-invariant** if

$$\begin{aligned}&\nu\big(\{\sigma^{(d)} \in \Sigma^d : (\sigma_1^{(d)},\ldots,\sigma_{d-1}^{(d)}) \in \Gamma\}\big) \\ &\qquad = \nu\big(\{\sigma^{(d)} \in \Sigma^d : (\sigma_2^{(d)},\ldots,\sigma_d^{(d)}) \in \Gamma\}\big), \quad \Gamma \in \mathcal{B}_{\Sigma^{d-1}}.\end{aligned}$$

Second, given $d \geq 2$ and $\mu \in \mathbf{M}_1(\Sigma^{d-1})$, we will denote by $\mu \otimes_d \Pi$ the element of $\mathbf{M}_1(\Sigma^d)$ defined by

$$\mu \otimes_d \Pi(\Gamma) = \int_{\Sigma^{d-1}} \left(\int_\Sigma \chi_\Gamma(\sigma^{(d-1)}, \tau)\, \Pi(\sigma^{(d-1)}_{d-1}, d\tau) \right) \mu(d\sigma^{(d-1)})$$

for $\Gamma \in \mathcal{B}_{\Sigma^d}$. Note that if $\pi^{(d)}_{d-1}$ denotes the mapping $\sigma^{(d)} \in \Sigma^d \longmapsto (\sigma^{(d)}_1, \dots, \sigma^{(d)}_{d-1}) \in \Sigma^{d-1}$, then $\mu \otimes_d \Pi$ is uniquely determined by the fact that $(\mu \otimes_d \Pi) \circ (\pi^{(d)}_{d-1})^{-1} = \mu$ together with the fact that

$$\sigma^{(d)} \in \Sigma^d \longmapsto \delta_{\pi^{(d)}_{d-1}(\sigma^{(d)})} \otimes_d \Pi(\sigma^{(d)}_{d-1}, \cdot)$$

is a r.c.p.d. of $\mu \otimes_d \Pi$ given $\mathcal{B}^{(d)}_{d-1} \equiv (\pi^{(d)}_{d-1})^{-1}(\mathcal{B}_{\Sigma^{d-1}})$.

4.4.9 Lemma. *Let Π be any transition probability on Σ, and define $J^{(d)}_\Pi$ accordingly. If $\nu \in \mathbf{M}_1(\Sigma^d)$, then*

$$J^{(d)}_\Pi(\nu) = \begin{cases} \mathbf{H}(\nu | \nu_{d-1} \otimes_d \Pi) & \text{if } \nu \text{ is shift-invariant} \\ \infty & \text{otherwise,} \end{cases} \tag{4.4.10}$$

where $\nu_{d-1} \equiv \nu \circ (\pi^{(d)}_{d-1})^{-1}$.

PROOF: First suppose that ν is not shift-invariant. Then there is a $\psi \in C_b(\Sigma^{d-1}; \mathbb{R})$ for which

$$\int_{\Sigma^d} \psi(\sigma^{(d)}_1, \dots, \sigma^{(d)}_{d-1})\, \nu(d\sigma^{(d)}) - \int_{\Sigma^d} \psi(\sigma^{(d)}_2, \dots, \sigma^{(d)}_d)\, \nu(d\sigma^{(d)}) \geq 1.$$

Thus, if

$$u_\alpha(\sigma^{(d)}) = \exp\left[\alpha\psi(\sigma^{(d)}_1, \dots, \sigma^{(d)}_{d-1})\right] \quad \text{for } \alpha > 0 \quad \text{and } \sigma^{(d)} \in \Sigma^d,$$

then $\log\left([\Pi^{(d)} u_\alpha](\sigma^{(d)})\right) = \alpha\psi(\sigma^{(d)}_2, \dots, \sigma^{(d)}_d)$, and so

$$J^{(d)}_\Pi(\nu) \geq -\int_{\Sigma^d} \log \frac{[\Pi^{(d)} u_\alpha]}{u_\alpha}\, d\nu \geq \alpha, \quad \alpha > 0,$$

which means that $J^{(d)}_\Pi(\nu) = \infty$.

We next suppose that ν is shift-invariant. Let $\sigma^{(d)} \longmapsto \nu^{(d-1)}(\sigma^{(d)}, \cdot)$ be a r.c.p.d. of ν given $\mathcal{B}^{(d)}_{d-1}$; and note that, by Lemma 4.4.7,

$$\mathbf{H}(\nu | \nu_{d-1} \otimes_d \Pi) = \int_{\Sigma^d} \mathbf{H}(\nu^{(d-1)}(\sigma^{(d)}, \cdot) | \delta_{\pi^{(d)}_{d-1}(\sigma^{(d)})} \otimes_d \Pi)\, \nu(d\sigma^{(d)}).$$

At the same time, by Lemma 3.2.13,

$$\int_{\Sigma^d} \mathbf{H}\big(\nu^{(d-1)}(\sigma^{(d)},\cdot)|\delta_{\pi^{(d)}_{d-1}(\sigma^{(d)})} \otimes_d \Pi\big)\,\nu(d\sigma^{(d)})$$

$$= \int_{\Sigma^d} \sup\bigg\{ \int_{\Sigma^d} \psi\big(\tau^{(d)}\big)\,\nu^{(d-1)}\big(\sigma^{(d)}, d\tau^{(d)}\big)$$

$$- \log\bigg(\int_{\Sigma} \exp\Big[\psi\big(\pi^{(d)}_{d-1}(\sigma^{(d)}),\tau\big)\Big]\,\Pi\big(\sigma^{(d)}_{d-1}, d\tau\big)\bigg)$$

$$: \psi \in C_{\mathrm{b}}\big(\Sigma^d;\mathbb{R}\big)\bigg\}\,\nu\big(d\sigma^{(d)}\big),$$

which obviously dominates

$$\int_{\Sigma^d} \psi\big(\sigma^{(d)}\big)\,\nu\big(d\sigma^{(d)}\big)$$

$$- \int_{\Sigma^d} \log\bigg(\int_{\Sigma} \exp\Big[\psi\big(\pi^{(d)}_{d-1}(\sigma^{(d)}),\tau\big)\Big]\,\Pi\big(\sigma^{(d)}_{d-1}, d\tau\big)\bigg)\nu\big(d\sigma^{(d)}\big)$$

for every $\psi \in C_{\mathrm{b}}\big(\Sigma^d;\mathbb{R}\big)$. But, by shift invariance, the preceding says that $\mathbf{H}\big(\nu\big|\nu_{d-1} \otimes_d \Pi\big)$ dominates

$$\sup\bigg\{ \int_{\Sigma^d} \psi\big(\sigma^{(d)}\big)\,\nu\big(d\sigma^{(d)}\big)$$

$$- \int_{\Sigma^d} \log\Big[\big[\Pi^{(d)} e^{\psi}\big]\big(\sigma^{(d)}\big)\Big]\,\nu\big(d\sigma^{(d)}\big) : \psi \in C_{\mathrm{b}}\big(\Sigma^d;\mathbb{R}\big)\bigg\} = J^{(d)}_{\Pi}(\nu).$$

Thus, we have now shown that $J^{(d)}_{\Pi}(\nu) \le \mathbf{H}\big(\nu\big|\nu_{d-1} \otimes_d \Pi\big)$. On the other hand, by Lemma 3.2.13, JENSEN's inequality, and shift-invariance:

$$\begin{aligned}
J^{(d)}_{\Pi}(\nu) &= \sup\bigg\{ \int_{\Sigma^d} \psi d\nu - \int_{\Sigma^d} \log\Big(\big[\Pi^{(d)} e^{\psi}\big]\Big)\,d\nu : \psi \in C_{\mathrm{b}}\big(\Sigma^d;\mathbb{R}\big)\bigg\} \\
&\ge \sup\bigg\{ \int_{\Sigma^d} \psi d\nu - \log\bigg(\int_{\Sigma^d} \big[\Pi^{(d)} e^{\psi}\big]\,d\nu\bigg) : \psi \in C_{\mathrm{b}}\big(\Sigma^d;\mathbb{R}\big)\bigg\} \\
&= \sup\bigg\{ \int_{\Sigma^d} \psi d\nu - \log\bigg(\int_{\Sigma^d} e^{\psi}\,d\big(\nu_{d-1} \otimes_d \Pi\big)\bigg) : \psi \in C_{\mathrm{b}}\big(\Sigma^d;\mathbb{R}\big)\bigg\} \\
&= \mathbf{H}\big(\nu\big|\nu_{d-1} \otimes_d \Pi\big);
\end{aligned}$$

and clearly this completes the proof. ∎

We are now ready to return to the problem, posed at the beginning of this section, of examining the large deviation theory for the $\mathbf{M}_1(\Omega)$-valued random variables in (4.4.1). Actually, as we are about to see, we are already quite close to having such a theory. Indeed, by **(ii)** in Exercise 3.2.22, we can identify $\mathbf{M}_1(\Omega)$ as the projective limit of the sequence $\{\mathbf{M}_1(\Sigma^d) : d \geq 2\}$. Furthermore, if π_d is the projection map

$$\omega \in \Omega \longmapsto \pi_d(\omega) = (\Sigma_0(\omega), \dots, \Sigma_{d-1}(\omega)) \in \Sigma^d,$$

then it is obvious that $\mu_{\sigma,n}^{(d)}$ is the distribution of $\omega \longmapsto \mathbf{R}_n(\omega) \circ (\pi_d)^{-1}$ under P_σ. Hence, just as in Exercise 2.1.21, if we set

$$J_\Pi^{(\infty)}(Q) \equiv \sup_{d\geq 2} J_\Pi^{(d)}(Q \circ (\pi_d)^{-1}), \quad Q \in \mathbf{M}_1(\Omega), \tag{4.4.11}$$

then we have the following uniform large deviation result as a consequence of Lemma 4.4.5.

4.4.12 Theorem. *Assume that* **(U)** *holds, and define* $J_\Pi^{(\infty)}$ *as in* (4.4.11). *Then* $J_\Pi^{(\infty)}$ *is a good rate function and*

$$\begin{aligned} -\inf_{\Gamma^\circ} J_\Pi^{(\infty)} &\leq \varliminf_{n\to\infty} \frac{1}{n} \log\left(\inf_{\sigma\in\Sigma} P_\sigma(\{\omega : \mathbf{R}_n(\omega) \in \Gamma\})\right) \\ &\leq \varlimsup_{n\to\infty} \frac{1}{n} \log\left(\sup_{\sigma\in\Sigma} P_\sigma(\{\omega : \mathbf{R}_n(\omega) \in \Gamma\})\right) \leq -\inf_{\overline{\Gamma}} J_\Pi^{(\infty)} \end{aligned} \tag{4.4.13}$$

for every $\Gamma \in \mathcal{B}_{\mathbf{M}_1(\Omega)}$. *(See* (4.4.16) *below for more information about* $J_\Pi^{(\infty)}$.*)*

Although Theorem 4.4.12 in conjunction with Lemma 4.4.9 provides a reasonably satisfactory description of the large deviation theory under consideration, it would be an even better theory if we could find a more direct method of computing $J_\Pi^{(\infty)}(Q)$. Unfortunately, in order to get a nicer expression for $J_\Pi^{(\infty)}$ it will be necessary for us to introduce some additional notation.

We will say that $Q \in \mathbf{M}_1(\Omega)$ is **shift-invariant** if $Q = Q \circ \theta_n^{-1}$ for every $n \geq 1$, and we will use $\mathbf{M}_1^{\mathrm{S}}(\Omega)$ to denote the set of all shift-invariant $Q \in \mathbf{M}_1(\Omega)$. Note that, by Lemma 4.4.9, we need only concern ourselves with the computation of $J_\Pi^{(\infty)}(Q)$ for $Q \in \mathbf{M}_1^{\mathrm{S}}(\Omega)$ since $J_\Pi^{(\infty)}(Q) = \infty$ for $Q \notin \mathbf{M}_1^{\mathrm{S}}(\Omega)$. Next, for $n \in \mathbb{Z}$, set $\mathbb{Z}_n = \mathbb{Z} \cap (-\infty, n]$, $\Omega_n^* = \Sigma^{\mathbb{Z}_n}$, and use $\Sigma_m(\omega^*)$ to denote the position of $\omega^* \in \Omega_n^*$ at time $m \in \mathbb{Z}_n$. Given $n \in \mathbb{Z}$ and $Q \in \mathbf{M}_1^{\mathrm{S}}(\Omega)$, one can use the KOLMOGOROV Extension Theorem to show that there is a unique $Q_n^* \in \mathbf{M}_1(\Omega_n^*)$ with the property that

$$\begin{aligned} Q_n^*\Big(\{\omega^* \in \Omega_n^* : (\Sigma_{-d+n}(\omega^*), \dots, \Sigma_n(\omega^*)) \in \Gamma\}\Big) \\ = Q\Big(\{\omega \in \Omega : (\Sigma_0(\omega), \dots, \Sigma_d(\omega)) \in \Gamma\}\Big) \end{aligned}$$

for all $d \geq 1$ and $\Gamma \in \Sigma^{d+1}$. Next, given $P \in \mathbf{M}_1(\Omega_0^*)$, define $P \otimes_0 \Pi$ to be the unique element of $\mathbf{M}_1(\Omega_1^*)$ satisfying

$$\int_{\Omega_1^*} \psi(\Sigma_{-d}(\omega^*), \dots, \Sigma_0(\omega^*), \Sigma_1(\omega^*)) \, (P \otimes_0 \Pi)(d\omega^*)$$
$$= \int_{\Omega_0^*} \left(\int_\Sigma \psi(\Sigma_{-d}(\omega^*), \dots, \Sigma_0(\omega^*), \tau) \, \Pi(\Sigma_0(\omega^*), d\tau) \right) P(d\omega^*)$$

for all $d \geq 1$ and $\psi \in B(\Sigma^{d+2}; \mathbb{R})$. Finally, for $n \in \mathbb{Z}$ and $k \leq \ell \leq n$, let $\mathcal{B}_{[k,\ell]}^{*(n)}$ denote the σ-algebra over Ω_n^* generated by the map

$$\omega^* \in \Omega_n^* \longmapsto \left(\Sigma_k(\omega^*), \dots, \Sigma_\ell(\omega^*) \right) \in \Sigma^{\ell - k};$$

and, for P, $P' \in \mathbf{M}_1(\Omega_n^*)$, let $\mathbf{H}_d^{(n)}(P' | P)$ denote the relative entropy of the restrictions to $\mathcal{B}_{[-d,n]}^{*(n)}$ of measures P and P'. It is then an easy matter to see that, for any $Q \in \mathbf{M}_1^{\mathrm{S}}(\Omega)$ and $d \geq 1$, Lemma 4.4.9 becomes the statement that

$$J_\Pi^{(d+2)}\left(Q \circ (\pi_{d+2})^{-1}\right) = \mathbf{H}_d^{(1)}\left(Q_1^* | Q_0^* \otimes_0 \Pi\right).$$

Hence,

$$J_\Pi^{(\infty)}(Q) = \sup_{d \geq 1} \mathbf{H}_d^{(1)}\left(Q_1^* | Q_0^* \otimes_0 \Pi\right), \quad Q \in \mathbf{M}_1^{\mathrm{S}}(\Omega). \tag{4.4.14}$$

In order to take advantage of the expression in (4.4.14), we will need the following simple continuity result for the relative entropy functional.

4.4.15 Lemma. *Let $(E, \mathcal{F})$ be a measurable space, suppose that $\{\mathcal{F}_n\}_{n=1}^\infty$ is a non-decreasing sequence of sub-σ-algebras of $\mathcal{F}$ such that $\bigcup_1^\infty \mathcal{F}_n$ generates $\mathcal{F}$. Then, for any pair of probability measures P and Q on $(E, \mathcal{F})$,*

$$\mathbf{H}\left(Q|_{\mathcal{F}_n} \big| P|_{\mathcal{F}_n}\right) \nearrow \mathbf{H}(Q|P) \text{ as } n \longrightarrow \infty.$$

PROOF: By the argument used to prove Lemma 3.2.13, we know that

$$\mathbf{H}\left(Q|_{\mathcal{F}_n} \big| P|_{\mathcal{F}_n}\right) = \sup \left\{ \int_E \psi \, dQ - \log \left(\int_E e^\psi \, dP \right) : \psi \in B\left(E, \mathcal{F}_n; \mathbb{R}\right) \right\},$$

where $B(E, \mathcal{F}_n; \mathbb{R})$ denotes the space of bounded, $\mathcal{F}_n$-measurable $\psi : E \longrightarrow \mathbb{R}$. At the same time,

$$\mathbf{H}(Q|P) = \sup \left\{ \int_E \psi \, dQ - \log \left(\int_E e^\psi \, dP \right) : \psi \in B\left(E, \mathcal{F}; \mathbb{R}\right) \right\}.$$

Hence, it is clear that $n \longmapsto \mathbf{H}\big(Q|_{\mathcal{F}_n}\big|P|_{\mathcal{F}_n}\big)$ is non-decreasing and that its limit does not exceed $\mathbf{H}(Q|P)$. On the other hand, the class of $\psi \in B(E,\mathcal{F};\mathbb{R})$ for which

$$\int_E \psi\, dQ - \log\left(\int_E e^{\psi}\, dP\right) \le \sup_{n\ge 1} \mathbf{H}\big(Q|_{\mathcal{F}_n}\big|P|_{\mathcal{F}_n}\big)$$

is closed under bounded, point-wise convergence and, obviously, contains $B\big(E,\mathcal{F}_n;\mathbb{R}\big)$ for all $n \ge 1$. ∎

Combining Lemma 4.4.15 with (4.4.14), we now see that

$$J_{\Pi}^{(\infty)}(Q) = \begin{cases} \mathbf{H}\big(Q_1^*\big|Q_0^* \otimes_0 \Pi\big) & \text{if } Q \in \mathbf{M}_1^{\mathrm{S}}(\Omega) \\ \infty & \text{otherwise.} \end{cases} \tag{4.4.16}$$

When (4.4.16) is put together with (4.4.12), we obtain a version of the process level large deviation result proved originally by DONSKER and VARADHAN [**36**].

Having dealt with the discrete time setting, we now want to see whether we cannot prove the analogous result in the continuous-time context. However, before we can do so, we must arrange that the sample space Ω at the beginning of Section 4.2 be itself amenable to a Polish structure. For this reason, we will assume that Ω is the space of right-continuous paths $\omega : [0,\infty) \longrightarrow \Sigma$ which have a left limit at each $t \in (0,\infty)$. Next, define $\Omega_T = D\big([0,T];\Sigma\big)$ for $T \in (0,\infty)$ to be the **Skorokhod space** of right-continuous $\omega_T : [0,T] \longrightarrow \Sigma$ which have a left limit at each $t \in (0,T]$ and are left-continuous at T, and endow Ω_T with the **Skorokhod topology**. That is, we give Ω_T the topology induced by the metric

$$\operatorname{dist}\big(\omega_T,\omega_T'\big) = \inf_{\lambda} \sup_{t\in[0,T]} \big\{\operatorname{dist}\big(\omega_T(t),\omega_T'(\lambda(t))\big) + |\lambda(t) - t|\big\}$$

for ω_T, $\omega_T' \in \Omega_T$, where λ runs over all increasing homeomorphisms of $[0,T]$ onto itself. The following facts about the SKOROKHOD topology on Ω_T are standard and will be important for our development below.

(**i**) The SKOROKHOD topology on Ω_T is Polish in the sense that it is separable and admits a complete metric.

(**ii**) The σ-algebra over Ω_T generated by the maps $\omega_T \in \Omega_T \longmapsto \omega_T(t) \in \Sigma$, $t \in [0,T]$, coincides with the BOREL field $\mathcal{B}_{\Omega_T}$.

Next, for $0 < T_1 < T_2 < \infty$, define $\pi_{T_2}^{(T_1)} : \Omega_{T_2} \longrightarrow \Omega_{T_1}$ so that

$$\pi_{T_2}^{(T_1)}\big(\omega_{T_2}\big)(t) = \begin{cases} \omega_{T_2}(t) & \text{if } t \in [0, T_1) \\ \lim_{t \nearrow T_1} \omega_{T_2}(t) & \text{if } t = T_1. \end{cases}$$

Unfortunately, although these natural restriction maps are, by **(ii)**, measurable, they are *not* continuous. Thus, it is not possible for us to simply define the topology on Ω as the projective limit of the topologies on the Ω_T's. However, we will postpone the consideration of this technicality until later on in our development and will, for now content ourselves with the introduction of the projective limit measurable structure on Ω; which, according to **(ii)**, is the one induced by the position maps $\omega \in \Omega \longrightarrow \Sigma_t(\omega)$, $t \in [0, \infty)$. Thus, we will use $\mathcal{B}_t$, $t \in [0, \infty)$, to stand for the σ-algebra over Ω generated by $\omega \in \Omega \longrightarrow \Sigma_s(\omega)$, $s \in [0, t]$, and we will use $\mathcal{B}$ to denote the smallest σ-algebra over Ω containing $\bigcup_{t \geq 0} \mathcal{B}_t$. Obviously, $(t, \omega) \in [0, \infty) \times \Omega \longmapsto \Sigma_t(\omega) \in \Sigma$ is $\{\mathcal{B}_t : t \in [0, \infty)\}$-progressively measurable. In addition, for each $T \in (0, \infty)$, the map $\pi_T : \Omega \longrightarrow \Omega_T$ defined by

$$\pi_T(\omega)(t) = \begin{cases} \Sigma_t(\omega) & \text{for } t \in [0, T) \\ \lim_{t \nearrow T} \Sigma_t(\omega) & \text{for } t = T, \end{cases}$$

is a measurable surjection; and, in fact, $\sigma\left(\bigcup_{t \in [0,T)} \mathcal{B}_t\right) = \pi_T^{-1}\big(\mathcal{B}_{\Omega_T}\big)$.

4.4.17 Warning.

Throughout the rest of this section, Ω will be the path-space just described, and we will be assuming that the transition probability function $P(t, \sigma, \cdot)$ permits us to realize the corresponding MARKOV family $\{P_\sigma : \sigma \in \Sigma\}$ on $(\Omega, \mathcal{B})$ with $\Sigma_t(\omega)$ being the position of $\omega \in \Omega$ at time $t \in [0, \infty)$. Furthermore, we will be assuming that

$$P_\sigma\Big(\big\{\omega : \Sigma_t(\omega) = \lim_{s \nearrow t} \Sigma_s(\omega)\big\}\Big) = 1, \quad (t, \sigma) \in (0, \infty) \times \Sigma. \tag{4.4.18}$$

In what follows, it will be convenient to have a notation for the "splice" of two paths. For this reason, if $T \in (0, \infty)$, $\omega_T \in \Omega_T$, and $\omega' \in \Omega$, define $\overline{\omega}_T \in \Omega$ by $\overline{\omega}_T(t) = \omega_T(t \wedge T)$, $t \in [0, \infty)$, and $\omega_T \otimes_T \omega' \in \Omega$ so that $\omega_T \otimes_T \omega' = \overline{\omega}_T$ if $\Sigma_0(\omega') \neq \omega_T(T)$ and

$$\Sigma_t(\omega_T \otimes_T \omega') = \begin{cases} \omega_T(t) & \text{for } t \in [0, T) \\ \omega'(t - T) & \text{for } t \in [T, \infty) \end{cases}$$

if $\Sigma_0(\omega') = \omega_T(T)$. Observe that the map $\big(\omega_T, \omega'\big) \in \Omega_T \times \Omega \longmapsto \omega_T \otimes_T \omega'$ is measurable. When $\omega, \omega' \in \Omega$, set $\omega \otimes_T \omega' = \big(\pi_T \omega\big) \otimes_T \omega'$. Finally, for

$\omega_T \in \Omega_T$ and $Q \in \mathbf{M}_1(\Omega)$ define $\delta_{\omega_T} \otimes_T Q \in \mathbf{M}_1(\Omega)$ so that

$$\int_\Omega \Psi\big(\omega'\big)\,\big(\delta_{\omega_T} \otimes_T Q\big)(d\omega') = \int_\Omega \Psi\big(\omega_T \otimes_T \omega'\big)\,Q(d\omega')$$

for $\Psi \in B\big((\Omega, \mathcal{B}); \mathbb{R}\big)$.

We will also need the **time-shift semigroup** $\{\theta_t : t \geq 0\}$ on Ω. Namely, for $t \in [0, \infty)$, define the **time-shift** $\theta_t : \Omega \longrightarrow \Omega$ by $\Sigma_s(\theta_t\omega) = \Sigma_{s+t}(\omega)$, $s \in [0, \infty)$; and note that $(t, \omega) \in [0, \infty) \times \Omega \longmapsto \theta_t\omega \in \Omega$ is measurable.

With these preliminaries taken care of, we can begin to formulate the problem which we want to study. To this end, let $\omega \in \Omega \longmapsto \mathbf{R}_t(\omega) \in \mathbf{M}_1\big((\Omega, \mathcal{B})\big)$ (the probability measures on the measurable space $(\Omega, \mathcal{B})$) be the map given by

$$\mathbf{R}_t(\omega) = \overline{\lambda}_{[0,t]} \circ \Big(\big(\theta_\cdot\omega\big)\big|_{[0,t]}\Big)^{-1}, \tag{4.4.19}$$

where $\overline{\lambda}_{[0,t]}$ denotes normalized LEBESGUE measure on $[0, t]$. (Cf. the comment in Remark 4.2.2 following the definition of $\mathbf{L}_t(\omega)$ for the appropriate expression when the paths are regular.) What we want to do is analyze the large deviations of $\{\mathbf{R}_t : t > 0\}$ under the measures P_σ. Of course, as yet, we do not even have a topological structure on $\mathbf{M}_1\big((\Omega, \mathcal{B})\big)$ and therefore are not really in a position to carry out such an analysis. Nonetheless, just as in the MARKOV chain case, our analysis will be actually accomplished at the level of the finite time-marginals of the $\mathbf{R}_t$'s; and this analysis we are ready to do. Unfortunately, although the ideas here are just as intrinsically simple as the ones in the discrete time setting, technicalities introduced by the continuity of time tend to make them appear more complicated than they really are.

Given $T \in (0, \infty)$ and $\omega_T \in \Omega_T$, define

$$P^{(T)}_{\omega_T} = \delta_{\omega_T} \otimes_T P_{\omega_T(T)}. \tag{4.4.20}$$

It is a relatively easy matter to check that $\omega_T \in \Omega_T \longmapsto P^{(T)}_{\omega_T} \in \mathbf{M}_1\big((\Omega, \mathcal{B})\big)$ is measurable. What is less obvious is that $\big\{P^{(T)}_{\omega_T} : \omega_T \in \Omega_T\big\}$ satisfies the MARKOV property described in the following.

4.4.21 Lemma. *For each $T \in (0, \infty)$, $\omega_T \in \Omega_T$, and $\omega' \in \Omega$,*

$$\omega_T = \lim_{t \searrow 0} \pi_T\big(\theta_t(\omega_T \otimes_T \omega')\big).$$

In addition, for each $T \in (0,\infty)$, $\omega_T \in \Omega_T$, $s \in [0,\infty)$, *and* $A \in \mathcal{B}_{s+T}$,

$$\begin{aligned}\int_A \Psi\big(\theta_s\omega'\big)\, P^{(T)}_{\omega_T}(d\omega') \\ = \int_A \left(\int_\Omega \Psi\big(\omega''\big)\, P^{(T)}_{\pi_T(\theta_s\omega')}\big(d\omega''\big)\right) P^{(T)}_{\omega_T}(d\omega')\end{aligned} \tag{4.4.22}$$

for every $\Psi \in B\big((\Omega,\mathcal{B});\mathbb{R}\big)$.

PROOF: To prove the first assertion, note that

$$\begin{aligned}&\operatorname{dist}\big(\omega_T, \pi_T\big(\theta_t(\omega_T \otimes_T \omega')\big)\big) \\ &\qquad \le \operatorname{dist}\big(\omega_T, \pi_T(\theta_t\overline{\omega}_T)\big) + \operatorname{dist}\big(\pi_{t+T}\overline{\omega}_T, \pi_{t+T}(\omega_T \otimes_T \omega')\big).\end{aligned}$$

To prove (4.4.22), set

$$A_\omega = \big\{\omega' : \omega_T \otimes_T \omega' \in A\big\},$$

and first suppose that $s \le T$. Then $\theta_s(\omega_T \otimes_T \omega') = \big(\theta_s\overline{\omega}_T\big) \otimes_{T-s} \omega'$, and so, by the MARKOV property for the P_σ 's,

$$\begin{aligned}&\int_A \Psi\big(\theta_s\omega'\big)\, P^{(T)}_{\omega_T}(d\omega') \\ &\quad = \int_{A_\omega} \Psi\Big(\big(\theta_s\overline{\omega}_T\big) \otimes_{T-s} \omega'\Big)\, P_{\omega_T(T)}(d\omega') \\ &\quad = \int A_\omega\left(\int_\Omega \Psi\Big(\big[\big(\theta_s\overline{\omega}_T\big) \otimes_{T-s} \omega'\big] \otimes_T \omega''\Big) P_{\omega'(s)}(d\omega'')\right) P_{\omega_T(T)}(d\omega') \\ &\quad = \int_A \left(\int_\Omega \Psi\Big(\big(\theta_s\omega'\big) \otimes_T \omega''\Big)\, P_{\Sigma_s(\omega')}(d\omega'')\right) P^{(T)}_{\omega_T}(d\omega') \\ &\quad = \int_A \left(\int_\Omega \Psi\big(\omega''\big)\, P^{(T)}_{\pi_T(\theta_s\omega')}\big(d\omega''\big)\right) P^{(T)}_{\omega_T}(d\omega'),\end{aligned}$$

since, by (4.4.18), $\Sigma_s(\omega') = \big[\pi_T\big(\theta_s(\omega_T \otimes_T \omega')\big)\big](T)$ for $P_{\omega_T(T)}$-almost every $\omega' \in \Omega$. When $s > T$, a similar argument, based on the identities

$$\theta_s\big(\omega_T \otimes_T \omega'\big) = \theta_{s-T}\omega' \quad \text{and} \quad \theta_{s-T}\big(\omega' \otimes_s \omega''\big) = \big[\theta_s\big(\omega_T \otimes_T \omega'\big)\big] \otimes_T \omega''$$

for $\omega' \in \Omega$ with $\Sigma_0(\omega') = \omega_T(T)$, yields the desired result. ∎

For $T \in (0,\infty)$ define $(t,\omega_T) \in (0,\infty) \times \Omega_T \longmapsto P^{(T)}(t,\omega_T,\cdot) \in \mathbf{M}_1(\Omega_T)$ so that

$$P^{(T)}(t,\omega_T,\Gamma) = P^{(T)}_{\omega_T}\big(\{\omega' : \pi_T\big(\theta_t\omega'\big) \in \Gamma\}\big), \quad \Gamma \in \mathcal{B}_{\Omega_T}.$$

Then, by (4.4.22), $P^{(T)}(t,\omega_T,\cdot)$ is a transition probability function on Ω_T. In addition, by the first part of Lemma 4.4.21, $P^{(T)}(t,\omega_T,\cdot) \Longrightarrow \delta_{\omega_T}$ as $t \searrow 0$. Finally, if $\mathcal{B}_t^{(T)} = \mathcal{B}_{t+T}$, $t \in [0,\infty)$, then the map

$$(t,\omega) \in [0,\infty) \times \Omega \longmapsto \Sigma_t^{(T)}(\omega) \equiv \pi_T(\theta_t\omega) \in \Omega_T$$

is $\{\mathcal{B}_t^{(T)} : t \in [0,\infty)\}$-progressively measurable and

$$P_{\omega_T}^{(T)}\left(\{\omega' : \Sigma_{s+t}^{(T)}(\omega') \in \Gamma\} \middle| \mathcal{B}_s^{(T)}\right) = P^{(T)}\left(t, \Sigma_s^{(T)}, \Gamma\right) \quad \left(\text{a.e.}, P_{\omega_T}^{(T)}\right)$$

for all s, $t \in [0,\infty)$ and every $\Gamma \in \mathcal{B}_{\Omega_T}$. Thus, we are in the situation treated in Section 4.2. In fact, if the original transition probability function $P(t,\sigma,\cdot)$ on Σ satisfies $(\tilde{\mathbf{U}})$, then it is clear that

$$(\tilde{\mathbf{U}}^{(\mathrm{T})}) \qquad \begin{aligned} &\int_{(0,1]} P^{(T)}\left(t+T,\omega_T,\cdot\right)\rho_1(dt) \\ &\leq M \int_{(0,1]} P^{(T)}\left(t+T,\omega_T',\cdot\right)\rho_2(dt), \quad \omega_T, \omega_T' \in \Omega_T. \end{aligned}$$

Thus (cf. Remark 4.2.8 as well as Exercise 4.2.61), the following statement is just an application of the results proved in Section 4.2.

4.4.23 Lemma. *Assume that $P(t,\sigma,\cdot)$ satisfies $(\tilde{\mathbf{U}})$, let $T \in (0,\infty)$ be given, and define*

$$\overline{J}_P^{(T)} \equiv \overline{J}_{P^{(T)}} : \mathbf{M}_1(\Omega_T) \longrightarrow [0,\infty]$$

in terms of $P^{(T)}(t,\omega_T,\cdot)$ in the same way as $\overline{J}_P$ is defined by (4.2.36) in terms of $P(t,\sigma,\cdot)$. Then: the level sets of $\overline{J}_P^{(T)}$ are strongly compact;

$$\overline{J}_P^{(T)}(\nu) = \sup_{h>0} \frac{1}{h} J_{\Pi_h^{(T)}}(\nu) = \lim_{h\searrow 0} \frac{1}{h} J_{\Pi_h^{(T)}}(\nu), \quad \nu \in \mathbf{M}_1(\Omega_T), \tag{4.4.24}$$

where $\Pi_h^{(T)}(\omega_T,\cdot) \equiv P^{(T)}(h,\omega_T,\cdot)$ and $J_{\Pi_h^{(T)}}$ is defined accordingly as in (4.1.38); and

$$\begin{aligned} -\inf_{\Gamma^{\circ\tau}} \overline{J}_P^{(T)} &\leq \varliminf_{t\to\infty} \frac{1}{t} \log\left(\inf_{\omega_T\in\Omega_T} P_{\omega_T}^{(T)}\left(\{\omega' : \mathbf{L}_t^{(T)}(\omega') \in \Gamma\}\right)\right) \\ &\leq \varlimsup_{t\to\infty} \frac{1}{t} \log\left(\sup_{\omega_T\in\Omega_T} P_{\omega_T}^{(T)}\left(\{\omega' : \mathbf{L}_t^{(T)}(\omega') \in \Gamma\}\right)\right) \leq -\inf_{\overline{\Gamma}^{\tau}} \overline{J}_P^{(T)} \end{aligned} \tag{4.4.25}$$

for $\Gamma \in \mathcal{B}_{\mathbf{M}_1(\Omega_T)}$, where

$$\mathbf{L}_t^{(T)}(\omega') \equiv \overline{\lambda}_{[0,t]} \circ \left(\Sigma_\cdot^{(T)}(\omega')\big|_{[0,t]}\right)^{-1}$$

and $\overline{\lambda}_{[0,t]}$ denotes normalized LEBESGUE *measure on $[0,t]$.*

The reason for our choosing to state (4.4.25) relative to the strong topology on $\mathbf{M}_1(\Omega_T)$ will become clear as we develop the theory for unbounded time intervals.

Noting that the distribution of

$$\omega' \in \Omega \longmapsto \big(\mathbf{R}_t(\omega')\big) \circ \pi_T^{-1} \in \mathbf{M}_1\big(\Omega_T\big)$$

under $P_{\omega_T(T)}$ coincides with that of

$$\omega' \in \Omega \longmapsto \mathbf{L}_t^{(T)}\big(\theta_T \omega'\big) \in \mathbf{M}_1\big(\Omega_T\big)$$

under $P_{\omega_T}^{(T)}$, we see that, as long as $\sigma = \omega_T(T)$,

$$\begin{aligned} &P_\sigma\big(\{\omega' : \mathbf{R}_t(\omega') \circ \pi_T^{-1} \in \Gamma\}\big) \\ &\quad = \int_{\Omega_T} P_{\omega''_T}^{(T)}\big(\{\omega' : \mathbf{L}_t^{(T)}(\omega') \in \Gamma\}\big)\, P^{(T)}\big(T, \omega_T, d\omega''_T\big), \quad \Gamma \in \mathcal{B}_{\mathbf{M}_1(\Omega_T)}, \end{aligned}$$

Moreover, having stated Lemma 4.4.23 in terms of the strong topology, we can circumvent the technical objection (raised after our initial discussion of the SKOROKHOD topology) to putting an inductive limit topology on Ω. That is, we will entirely avoid putting a topology on Ω itself and will, instead, go directly to the **projective limit strong topology** on $\mathbf{M}_1\big((\Omega, \mathcal{B})\big)$. To be precise, we consider the topology on $\mathbf{M}_1\big((\Omega, \mathcal{B})\big)$ for which the sets

$$\left\{Q' : \left|\int_\Omega \Phi\, dQ' - \int_\Omega \Phi\, dQ\right| < \epsilon\right\}, \quad \epsilon \in (0, \infty), \tag{4.4.26}$$

form a neighborhood basis at Q as Φ runs over the bounded functions which are $\mathcal{B}_T$-measurable for some $T \in [0, \infty)$. Clearly this is the projective limit, under the maps

$$Q \in \mathbf{M}_1\big((\Omega, \mathcal{B})\big) \longrightarrow Q \circ \pi_T^{-1} \in \mathbf{M}_1(\Omega_T)$$

of the strong topology on the spaces $\mathbf{M}_1(\Omega_T)$. In particular,

$$K \subset\subset \mathbf{M}_1\big((\Omega, \mathcal{B})\big) \quad \text{if and only if} \quad K = \bigcap_{d=1}^{\infty} \big\{Q : Q \circ \pi_d^{-1} \in K_d\big\},$$

where K_d is a strongly compact subset of $\mathbf{M}_1(\Omega_d)$ for each $d \in \mathbb{Z}^+$. Finally, we will say that $\Gamma \subseteq \mathbf{M}_1\big((\Omega, \mathcal{B})\big)$ is **measurable** if it is an element of the σ-algebra over $\mathbf{M}_1\big((\Omega, \mathcal{B})\big)$ generated by the sets in (4.4.26). In case there is any doubt about it, we point out that this notion of measurability will, in most cases, be much more restrictive than the one determined by the BOREL structure associated with the projective limit strong topology.

Having made these preparations, we can now prove the following large deviation principle.

4.4.27 Theorem. *Assume that $P(t,\sigma,\cdot)$ satisfies $(\tilde{\mathbf{U}})$ and define the function $\overline{J}_P^{(\infty)} : \mathbf{M}_1\big((\Omega,\mathcal{B})\big) \longrightarrow [0,\infty]$ by*

$$\overline{J}_P^{(\infty)}(Q) = \sup\Big\{\overline{J}_P^{(T)}\big(Q\circ\pi_T^{-1}\big) : T\in(0,\infty)\Big\}, \tag{4.4.28}$$

for each $Q\in\mathbf{M}_1\big((\Omega,\mathcal{B})\big)$. (See (4.4.38) below for more information about $\overline{J}_P^{(\infty)}$.) Then the level sets of $\overline{J}_P^{(\infty)}$ are compact; and, for every measurable $\Gamma\subseteq\mathbf{M}_1\big((\Omega,\mathcal{B})\big)$,

$$\begin{aligned} -\inf_{\Gamma^\circ}\overline{J}_P^{(\infty)} &\le \varliminf_{t\to\infty}\frac{1}{t}\log\left(\inf_{\sigma\in\Sigma}P_\sigma\big(\{\omega:\mathbf{R}_t(\omega)\in\Gamma\}\big)\right)\\ &\le \varlimsup_{t\to\infty}\frac{1}{t}\log\left(\sup_{\sigma\in\Sigma}P_\sigma\big(\{\omega:\mathbf{R}_t(\omega)\in\Gamma\}\big)\right) \le -\inf_{\overline{\Gamma}}\overline{J}_P^{(\infty)}. \end{aligned} \tag{4.4.29}$$

PROOF: In view of (4.4.26) and (4.4.25), the only part of this statement which requires comment is the last inequality in (4.4.29). The main difficulty in the proof stems from the fact that the strong topology on $\mathbf{M}_1(\Omega_T)$ is not first countable. In particular, it is not immediately clear whether strongly compact sets $\Big\{\nu\in\mathbf{M}_1(\Omega_T):\overline{J}_P^{(T)}(Q)\le L\Big\}$ are sequentially compact in the strong topology. To see that they are, let $\{\nu_n\}_1^\infty\subseteq\mathbf{M}_1(\Omega_T)$ satisfying $\overline{J}_P^{(T)}(\nu_n)\le L<\infty$ for all $n\in\mathbb{Z}^+$ be given. Then, because $\Big\{\nu\in\mathbf{M}_1(\Omega_T):\overline{J}_P^{(T)}(\nu)\le L\Big\}$ is weakly compact, we can choose a subsequence $\{\nu_{n'}\}$ which converges weakly to some $\nu\in\mathbf{M}_1(\Omega_T)$. But $\{\nu\in\mathbf{M}_1(\Omega_T):\overline{J}_P^{(T)}(\nu)\le L\}$ is also strongly compact, and therefore $\nu_{n'}\longrightarrow\nu$ in the strong topology.

Once one has the preceding, it becomes an easy matter to check that if

$$\{Q_d\}_{d=1}^\infty\subseteq\mathbf{M}_1\big((\Omega,\mathcal{B})\big)\ \text{and}\ \sup_{d\ge1}\overline{J}_P^{(d)}\big(Q_d\circ\pi_d^{-1}\big)<\infty,$$

then there is a subsequence $\{Q_{d_n}\}$ which converges in $\mathbf{M}_1\big((\Omega,\mathcal{B})\big)$.

Now suppose that Γ is a measurable subset of $\mathbf{M}_1\big((\Omega,\mathcal{B})\big)$ and that $F\supseteq\Gamma$ is a closed subset of $\mathbf{M}_1\big((\Omega,\mathcal{B})\big)$. Choose $\{F_d\}_{d=1}^\infty$ so that $F=\bigcap_{d=1}^\infty\{Q:Q\circ\pi_d^{-1}\in F_d\}$ and each F_d is a strongly closed set in $\mathbf{M}_1(\Omega_d)$. We then know, from (4.4.25), that

$$\varlimsup_{t\to\infty}\frac{1}{t}\log\left(\sup_{\sigma\in\Sigma}P_\sigma\big(\{\omega:\mathbf{R}_t(\omega)\in\Gamma\}\big)\right)\le-\sup_{d\in\mathbb{Z}^+}\inf_{F_d}\overline{J}_P^{(d)}.$$

Next, suppose that $\ell\equiv\sup_{d\in\mathbb{Z}^+}\inf_{F_d}\overline{J}_P^{(d)}<\infty$. Then we can choose $\{Q_d\}_{d=1}^\infty\subseteq F$ so that $\overline{J}_P^{(d)}\big(Q_d\circ\pi_d^{-1}\big)=\inf_{F_d}\overline{J}_P^{(d)}\le\ell$. Thus, by the preceding paragraph, we can find a subsequence $\{Q_{d_n}\}$ and a $Q \in \mathbf{M}_1 \in ((\Omega, \mathcal{B}))$ so that $Q_{d_n} \longrightarrow Q$ in $\mathbf{M}_1((\Omega, \mathcal{B}))$. Since F is closed, $Q \in F$, and clearly $\overline{J}_P^{(\infty)}(Q) \leq \ell$. ∎

Once again, we want to develop a better expression for our rate function. Our development will turn on (4.4.24) combined with the sort of reasoning with which we solved the analogous problem in the case of MARKOV chains.

In what follows, it will be handy to have some more notation. In the first place, if $\nu \in \mathbf{M}_1(\Omega_T)$ for some $T \in (0, \infty)$ or if $Q \in \mathbf{M}_1((\Omega, \mathcal{B}))$, then we will use ν_t, $t \in [0, T]$, or Q_t, $t \in [0, \infty)$, to denote $\nu \circ (\pi_t^{(T)})^{-1}$ or $Q \circ \pi_t^{-1}$, respectively. Secondly, for $T \in (0, \infty)$ and $h \in [0, T]$, define $\theta_h^{(T)} : \Omega_T \longrightarrow \Omega_T$ by $\theta_h^{(T)} \omega_T = \pi_T(\theta_h \overline{\omega}_T)$ and say that $\nu \in \mathbf{M}_1(\Omega_T)$ is **shift-invariant** if $[\nu \circ (\theta_h^{(T)})^{-1}]_{T-h} = \nu_{T-h}$ for all $h \in [0, T]$. Similarly, we say that $Q \in \mathbf{M}_1((\Omega, \mathcal{B}))$ is **shift-invariant** if $Q \circ \theta_h = Q$ for every $h \in [0, \infty)$; and we will use $\mathbf{M}_1^{\mathrm{S}}(\Omega_T)$ and $\mathbf{M}_1^{\mathrm{S}}((\Omega, \mathcal{B}))$ to denote the set of shift-invariant elements of $\mathbf{M}_1(\Omega_T)$ and $\mathbf{M}_1((\Omega, \mathcal{B}))$, respectively.

4.4.30 Lemma. *$Q \in \mathbf{M}_1((\Omega, \mathcal{B}))$ is shift-invariant if and only if $Q_T \in \mathbf{M}_1^{\mathrm{S}}(\Omega_T)$ for every $T \in (0, \infty)$. Moreover, if $\nu \in \mathbf{M}_1^{\mathrm{S}}(\Omega_T)$, then*

$$\nu\Big(\Big\{\omega_T : \omega_T(t) = \lim_{s \nearrow t} \omega_T(s)\Big\}\Big) = 1 \quad \text{for } t \in (0, T].$$

In particular, if $Q \in \mathbf{M}_1^{\mathrm{S}}((\Omega, \mathcal{B}))$, then, for each $t \in (0, \infty)$, $\Sigma_t(\omega) = \lim_{s \nearrow t} \Sigma_s(\omega)$ for Q-almost every $\omega \in \Omega$.

PROOF: Obviously it suffices to prove the second assertion. To this end, define

$$\Gamma(\omega_T) = \Big\{t \in (0, T] : \omega_T(t) \neq \lim_{s \nearrow t} \omega_T(s)\Big\}, \quad \omega_T \in \Omega_T.$$

Because each ω_T has at most countably many discontinuities,

$$\overline{\lambda}_{[0,T]}\big(\Gamma(\omega_T)\big) = 0, \quad \omega_T \in \Omega_T,$$

where $\overline{\lambda}_{[0,T]}$ denotes normalized LEBESGUE measure on $[0, T]$. On the other hand, if $\nu \in \mathbf{M}_1^{\mathrm{S}}(\Omega_T)$, then

$$\nu\big(\big\{\omega_T : \omega_T(t) \neq \lim_{s \nearrow t} \omega_T(s)\big\}\big)$$

is independent of $t \in (0, T]$; and therefore, by FUBINI's Theorem, we get the desired result. ∎

For $T \in [0, \infty)$ and $\nu \in \mathbf{M}_1(\Omega_T)$, we set

$$\nu \otimes_T P_* = \int_{\Omega_T} P^{(T)}_{\omega_T} \, \nu(d\omega_T); \tag{4.4.31}$$

and when $Q \in \mathbf{M}_1\big((\Omega, \mathcal{B})\big)$, we use $Q \otimes_T P_*$ instead of $Q_T \otimes_T P_*$. Note that $\nu \otimes_T P_*$ is the unique $Q \in \mathbf{M}_1\big((\Omega, \mathcal{B})\big)$ with the properties that $Q_T = \nu$ and $\omega \in \Omega \longmapsto P^{(T)}_{\pi_T(\omega)}$ is a r.c.p.d. of Q given $\pi_T^{-1}(\mathcal{B}_{\Omega_T})$. In addition, by (4.4.18) and the MARKOV property for $\{P_\sigma : \sigma \in \Sigma\}$, one can easily check that

$$\nu \otimes_{T_1} P_* = \big(\nu \otimes_{T_1} P_*\big) \otimes_{T_2} P_* \tag{4.4.32}$$

for $0 \le T_1 < T_2 < \infty$ and $\nu \in \mathbf{M}_1\big(\Omega_{T_1}\big)$.

4.4.33 Lemma. *Let $T \in (0, \infty)$ and $\nu \in \mathbf{M}_1(\Omega_T)$ be given. If ν is not shift-invariant, then*

$$\sup_{h \in (0,T)} J_{\Pi_h^{(T)}}(\nu) = \infty.$$

On the other hand, if $\nu \in \mathbf{M}_1^{\mathrm{S}}(\Omega_T)$, then, for $h \in (0, T)$,

$$\begin{aligned} J_{\Pi_h^{(T)}}(\nu) &= \mathbf{H}\big(\nu \big| (\nu \otimes_{T-h} P_*)_T\big) \\ &= \int_{\Omega_T} \mathbf{H}\left(\nu^{(T-h)}(\omega_T, \cdot) \middle| \left(P^{(T-h)}_{\pi^{(T)}_{T-h}(\omega_T)}\right)_T\right) \nu(d\omega_T), \end{aligned} \tag{4.4.34}$$

where $\omega_T \in \Omega_T \longmapsto \nu^{(T-h)}(\omega_T, \cdot)$ is a r.c.p.d. of ν given $\big(\pi^{(T)}_{T-h}\big)^{-1}\big(\mathcal{B}_{\Omega_{T-h}}\big)$. In particular, if $\nu \in \mathbf{M}_1^{\mathrm{S}}(\Omega_T)$ and s, $t > 0$ satisfy $s + t < T$, then

$$J_{\Pi_{s+t}^{(T)}}(\nu) = J_{\Pi_t^{(T-s)}}(\nu_{T-s}) + J_{\Pi_s^{(T)}}(\nu). \tag{4.4.35}$$

PROOF: First note that

$$\begin{aligned} \big[\Pi_h^{(T)} \psi\big](\omega_T) &= \int_{\Omega_T} \psi\big(\pi_T \circ \theta_h(\omega')\big) \, P^{(T)}_{\omega_T}(d\omega') \\ &= \int_{\Omega} \psi\big(\pi_T(\omega')\big) \, P^{(T-h)}_{\pi_{T-h}(\theta_h \overline{\omega}_T)}(d\omega') \end{aligned}$$

for any $\psi \in B(\Omega_T; \mathbb{R})$. In particular, if ψ is $\big(\pi^{(T)}_{T-h}\big)^{-1}(\Omega_{T-h})$-measurable, then $\big[\Pi_h^{(T)} \psi\big] = \psi \circ \theta_h^{(T)}$; and if $\nu \in \mathbf{M}_1^{\mathrm{S}}(\Omega_T)$, then

$$\int_{\Omega_T} \big[\Pi_h^{(T)} \psi\big](\omega_T) \, \nu(d\omega_T) = \int_{\Omega_T} \psi\big(\pi_T(\omega')\big) \big(\nu_{T-h} \otimes_{T-h} P_*\big)(d\omega').$$

With these preliminaries, the argument used to prove Lemma 4.4.9 can be easily adapted to prove the first assertion of the present lemma as well as (4.4.34). Finally, by combining Lemma 4.4.7 with (4.4.32), we see that

$$\begin{aligned}&\mathbf{H}\big(\nu\big|\big(\nu_{T-(s+t)}\otimes_{T-(s+t)}P_*\big)_T\big)\\&\quad=\mathbf{H}\Big(\nu_{T-s}\Big|\big(\nu_{T-s-t}\otimes_{T-s-t}P_*\big)_{T-s}\Big)\\&\qquad\qquad+\int_{\Omega_T}\mathbf{H}\left(\nu^{(T-s-t)}(\omega_T,\cdot)\middle|\left(P^{(T-s-t)}_{\pi^{(T)}_{T-s-t}(\omega_T)}\right)_T\right)\nu(d\omega_T).\end{aligned}$$

Thus, if ν is shift-invariant, then (4.4.35) follows from (4.4.34). ∎

As was the case in the MARKOV chain setting, in order to complete our program it will be convenient to move our measures to the left half-line. Thus, for $T\in[0,\infty)$, let Ω_T^* be the space of right-continuous paths $\omega_T^*:(-\infty,T]\longrightarrow\Sigma$ which have a left limit at each $t\in(-\infty,T]$ and are left-continuous at T. For $-\infty<s\le t\le T<\infty$, denote by $\mathcal{B}^{(T)}_{[s,t]}$ the σ-algebra over Ω_T^* generated by the maps $\omega_T^*\in\Omega_T^*\longmapsto\omega_T^*(\tau)\in\Sigma$ for $\tau\in[s,t]$; and use $\mathcal{B}^{(T)}$ to stand for the smallest σ-algebra over Ω_T^* which contains $\mathcal{B}^{(T)}_{[s,T]}$ for all $s\in(-\infty,T]$.

4.4.36 Lemma. *Let $Q\in\mathbf{M}_1^{\mathrm{S}}\big((\Omega,\mathcal{B})\big)$ be given. Then, for every $T\in[0,\infty)$, there is a unique $Q_T^*\in\mathbf{M}_1^{\mathrm{S}}\big((\Omega_T^*,\mathcal{B}^{(T)})\big)$ with the property that*

$$\begin{aligned}&Q_T^*\Big(\big\{\omega_T^*:\big(\omega_T^*(t_1),\dots,\omega_T^*(t_n)\big)\in\Gamma\big\}\Big)\\&\qquad=Q\big(\big\{\omega:\big(\omega(0),\dots,\omega(t_n-t_1)\big)\in\Gamma\big\}\big)\end{aligned}$$

for every $n\in\mathbb{Z}^+$, $-\infty<t_1<\cdots<t_n\le T$, and $\Gamma\in\mathcal{B}_{\Sigma^n}$.

PROOF: The uniqueness assertion is obvious; and clearly it suffices to prove existence in the case when $T=0$.

For $d\in\mathbb{Z}^+$, let $\Omega^*_{[-d,0]}$ be the space of right-continuous paths $\omega^*_{[-d,0]}:[-d,0]\longrightarrow\Sigma$ which have a left limit at each $t\in[-d,0]$ and are continuous at each $t\in[-d,0]$ for which $-t\in\mathbb{Z}$. Then (cf. Exercise 4.4.40 below), $\Omega^*_{[-d,0]}$ becomes a Polish space when it is given the topology determined by the SKOROKHOD metric in which the homeomorphisms $\lambda:[-d,0]\longrightarrow[-d,0]$ have the property that $\lambda(t)=t$ for every $t\in[-d,0]\cap\mathbb{Z}$. Also, it is then easy to see that the natural restriction mapping taking $\Omega^*_{[-d-1,0]}$ onto $\Omega^*_{[-d,0]}$ is continuous for each $d\in\mathbb{Z}^+$; and, clearly, the projective limit of $\big\{\Omega^*_{[-d,0]}:d\in\mathbb{Z}^+\big\}$ can be identified with the space $\Omega^*_{(\infty,0]}$ consisting of those paths $\omega_0^*\in\Omega_0^*$ which are continuous at $-n$ for every $n\in\mathbb{N}$.

Now let $Q \in \mathbf{M}_1^{\mathrm{S}}((\Omega,\mathcal{B}))$ be given. Then, by Lemma 4.4.30, there is, for each $d \in \mathbb{Z}^+$, a unique $Q^*_{[-d,0]} \in \mathbf{M}_1^{\mathrm{S}}(\Omega^*_{[-d,0]})$ such that

$$\begin{aligned} Q^*_{[-d,0]}\Big(\big\{\omega^*_{[-d,0]} : \big(\omega^*_{[-d,0]}(t_1-d),\dots,\omega^*_{[-d,0]}(t_n-d)\big)\in\Gamma\big\}\Big) \\ = Q\big(\{\omega : \big(\omega(t_1),\dots,\omega(t_n)\big)\in\Gamma\}\big) \end{aligned}$$

for all $n \in \mathbb{Z}^+$, $0 \le t_1 < \cdots < t_n \le d$, and $\Gamma \in \mathcal{B}_{\Sigma^n}$. Moreover, the family $\{Q^*_{[-d,0]} : d \in \mathbb{Z}^+\}$ is consistently defined on the spaces $\{\Omega^*_{[-d,0]} : d \in \mathbb{Z}^+\}$. Hence, by KOLMOGOROV's Extension Theorem, there is a unique $Q^*_0 \in \mathbf{M}_1^{\mathrm{S}}((\Omega^*_0,\mathcal{B}^{(0)}))$ which extends all the $Q^*_{[-d,0]}$'s; and clearly this is the measure which we were seeking. ∎

Given $T \in [0,\infty)$, $\omega_T \in \Omega_T$, and $\omega^*_0 \in \Omega^*_0$, define $\omega^*_0 \otimes_0 \omega_T \in \Omega^*_T$ so that $\big(\omega^*_0 \otimes_0 \omega_T\big)(t) = \omega^*_0(t\wedge 0)$ if $\omega^*_0(0) \ne \omega_T(0)$ and

$$\big(\omega^*_0 \otimes_0 \omega_T\big)(t) = \begin{cases} \omega^*_0(t) & \text{for } t \in (-\infty,0] \\ \omega_T(t) & \text{for } t \in (0,T] \end{cases}$$

if $\omega^*_0(0) = \omega_T(0)$. It is then an easy matter to check that

$$\big(\omega^*_0,\omega_T\big) \in \Omega^*_0 \times \Omega_T \longmapsto \omega^*_0 \otimes_0 \omega_T \in \Omega^*_T$$

is measurable. Thus, for $Q \in \mathbf{M}_1^{\mathrm{S}}((\Omega,\mathcal{B}))$ and $T \in [0,\infty)$, we can determine $\big(Q^*_0 \otimes_0 P_*\big)_T \in \mathbf{M}_1\big((\Omega^*_T,\mathcal{B}^{(T)})\big)$ by

$$\big(Q^*_0 \otimes_0 P_*\big)_T(\Gamma) = \int_{\Omega^*_0}\left(\int_\Omega \chi_\Gamma\big(\omega^*_0 \otimes_0 \pi_T(\omega)\big)\,P_{\omega^*_0(0)}(d\omega)\right)Q^*_0(d\omega^*_0)$$

for all $\Gamma \in \mathcal{B}^{(T)}$. Finally, for $T \in [0,\infty)$, $s \in (-\infty,T]$, and $\mu^*_T, \nu^*_T \in \mathbf{M}_1\big((\Omega^*_T,\mathcal{B}^{(T)})\big)$, we will set

$$\mathbf{H}^{(T)}_s\big(\nu^*_T\big|\mu^*_T\big) = \mathbf{H}\left(\nu^*_T\big|_{\mathcal{B}^{(T)}_{[s,T]}}\Big|\mu^*_T\big|_{\mathcal{B}^{(T)}_{[s,T]}}\right).$$

After one reconciles the notation just introduced with our earlier notation, one finds that (4.4.34) says that, for all $0 < h < T$,

$$J_{\Pi^{(T)}_h}(Q_T) = \mathbf{H}^{(h)}_{-T+h}\Big(Q^*_h\big|\big(Q^*_0 \otimes_0 P_*\big)_h\Big), \quad Q \in \mathbf{M}_1^{\mathrm{S}}((\Omega,\mathcal{B})); \tag{4.4.37}$$

and, as we are about to see, (4.4.37) is the key to the last step in our identification of $\overline{J}^{(\infty)}_P$.

4.4.38 Theorem. *Let $Q \in \mathbf{M}_1\big((\Omega, \mathcal{B})\big)$ be given. Then, for any $h > 0$,*

$$\overline{J}_P^{(\infty)}(Q) = \lim_{T \nearrow \infty} \lim_{h \searrow 0} \frac{1}{h} J_{\Pi_h^{(T)}}(Q_T) \tag{4.4.39}$$

$$= \begin{cases} \mathbf{H}\left(Q_1^* \middle| \left(Q_0^* \otimes_0 P_*\right)_1\right) & \text{if } Q \in \mathbf{M}_1^{\mathrm{S}}\big((\Omega, \mathcal{B})\big) \\ \infty & \text{otherwise.} \end{cases}$$

PROOF: If $Q \notin \mathbf{M}_1^{\mathrm{S}}\big((\Omega, \mathcal{B})\big)$, then, by Lemma 4.4.30 and Lemma 4.4.33, $\overline{J}_P^{(\infty)}(Q) = \infty$. Thus, we will now assume that $Q \in \mathbf{M}_1^{\mathrm{S}}\big((\Omega, \mathcal{B})\big)$.

Set $f(h,T) = J_{\Pi_h^{(T)}}(Q_T)$ for $0 < h < T < \infty$. Then, $f(h, \cdot)$ is non-decreasing on (h, ∞) for each $h \in (0, \infty)$; and, by (4.4.35), $f(s+t, T) = f(t, T-s) + f(s, T)$ as long as $s + t < T$. In particular, if $h \in (0, \infty)$ and $T \in (1, \infty)$ and $n \in \mathbb{Z}^+$, then by induction on $0 \le \ell \le n$:

$$f\left(\frac{\ell}{n}, T\right) = \sum_{k=0}^{\ell-1} f\left(\frac{1}{n}, T - \frac{k}{n}\right);$$

and so

$$nf\left(\frac{h}{n}, T\right) \ge f(h,T) \ge nf\left(\frac{h}{n}, T-1\right), \quad T \in (2, \infty) \text{ and } n \in \mathbb{Z}^+.$$

Consequently,

$$f(h,T) \le nf\left(\frac{h}{n}, T\right) \le f(h, T+1), \quad T \in (2h, \infty),$$

for every $n \in \mathbb{Z}^+$; and therefore, by (4.4.24),

$$f(h,T) \le h\overline{J}_P^{(T)}(Q_T) \le f(h, T+1), \quad T \in (2h, \infty);$$

and clearly the desired result now follows immediately from (4.4.37) and LEMMA 4.4.7. ∎

In conjunction with Theorem 4.4.38, Theorem 4.4.27 becomes a version of the DONSKER and VARADHAN's result on this subject [**36**].

4.4.40 Exercise.

Working with the SKOROKHOD topology is notoriously unpleasant; and, in order not to burden the presentation with even more technicalities, we have swept some annoying details under the rug. What follows is a selection of some points which we have used without proof.

(**i**) Show that, for each $T \in (0,\infty)$ and $t \in [0,T]$, the map $\omega_T \in \Omega_T \longmapsto \omega_T(t) \in \Sigma$ is $\mathcal{B}_{\Omega_T}$-measurable. This fact, which is well-known when $\Sigma = \mathbb{R}$, can be proved for general Σ's by using the fact that every Polish space may be continuously embedded as a $\mathfrak{G}_\delta$ in $[0,1]^{\mathbb{Z}^+}$ and applying the $\Sigma = \mathbb{R}$ result to each of the coordinates of the embedding.

(**ii**) In the proof of Lemma 4.4.36, we tacitly used the fact that if $d \in \mathbb{Z}^+$ and we define the SKOROKHOD distance $\operatorname{dist}\big(\omega^*_{[-d,0]}, \overline{\omega}^*_{[-d,0]}\big)$ between paths $\omega^*_{[-d,0]}, \overline{\omega}^*_{[-d,0]} \in \Omega^*_{[-d,0]}$ by

$$\inf_\lambda \sup_{t\in[-d,0]} \Big\{\operatorname{dist}\Big(\omega^*_{[-d,0]}\big(\lambda(t)\big), \overline{\omega}^*_{[-d,0]}(t)\Big) + \big|\lambda(t) - t\big|\Big\},$$

where λ runs over increasing homeomorphisms of $[-d,0]$ satisfying $\lambda(t) = t$ for $t \in [-d,0] \cap \mathbb{Z}$, then the resulting metric makes $\Omega^*_{[-d,0]}$ into a Polish space and the natural restriction maps from $\Omega^*_{[-d-1,0]}$ onto $\Omega^*_{[-d,0]}$ continuous. Check this fact.

4.4.41 Exercise.

A remarkable dividend of looking at large deviations at the level of processes is that the rate functions $J_\Pi^{(\infty)}$ and $\overline{J}_P^{(\infty)}$ have the pleasing property that they are affine on the space of shift-invariant probability measures. (As we will see in Section 5.3 below, this fact can be made to play an extremely important role in the derivation of process-level large deviation results.) In this exercise, we outline a simple way to see this fact for $\overline{J}_P^{(\infty)}$; an analogous approach leads to the same fact for $J_\Pi^{(\infty)}$.

What we want to show is that, for Q, $Q' \in \mathbf{M}_1^{\mathrm{S}}(\Omega)$,

$$\text{(4.4.42)} \quad \overline{J}_P^{(\infty)}\big(\alpha Q + (1-\alpha)Q'\big) = \alpha \overline{J}_P^{(\infty)}(Q) + (1-\alpha)\overline{J}_P^{(\infty)}(Q'), \quad \alpha \in (0,1).$$

Since we already know that $\overline{J}_P^{(\infty)}$ is convex, all that we need to do is check that the right hand side of (4.4.42) is dominated by the left. The first step will be to develop yet another expression (cf. (4.4.43) below) for $\overline{J}_P^{(\infty)}$.

(**i**) Given $\nu \in \mathbf{M}_1(\Sigma)$, set

$$P_\nu = \int_\Sigma P_\sigma\, \nu(d\sigma).$$

Using (4.4.8) and (4.4.34), show that for any $Q \in \mathbf{M}_1^{\mathrm{S}}(\Omega)$, $\nu \in \mathbf{M}_1(\Sigma)$, and $T \in [0,\infty)$:

$$\mathbf{H}\big(Q_{T+h}\big|(P_\nu)_{T+h}\big) = \mathbf{H}\big(Q_T\big|(P_\nu)_T\big) + J_{\Pi_h^{(T+h)}}\big(Q_{T+h}\big), \quad h \in (0,\infty).$$

Starting from the preceding and using (4.4.39), conclude that

$$\overline{J}_P^{(\infty)}(Q) = \lim_{T\to\infty} \frac{1}{T}\mathbf{H}\big(Q_T\big|(P_\nu)_T\big) \text{ if } \mathbf{H}\big(Q_0\big|\nu\big) < \infty. \tag{4.4.43}$$

(**ii**) To complete the proof of (4.4.42), prove that

$$\big(\alpha a + (1-\alpha)b\big)\log\big(\alpha a + (1-\alpha)b\big) \geq \alpha a \log a + (1-\alpha)b\log b - \frac{|b-a|}{e}$$

for every $\alpha \in (0,1)$ and all $a,\, b \in [0,\infty)$. Now suppose that $Q,\, Q' \in \mathbf{M}_1^{\mathrm{S}}(\Omega)$ and $\alpha \in (0,1)$ are given, set $\nu = \alpha Q_0 + (1-\alpha)Q_0'$, and use the preceding together with (4.4.43) to conclude that

$$\overline{J}_P^{(\infty)}\big(\alpha Q + (1-\alpha)Q'\big) \geq \alpha\overline{J}_P^{(\infty)}(Q) + (1-\alpha)\overline{J}_P^{(\infty)}(Q').$$

(**iii**) The equation (4.4.43) is interesting in its own right. Indeed, it expresses $\overline{J}_P^{(\infty)}(Q)$ as a **specific relative entropy**. This expression becomes particularly interesting in the case when one knows (as one does if $P(t,\sigma,\cdot)$ satisfies $(\tilde{\mathbf{U}})$) *a priori* that there is a $\{P_t : t > 0\}$-invariant $\mu \in \mathbf{M}_1(\Sigma)$ with the property that $\mathbf{H}\big(Q_0\big|\mu\big) < \infty$ for every $Q \in \mathbf{M}_1^{\mathrm{S}}(\Omega)$ with $\overline{J}_P^{(\infty)}(Q) < \infty$. Indeed, show that, in this case, one can replace (4.4.43) by

$$\overline{J}_P^{(\infty)}(Q) = \lim_{T\to\infty} \frac{1}{T}\mathbf{H}\big(Q_T\big|(P_\mu)_T\big), \quad Q \in \mathbf{M}_1^{\mathrm{S}}(\Omega). \tag{4.4.44}$$

4.4.45 Exercise.

Let Π be a transition probability on Σ, and define $J_\Pi : \mathbf{M}_1(\Sigma) \longrightarrow [0,\infty]$ accordingly (as in (4.1.38)). Also, for given $\nu \in \mathbf{M}_1(\Sigma)$, let $\mathbf{M}_1^{(\nu)}(\Sigma^2)$ denote the space of $\mu \in \mathbf{M}_1(\Sigma^2)$ with the property that $\mu \circ \pi_1^{-1} = \nu = \mu \circ \pi_2^{-1}$, where π_i, $i \in \{1,2\}$, is used here to denote the ith projection from Σ^2 into Σ.

(**i**) Assume that Π satisfies the condition $(\mathbf{U})$ of Section 4.1, and use the results in this section together with those in Section 4.1 to prove the equality

$$J_\Pi(\nu) = \inf\Big\{\mathbf{H}\big(\mu\big|\nu \otimes_2 \Pi\big) : \mu \in \mathbf{M}_1^{(\nu)}(\Sigma^2)\Big\} \tag{4.4.46}$$

as an application of the last part of Lemma 2.1.4. Conclude, in particular, that if $J_\Pi(\nu) < \infty$, then there must exist a $\mu \in \mathbf{M}_1^{(\nu)}(\Sigma^2)$ such that $J_\Pi(\nu) = \mathbf{H}\big(\mu\big|\nu \otimes_2 \Pi\big)$.

(ii) Half of (4.4.46) is trivial and depends in no way on the condition (**U**). Namely, to see that the left hand side of (4.4.46) is always dominated by the right, check directly from the definitions of J_Π and $J_\Pi^{(2)}$ (cf. (4.4.4)) that $J_\Pi(\nu) \le J_\Pi^{(2)}(\mu)$ for every $\mu \in \mathbf{M}_1^{(\nu)}(\Sigma^2)$, and then apply Lemma 4.4.9.

(iii) Even when Π satisfies (**U**), a direct proof that the left hand side of (4.4.46) dominates the right is not so easy. Thus, all that we will attempt to do here is explain how the existence of a $\mu \in \mathbf{M}_1^{(\nu)}(\Sigma^2)$ satisfying $J_\Pi(\nu) = \mathbf{H}(\mu|\nu \otimes_2 \Pi)$ is related to the functions $u \in B(\Sigma; [1,\infty))$ in terms of which $J_\Pi(\nu)$ is defined.

Given a $u \in B(\Sigma; [1,\infty))$, consider the transition probability defined by

$$\Pi_u(\sigma, d\tau) = \frac{u(\tau)}{[\Pi u](\sigma)} \Pi(\sigma, d\tau).$$

(Note that, in the notation of Section 4.1, the Π_u above would have been denoted there by Π_V with $V = \log \frac{u}{\Pi u}$.) Next, define $\mu_u = \nu \otimes_2 \Pi_u$, check that

$$\mathbf{H}(\mu_u|\nu \otimes_2 \Pi) = \int_{\Sigma^2} \log u(\tau)\, \mu_u(d\sigma \times d\tau) - \int_{\Sigma^2} \log\big([\Pi u](\sigma)\big)\, \mu_u(d\sigma \times d\tau),$$

and conclude that

$$J_\Pi(\nu) = -\int_\Sigma \log \frac{[\Pi u]}{u}\, d\nu = \mathbf{H}(\mu_u|\nu \otimes_2 \Pi). \tag{4.4.47}$$

for $\mu_u \in \mathbf{M}_1^{(\nu)}(\Sigma^2)$ Conversely, use Lemma 3.2.13 to check that

$$\mathbf{H}(\nu|\mu_u \circ \pi_2^{-1}) \le J_\Pi(\nu) + \int_\Sigma \log \frac{[\Pi u]}{u}\, d\nu,$$

and conclude that

$$J_\Pi(\nu) = -\int_\Sigma \log \frac{[\Pi u]}{u}\, d\nu \text{ implies } \mu_u \in \mathbf{M}_1^{(\nu)}(\Sigma^2).$$

Summarizing, we now see that $J_\Pi(\nu) = -\int_\Sigma \log \frac{[\Pi u]}{u}\, d\nu$ if and only if $\mu_u \in \mathbf{M}_1^{(\nu)}(\Sigma^2)$, in which case $J_\Pi(\nu) = \mathbf{H}(\mu_u|\nu \otimes_2 \Pi)$. The problem is, of course, that one cannot expect, in general, that there will exist a $u \in B(\Sigma; [0,\infty))$ for which $J_\Pi(\nu) = -\int_\Sigma \log \frac{[\Pi u]}{u}\, d\nu$.

4.4.48 Exercise.

It is no accident that the rate function governing the large deviations of the empirical process is infinite off of the space of shift-invariant measures. To see this, let $\Omega = \Sigma^{\mathbb{N}}$, define $\omega \in \Omega \longmapsto \mathbf{R}_n(\omega) \in \mathbf{M}_1(\Omega)$ as in (4.4.1), and suppose that $P \in \mathbf{M}_1(\Omega)$ and $I : \Omega \longrightarrow [0, \infty]$ satisfy

$$\varliminf_{n\to\infty} \frac{1}{n} \log\left(P\left(\left\{\omega : \mathbf{R}_n(\omega) \in G\right\}\right)\right) \geq -I(Q)$$

for every open G in $\mathbf{M}_1(\Omega)$ and $Q \in G$. Show that I must be identically infinite off of $\mathbf{M}_1^{\mathrm{S}}(\Omega)$.

Hint: First check that $\mathbf{M}_1^{\mathrm{S}}(\Omega)$ is a closed subset of $\mathbf{M}_1(\Omega)$; and, second, note that, for any $\epsilon > 0$, there is an $N \in \mathbb{Z}^+$ such that the LÉVY distance between elements Q and Q' of $\mathbf{M}_1(\Omega)$ is less than ϵ if

$$\left\|Q \circ \pi_{[0,N]}^{-1} - Q' \circ \pi_{[0,N]}^{-1}\right\|_{\mathrm{var}} < \frac{\epsilon}{2}.$$

(The map $\pi_{[0,N]}$ is the projection of Ω onto Σ^N obtained by restricting a path $\omega \in \Omega$ to $\mathbb{N} \cap [0, N]$.) Finally, for any $\omega \in \Omega$ and $n \in \mathbb{Z}^+$, let $\tilde{\omega}_n \in \Omega$ be the path determined by

$$\Sigma_{kn+\ell}\left(\tilde{\omega}_n\right) = \Sigma_\ell(\omega) \text{ for } k \in \mathbb{N} \text{ and } 1 \leq \ell < n;$$

and show both that $\mathbf{R}_n\left(\tilde{\omega}_n\right) \in \mathbf{M}_1^{\mathrm{S}}(\Omega)$ and that

$$\left\|\mathbf{R}_n(\omega) \circ \pi_{[0,N]}^{-1} - \mathbf{R}_n\left(\tilde{\omega}_n\right) \circ \pi_{[0,N]}^{-1}\right\|_{\mathrm{var}} \leq \frac{2n}{N}.$$

V Non-Uniform Results

5.1 Generalities about the Upper Bound

We begin by restating Theorem 2.2.4 for the setting in which we will be working. Namely, let Ω be a Polish space and suppose that $\{Q_\epsilon : \epsilon > 0\}$ is a family of probability measures on $\mathbf{M}_1(\Omega)$ with the property that

$$\Lambda(V) \equiv \lim_{\epsilon \to 0} \epsilon \log \left(\int_{\mathbf{M}_1(\Omega)} \exp\left[\frac{1}{\epsilon} \int_\Omega V(\omega)\, \mu(d\omega) \right] Q_\epsilon(d\mu) \right) \tag{5.1.1}$$

exists for every $V \in C_\mathrm{b}(\Omega; \mathbb{R})$. We then know that

$$\overline{\lim_{\epsilon \to 0}}\, \epsilon \log\bigl(Q_\epsilon(C)\bigr) \le - \inf_C \Lambda^* \tag{5.1.2}$$

for $C \subset\subset \mathbf{M}_1(\Omega)$, where $\Lambda^* : \mathbf{M}_1(\Omega) \longrightarrow [0, \infty]$, given by

$$\Lambda^*(\mu) \equiv \sup\left\{ \int_\Omega V\, d\mu - \Lambda(V) : V \in C_\mathrm{b}(\Omega; \mathbb{R}) \right\}, \tag{5.1.3}$$

is the LEGENDRE transform of Λ. Our goal in this section is to find out when we can remove the restriction that the C in (5.1.2) be compact.

Given a function $\Phi : C_\mathrm{b}(\Omega; \mathbb{R}) \longrightarrow \mathbb{R}$, we will say that Φ is **non-decreasing** if $\Phi(V_1) \le \Phi(V_2)$ whenever $V_1 \le V_2$; and we will say that Φ is **tight** if for each $M \in (0, \infty)$ there is a $K(M) \subset\subset \Omega$ such that $\Phi(V) \le 1$ whenever V is an element of $C_\mathrm{b}(\Omega; \mathbb{R})$ which vanishes on $K(M)$ and is bounded by M.

5.1.4 Lemma. *Let $\Phi : C_b(\Omega;\mathbb{R}) \longrightarrow \mathbb{R}$ be a non-decreasing, convex function with the property that $\Phi(c\mathbf{1}) = c$, $c \in \mathbb{R}$. Then $\left|\Phi(V_2) - \Phi(V_1)\right| \leq \|V_2 - V_1\|_B$ for all V_1, $V_2 \in C_b(\Omega;\mathbb{R})$. Moreover, if, in addition, Φ is tight, then for every $\epsilon > 0$ and $M \in (0,\infty)$ there is a $K(\epsilon, M) \subset\subset \Omega$ such that $\left|\Phi(V_2) - \Phi(V_1)\right| \leq \epsilon$ for all V_1, $V_2 \in C_b(\Omega;\mathbb{R})$ with the properties that $V_1 = V_2$ on $K(\epsilon, M)$ and $\|V_1\|_B \vee \|V_2\|_B \leq M$.*

PROOF: First, note that $\Phi(V) \leq \Phi\big(\|V\|_B\mathbf{1}\big) = \|V\|_B$ and that

$$0 = \Phi\left(\frac{V - V}{2}\right) \leq \frac{\Phi(V) + \Phi(-V)}{2}.$$

Thus, $|\Phi(V)| \leq \|V\|_B$, $V \in C_b(\Omega;\mathbb{R})$. Second, using convexity and writing

$$V_2 = (1-\theta)V_1 + \frac{\theta}{2}\left(\frac{2(V_2 - V_1)}{\theta}\right) + \frac{\theta}{2}(2V_1), \quad \theta \in (0,1),$$

one sees that

$$\text{(5.1.5)} \qquad \Phi(V_2) - \Phi(V_1) \leq \frac{\theta}{2}\Phi\left(\frac{2(V_2 - V_1)}{\theta}\right) + \frac{\theta\big(\Phi(2V_1) - 2\Phi(V_1)\big)}{2}$$

for $\theta \in (0,1)$.

From (5.1.5) and the remark preceding it, we have that

$$\Phi(V_2) - \Phi(V_1) \leq \|V_2 - V_1\|_B + 2\theta\|V_1\|_B$$

for all $\theta \in (0,1)$; and, therefore, after letting $\theta \searrow 0$ and reversing the roles of V_1 and V_2, one gets the first assertion. To prove the second assertion, let $\epsilon > 0$ and $M \in (0,\infty)$ be given and use (5.1.5) to see that

$$\Phi(V_2) - \Phi(V_1) \leq \frac{\theta}{2}\left[\Phi\left(\frac{2(V_2 - V_1)}{\theta}\right) + 4M\right], \quad \theta \in (0,1),$$

as long as $\|V_1\|_B \vee \|V_2\|_B \leq M$. Finally, define $\theta \in (0,1)$ so that

$$\frac{\theta}{2}(1 + 4M) = \epsilon \wedge \frac{1}{2},$$

and set $K(\epsilon, M) = K(4M/\theta)$, where $\{K(M) : M \in (0,\infty)\}$ is the family of compact sets which appears in the definition of tightness for Φ. After reversing the roles of V_1 and V_2, one then arrives at the desired conclusion. ∎

Before presenting the next result, we need to introduce some notation. Let $\hat{\rho}$ be a compatible metric on Ω with the property that $(\Omega, \hat{\rho})$ is totally bounded, and denote by $\hat{\Omega}$ the completion of Ω with respect to $\hat{\rho}$. Obviously,

$\hat{\Omega}$ is compact and, because it is Polish, Ω can be thought of as a dense $\mathfrak{G}_\delta$ subset of $\hat{\Omega}$. In particular, we will identify $\mathbf{M}_1(\Omega)$ with the subset of those $\hat{\mu} \in \mathbf{M}_1(\hat{\Omega})$ for which $\hat{\mu}(\hat{\Omega} \setminus \Omega) = 0$. In addition, if $\hat{C}_{\mathrm{b}}(\Omega;\mathbb{R})$ denotes the space of bounded, $\hat{\rho}$-uniformly continuous functions on Ω, then $\hat{\phi} \in C(\hat{\Omega};\mathbb{R}) \longmapsto \hat{\phi}\big|_\Omega \in \hat{C}_{\mathrm{b}}(\Omega;\mathbb{R})$ is a surjective isometry.

What the following theorem turns on is the observation that "tightness" allows one to work on the compact space $\hat{\Omega}$ and then transfer one's conclusions there back to Ω itself.

5.1.6 Theorem. *Let $\Phi : C_{\mathrm{b}}(\Omega;\mathbb{R}) \longrightarrow \mathbb{R}$ be a non-decreasing, convex function with the property that $\Phi(c\mathbf{1}) = c$, $c \in \mathbb{R}$; and define Ψ on $\mathbf{M}_1(\Omega)$ by*

$$\Psi(\mu) = \sup\left\{\int_\Omega V\,d\mu - \Phi(V) : V \in C_{\mathrm{b}}(\Omega;\mathbb{R})\right\}, \quad \mu \in \mathbf{M}_1(\Omega). \tag{5.1.7}$$

Then Ψ is convex rate function. Moreover, if Φ is tight, then Ψ is good, there is a $\mu_0 \in \mathbf{M}_1(\Omega)$ at which Ψ vanishes,

$$\Phi(V) = \sup\left\{\int_\Omega V\,d\mu - \Psi(\mu) : \mu \in \mathbf{M}_1(\Omega)\right\}, \quad V \in C_{\mathrm{b}}(\Omega;\mathbb{R}), \tag{5.1.8}$$

and

$$\hat{\Psi}(\hat{\mu}) = \begin{cases} \Psi(\hat{\mu}) & \text{if } \hat{\mu} \in \mathbf{M}_1(\Omega) \\ \infty & \text{if } \hat{\mu} \in \mathbf{M}(\hat{\Omega}) \setminus \mathbf{M}_1(\Omega) \end{cases} \tag{5.1.9}$$

where $\hat{\Psi}$ is defined on $\mathbf{M}(\hat{\Omega})$ by

$$\hat{\Psi}(\hat{\mu}) = \sup\left\{\int_{\hat{\Omega}} \hat{V}\,d\hat{\mu} - \Phi\big(\hat{V}\big|_\Omega\big) : \hat{V} \in C(\hat{\Omega};\mathbb{R})\right\}, \quad \hat{\mu} \in \mathbf{M}(\hat{\Omega}). \tag{5.1.10}$$

Conversely, suppose that $\Psi : \mathbf{M}_1(\Omega) \longrightarrow [0,\infty]$ is a convex rate function which vanishes at some $\mu_0 \in \mathbf{M}_1(\Omega)$; and define Φ on $C_{\mathrm{b}}(\Omega;\mathbb{R})$ by (5.1.8). *Then Φ is a non-decreasing, convex function which satisfies $\Phi(c\mathbf{1}) = c$, $c \in \mathbb{R}$; Ψ can be recovered from Φ via* (5.1.7); *and Ψ is good if and only if Φ is tight.*

Proof: Let Φ be a function of the sort described in the first part of the theorem, and define Ψ accordingly by (5.1.7). Obviously, Ψ is lower semicontinuous and convex. In addition, since $\Phi(0) = 0$, it is clear that $\Psi \geq 0$.

Next, add the assumption that Φ is tight. To see that Ψ is good, let $\{K(M) : M \in (0,\infty)\}$ be the compact subsets of Ω described in tightness property for Φ. If $\Psi(\mu) \leq L$, then

$$\int_\Omega V\,d\mu \leq \Phi(V) + L \leq 1 + L$$

for all $V \in C_b(\Omega;\mathbb{R})$ satisfying $\|V\|_B \le M$ and $V = 0$ on $K(M)$. Hence, $\Psi(\mu) \le L$ implies that $\mu(K(M)^c) \le \frac{1+L}{M}$ for all $M \in (0,\infty)$; and therefore

$$\{\mu : \Psi(\mu) \le L\} \subseteq \left\{\mu : \mu\big(K(M)^c\big) \le \frac{1+L}{M} \text{ for all } M \in (0,\infty)\right\}$$

is compact in $\mathbf{M}_1(\Omega)$. We next turn to the proof of (5.1.9). To see that $\hat{\Psi}(\hat{\mu}) = \infty$ unless $\hat{\mu} \in \mathbf{M}_1(\Omega)$, suppose that $\hat{\mu} \in \mathbf{M}(\hat{\Omega})\setminus\mathbf{M}_1(\Omega)$. If $\hat{\mu}$ is not a probability measure, then $\hat{\Psi}(\hat{\mu}) = \infty$ follows easily from $\Phi(c\mathbf{1}) = c$, $c \in \mathbb{R}$. Thus, suppose that $\hat{\mu} \in \mathbf{M}_1(\hat{\Omega}) \setminus \mathbf{M}_1(\Omega)$. Then $\hat{\mu} = \theta\mu + (1-\theta)\hat{\nu}$, where $\mu \in \mathbf{M}_1(\Omega)$, $\hat{\nu} \in \mathbf{M}_1(\hat{\Omega})$ with $\hat{\nu}(\Omega) = 0$, and $\theta \in [0,1)$. Since Ω is a $\mathfrak{G}_\delta$ subset of $\hat{\Omega}$, $\hat{\Omega} \setminus \Omega$ can be written as the countable union of compact subsets of $\hat{\Omega}$. Hence, there exists a compact $\hat{K} \subseteq \hat{\Omega}\setminus\Omega$ for which $\hat{\nu}(\hat{K}) \ge \frac{1}{2}$. Now let $M \in (0,\infty)$ be given and use the TIETZE Extension Theorem to construct a $\hat{V}_M \in C\big(\hat{\Omega};[0,M]\big)$ with the properties that $\hat{V}_M = 0$ on $K(M)$ and $\hat{V}_M = M$ on $\hat{K}$. We then have that

$$\hat{\Psi}(\hat{\mu}) \ge \int_{\hat{\Omega}} \hat{V}_M \, d\hat{\mu} - \Phi\big(\hat{V}_M|_\Omega\big) \ge \frac{(1-\theta)M}{2} - 1, \quad M \in (0,\infty);$$

and this shows that $\hat{\Psi}(\hat{\mu}) = \infty$. To complete the proof of (5.1.9), we must still check that $\hat{\Psi}(\mu) = \Psi(\mu)$ for $\mu \in \mathbf{M}_1(\Omega)$. Obviously, $\hat{\Psi}(\mu) \le \Psi(\mu)$, and so it suffices to check that $\int_\Omega V\,d\mu - \Phi(V) \le \hat{\Psi}(\mu)$ for all $V \in C_b(\Omega;\mathbb{R})$. Given $V \in C_b(\Omega;\mathbb{R})$ and $\epsilon > 0$, set $M = \|V\|_B$, choose $K(\epsilon, M) \subset\subset \Omega$ as in the last part of Lemma 5.1.4, and take $K \subset\subset \Omega$ so that $K \supseteq K(\epsilon, M)$ and $\mu(K^c) < \epsilon/(M+1)$. Now use the TIETZE Extension Theorem to construct a $\hat{V} \in C(\hat{\Omega};\mathbb{R})$ so that $\hat{V} = V$ on K and $\|\hat{V}\|_B \le \|V\|_B$. Then

$$\int_\Omega V\,d\mu - \Phi(V) \le \int_\Omega \hat{V}\,d\mu - \Phi\big(\hat{V}|_\Omega\big) + 3\epsilon \le \hat{\Psi}(\mu) + 3\epsilon.$$

Continuing in the setting of the preceding paragraph, we next want to derive (5.1.8). To this end, first observe that, because of (5.1.10), (5.1.9), and the fact that $\mathbf{M}(\hat{\Omega})$ is the dual of $C(\hat{\Omega};\mathbb{R})$, Theorem 2.2.15 implies (5.1.8) for $V \in \hat{C}_b(\Omega;\mathbb{R})$. Also, it is clear that for all $V \in C_b(\Omega;\mathbb{R})$ the left hand side of (5.1.8) dominates the right. With these preliminaries in mind, let $V \in C_b(\Omega;\mathbb{R})$ and $0 < \epsilon \le 1$ be given. Set $M = \|V\|_B$ and (recalling that we already know that Ψ is good) choose $K \subset\subset \Omega$ so that $K \supseteq K(\epsilon, M)$ and $\mu(K^c) < \epsilon/(M+1)$ whenever $\Psi(\mu) \le 2M+1$. Next, construct $W \in \hat{C}_b(\Omega;\mathbb{R})$ so that $\|W\|_B \le M$ and $W = V$ on K, and choose $\mu \in \mathbf{M}_1(\Omega)$ so that $\Phi(W) \le \int_\Omega W\,d\mu - \Psi(\mu) + \epsilon$. Then, $\Psi(\mu) \le 2M+1$, and so

$$\Phi(V) \le \Phi(W) + \epsilon \le \int_\Omega W\,d\mu - \Psi(\mu) + 2\epsilon \le \int_\Omega V\,d\mu - \Psi(\mu) + 3\epsilon.$$

In other words, (5.1.8) is now proved. Finally, by taking $V = 1$ in (5.1.8), we see that $\inf_{\mathbf{M}_1(\Omega)} \Psi = 0$; and therefore, by Lemma 2.1.2, there is a μ_0 at which Ψ vanishes.

It remains to prove the converse assertions. Let Ψ be given as in the second part of the theorem, and define Φ by (5.1.8). It is then an easy matter to check that Φ is a non-decreasing, convex function for which $\Phi(c\mathbf{1}) = c$, $c \in \mathbb{R}$. Moreover, the ability to recover Ψ via (5.1.7) is a simple application of Theorem 2.2.15. In particular, by the first part of this theorem, Ψ is good if Φ is tight. Finally, to see that Φ is tight if Ψ is good, let $M \in (0, \infty)$ be given; and choose $K \subset\subset \Omega$ so that $\mu(K^c) < 1/M$ whenever $\Psi(\mu) \le M$. Then the right hand side of (5.1.8) is dominated by 1 for all $V \in C_b(\Omega; \mathbb{R})$ which vanish on K and satisfy $\|V\|_B \le M$. ∎

5.1.11 Corollary. *Let $\{Q_\epsilon : \epsilon > 0\}$ be a family of probability measures on $\mathbf{M}_1(\Omega)$ and assume that the limit $\Lambda(V)$ in* (5.1.1) *exists for each $V \in C_b(\Omega; \mathbb{R})$. Then Λ is a non-decreasing, convex function with the property that $\Lambda(c\mathbf{1}) = c$, $c \in \mathbb{R}$. Moreover, if Λ is tight, then the function Λ^* in* (5.1.3) *is good and* (5.1.2) *holds for every closed set $C \subseteq \mathbf{M}_1(\Omega)$.*

PROOF: The only assertion which is not an immediate consequence of Theorem 5.1.6 is the final one. To handle this one, denote by $\hat{Q}_\epsilon$ the measure on $\mathbf{M}_1(\hat{\Omega})$ induced from Q_ϵ by the inclusion $\mathbf{M}_1(\Omega) \subseteq \mathbf{M}_1(\hat{\Omega})$. Then

$$\Lambda(\hat{V}|_\Omega) = \lim_{\epsilon \to 0} \epsilon \log \left(\int_{\mathbf{M}_1(\hat{\Omega})} \exp \left[\int_{\hat{\Omega}} \hat{V}(\hat{\omega})\, \hat{\mu}(d\hat{\omega}) \right] \hat{Q}_\epsilon(d\hat{\mu}) \right)$$

for $\hat{V} \in C(\hat{\Omega}; \mathbb{R})$. Thus, if $\widehat{\Lambda^*}$ is defined in terms of Λ as in (5.1.10), then

$$\overline{\lim_{\epsilon \to 0}}\, \epsilon \log\big(\hat{Q}_\epsilon(\hat{C})\big) \le -\inf_{\hat{C}} \widehat{\Lambda^*}$$

for all closed $\hat{C} \subseteq \mathbf{M}_1(\hat{\Omega})$. At the same time, if Λ is tight, then, by (5.1.9), $\inf_{\hat{C}} \widehat{\Lambda^*} = \inf_{\hat{C} \cap \mathbf{M}_1(\Omega)} \Lambda^*$; and clearly this shows that (5.1.2) holds for every closed C. ∎

5.1.12 Exercise.

It turns out that there is no need to know that the limit $\Lambda(V)$ in (5.1.1) exists in order to get an upper bound. Indeed, let $\{Q_\epsilon : \epsilon > 0\} \subseteq \mathbf{M}_1\big(\mathbf{M}_1(\Sigma)\big)$, suppose that $\Phi : C_b(\Sigma; \mathbb{R}) \longrightarrow \mathbb{R}$ is a function which dominates

$$\text{(5.1.13)} \qquad \tilde{\Lambda}(V) \equiv \overline{\lim_{\epsilon \to 0}}\, \epsilon \log \left(\int_{\mathbf{M}_1(\Sigma)} \exp \left[\frac{1}{\epsilon} \int_\Sigma V(\sigma) \mu(d\sigma) \right] Q_\epsilon(d\mu) \right)$$

for $V \in C_b(\Sigma; \mathbb{R})$; and let $\Psi : \mathbf{M}_1(\Sigma) \longrightarrow \mathbb{R}$ be defined as in (5.1.7).

(i) Show that

$$\overline{\lim_{\epsilon\to 0}}\,\epsilon \log\left(Q_\epsilon(C)\right) \le -\inf_C \Psi \tag{5.1.14}$$

for all $C \subset\subset \mathbf{M}_1(\Sigma)$. Next, show that $\tilde{\Lambda}$ is a non-decreasing, convex function which satisfies $\tilde{\Lambda}(c\mathbf{1}) = c$, $c \in \mathbb{R}$; and conclude that (5.1.14) continues to hold for all closed $C \subseteq \mathbf{M}_1(\Sigma)$ if $\tilde{\Lambda}$ is tight. In particular, these considerations apply when $\Phi = \tilde{\Lambda}$; in which case we will use $\tilde{\Lambda}^*$ to denote the corresponding Ψ.

(ii) Suppose that there exists a function $F : \Sigma \longrightarrow \mathbb{R}$ with the properties that F is bounded below, $\{\sigma : F(\sigma) \le M\} \subset\subset \Sigma$ for every $M \in [0,\infty)$, and

$$\sup_{\epsilon>0}\left(\int_{\mathbf{M}_1(\Sigma)} \exp\left[\frac{1}{\epsilon}\int_\Sigma F(\sigma)\, d\mu(\sigma)\right] Q_\epsilon(d\mu)\right)^\epsilon < \infty. \tag{5.1.15}$$

Show that $\tilde{\Lambda}$ is then tight; and conclude that $\tilde{\Lambda}^*$ is good, that (5.1.14) holds with $\Psi = \tilde{\Lambda}^*$ for every closed $C \subseteq \mathbf{M}_1(\Sigma)$, and that

$$\tilde{\Lambda}(V) = \sup\left\{\int_\Sigma V\, d\mu - \tilde{\Lambda}^*(\mu) : \mu \in \mathbf{M}_1(\Sigma)\right\} \tag{5.1.16}$$

for every $V \in C_b(\Sigma;\mathbb{R})$.

5.1.17 Exercise.

Return to the setting of Remark 4.2.2 in Section 4.2, and define $\tilde{\Lambda}_P(V)$ to be

$$\sup_{\sigma\in\Sigma}\,\overline{\lim_{t\to\infty}}\,\frac{1}{t}\log\left(\int_\Omega \exp\left[\int_0^t V\left(\Sigma_s(\omega)\right) ds\right] P_\sigma(d\omega)\right) \tag{5.1.18}$$

for $V \in C_b(\Sigma;\mathbb{R})$.

(i) Check that $\tilde{\Lambda}_P$ is non-decreasing, convex, and satisfies $\tilde{\Lambda}_P(c\mathbf{1}) = c$, $c \in \mathbb{R}$. Thus, if (5.1.7) is used to define $\tilde{\Lambda}_P^*$ from $\tilde{\Lambda}_P$, then

$$\sup_{\sigma\in\Sigma}\,\overline{\lim_{t\to\infty}}\,\frac{1}{t}\log\Big(P_\sigma\big(\{\omega : \mathbf{L}_t(\omega) \in C\}\big)\Big) \le -\inf_C \tilde{\Lambda}_P^* \tag{5.1.19}$$

(cf. Remark 4.2.2) holds always for $C \subset\subset \mathbf{M}_1(\Sigma)$ and will hold for every closed $C \subseteq \mathbf{M}_1(\Sigma)$ if $\tilde{\Lambda}_P$ is tight.

(ii) Show that if $F : \Sigma \longrightarrow \mathbb{R}$ is a function which is bounded below and has the properties that $\{\sigma : F(\sigma) \leq M\} \subset\subset \Sigma$, $M \in [0, \infty)$, and

$$\sup_{\sigma\in\Sigma} \overline{\lim_{t\to\infty}} \left(\int_\Omega \exp\left[\int_0^t F\big(\Sigma_s(\omega)\big)\, ds \right] P_\sigma(d\omega) \right)^{1/t} < \infty, \tag{5.1.20}$$

then $\tilde{\Lambda}_P$ is tight.

(iii) Let $F : \Sigma \longrightarrow \mathbb{R}$ be a lower semi-continuous function which is bounded below, and suppose that there is a measurable $u : [0, \infty) \times \Sigma \longrightarrow [0, \infty)$ which satisfies (see the paragraph preceding Lemma 4.2.23)

$$u(t, \sigma) = \big[P_t u(0, \cdot)\big](\sigma) + \int_0^t \big[P_{t-s}\big(F u(s, \cdot)\big)\big](\sigma)\, ds.$$

Show that

$$u(t, \sigma) = \int_\Omega u\big(0, \Sigma_t(\omega)\big) \exp\left[\int_0^t F\big(\Sigma_s(\omega)\big)\, ds \right] P_\sigma(d\omega).$$

Finally, if $\{\sigma : F(\sigma) \leq M\} \subset\subset \Sigma$, $M \in [0, \infty)$, u is uniformly positive, and

$$\sup_{\sigma\in\Sigma} \overline{\lim_{t\to\infty}} \frac{1}{t} \log\big(u(t, \sigma)\big) < \infty,$$

conclude that $\tilde{\Lambda}_P$ is tight.

At least when dealing with processes whose paths are continuous, one often finds the function u by a localization procedure. Namely, one starts with a function F with compact level sets and seeks a non-decreasing, locally bounded sequence of functions $u_n \in \mathbf{D}$ which satisfy $u_n \geq 1$ and $L u_n = -F u_n$ on a sequence of open sets U_n which exhaust Σ; and one then takes u to be the limit of the u_n 's.

(iv) It is clear that $\tilde{\Lambda}_P \leq \Lambda_P$ (where Λ_P is defined in (4.2.21)) and therefore that $\Lambda_P^*(\nu) \leq \tilde{\Lambda}_P^*(\nu)$ and also that

$$\overline{\Lambda}_P^* \leq \overline{\tilde{\Lambda}}_P^*(\nu) \equiv \sup\left\{ \int_\Sigma V\, d\nu - \tilde{\Lambda}_P(V) : V \in B(\Sigma; \mathbb{R}) \right\}$$

for all $\nu \in \mathbf{M}_1(\Sigma)$ (cf. (4.2.22) and (4.2.37) for the notation here). Thus (cf. (4.2.36) and (4.2.38)), we see that $\overline{J}_P \leq \overline{\tilde{\Lambda}}_P^*$ and that, when $P(t, \sigma, \cdot)$ is FELLER-continuous, $J_P \leq \tilde{\Lambda}_P^*$. Check that the following line of reasoning leads to $\tilde{\Lambda}_P^* \leq \overline{J}_P$ and thence to

$$\tilde{\Lambda}_P^* = \overline{J}_P \quad \text{if } \overline{\tilde{\Lambda}}_P^* = \tilde{\Lambda}_P^*. \tag{5.1.21}$$

Let $V \in C_b(\Sigma;\mathbb{R})$ be given and define $\{P_t^V : t > 0\}$ accordingly as in Lemma 4.2.23. Given $\lambda > \tilde{\Lambda}_P(V)$, define

$$u_n(\sigma) = \int_0^n e^{-\lambda t}\,[P_t^V 1](\sigma)\,dt \quad \text{for } n \in \mathbb{Z}^+ \text{ and } \sigma \in \Sigma.$$

Show that $\inf_{n\in\mathbb{Z}^+} \inf_{\sigma\in\Sigma} u_n(\sigma) > 0$, $u_n \in \mathbf{D}$ (cf. the discussion preceding Lemma 4.2.31), and that

$$\lambda u_n - V u_n - L u_n = 1 - v_n \quad \text{where } v_n \equiv e^{-\lambda n}\big[P_n^V 1\big].$$

Next, check that

$$u_n \geq e^{-\lambda n} \int_{n-1}^n e^{-\lambda(t-n)}\big[P_t^V 1\big]\,dt \geq \exp\big[-(\|V\|_B + |\lambda|)\big] e^{-\lambda n}\big[P_n^V 1\big],$$

and therefore that $\sup_{n\in\mathbb{Z}^+}\big\|v_n/u_n\big\|_B < \infty$. Since $\lambda > \tilde{\Lambda}_P(V)$, conclude that $v_n/u_n \longrightarrow 0$ boundedly. After combining this with the preceding, one is led to

$$\overline{J}_P(\nu) \geq -\varliminf_{n\to\infty} \int_\Sigma \frac{Lu_n}{u_n}\,d\nu \geq \int_\Sigma V\,d\nu - \lambda, \quad \nu \in \mathbf{M}_1(\Sigma);$$

and from here it is an easy step to the desired conclusion.

Finally, by the same reasoning which just led to (5.1.21), prove that

$$\Lambda_P^* = \tilde{\Lambda}_P^* = J_P \quad \text{when } P(t,\sigma,\cdot) \text{ is FELLER-continuous.} \tag{5.1.22}$$

(iv) Formulate and verify the results in **(i)** through **(iv)** for the discrete-time setting.

5.2 A Little Ergodic Theory

Before attempting to develop lower bounds which will complement the upper bounds obtained in Section 5.1, we make a digression in which we will discuss a few essential facts from ergodic theory. Because it is not so readily available in standard texts, we will work in the continuous parameter setting.

We begin our discussion with the lovely Sunrise Lemma of F. RIESZ [**91**]. To understand both the name as well as the intuition behind what is going on, think about the distribution of light and shade in a (one-dimensional) mountainous region at precisely the moment when the sun comes up over the horizon. In the lemma, the sun is on the right, the set E is the region in the shade, and "$F(s)$ is the altitude at s."

5.2.1 Lemma. *Let $I = [a,b]$ be a non-empty compact interval and $F : I \longrightarrow \mathbb{R}$ a continuous function. Denote by E the set of $s \in I^\circ$ with the property that $F(t) > F(s)$ for some $t \in (s,b)$. Then E is an open subset of $\mathbb{R}$; and if $E \neq \emptyset$, then it is the union of countably many mutually disjoint open intervals (α,β) each of which has the property that $F(\beta) \geq F(\alpha)$.*

PROOF: Clearly, E is open in $\mathbb{R}$, and therefore all that we have to do is check that if (α,β) is a non-empty connected component of E then $F(\beta) \geq F(\alpha)$. To this end, suppose that $F(\beta) < F(\alpha)$ and set $A = \big(F(\alpha)+F(\beta)\big)/2$. Then $C \equiv \{s \in (\alpha,\beta) : F(s) = A\}$ is a non-empty, compact subset of (α,β). Let $\gamma = \max\{s : s \in C\}$, and observe that $F(t) < A$ for all $t \in (\gamma,\beta]$. In addition, since $\beta \notin E$, $F(t) \leq F(\beta) < A$ for every $t \in (\beta,b)$. Hence, $F(t) < A = F(\gamma)$ for all $t \in (\gamma,b)$, and therefore $\gamma \notin E$. However, $\gamma \in (\alpha,\beta) \subseteq E$; and so we have a contradiction. ∎

As a direct consequence of Lemma 5.2.1, we get the following sharp form of the HARDY–LITTLEWOOD Maximal Inequality [**58**].

5.2.2 Theorem. *Given a function $f \in L^1(\mathbb{R})$, define*

$$\tilde{f}(s) = \sup_{t>s} \frac{1}{t-s} \int_s^t \big|f(\xi)\big|\,d\xi, \quad s \in \mathbb{R}. \tag{5.2.3}$$

Then $s \in \mathbb{R} \longmapsto \tilde{f}(s) \in [0,\infty)$ is lower semi-continuous and

$$\Big|\big\{s : \tilde{f}(s) \geq \lambda\big\}\Big| \leq \frac{1}{\lambda} \int_{\{s:\tilde{f}(s)\geq\lambda\}} \big|f(s)\big|\,ds \leq \frac{1}{\lambda}\big\|f\big\|_{L^1(\mathbb{R})} \tag{5.2.4}$$

for all $\lambda \in (0,\infty)$. (We use $|\Gamma|$ to denote the LEBESGUE *measure of $\Gamma \subseteq \mathbb{R}$.) In particular, for all $p \in (1,\infty]$,*

$$\big\|\tilde{f}\big\|_{L^p(\mathbb{R})} \leq \frac{p}{p-1}\big\|f\big\|_{L^p(\mathbb{R})}. \tag{5.2.5}$$

PROOF: Without loss of generality, we will assume that $f \geq 0$.

Given $n \in \mathbb{Z}^+$ and $\lambda \in (0,\infty)$, set $I_n = [-n,n]$ and define

$$\tilde{f}_n(s) = \sup_{t\in(s,n]} \frac{1}{t-s}\int_s^t f(\xi)\,d\xi$$

and

$$F_{n,\lambda}(s) = \int_{-n}^{s} |f(\xi)|\,d\xi - \lambda(s+n)$$

for $s \in [-n,n)$. Clearly, $\{s \in I_n^\circ : \tilde{f}_n(s) > \lambda\}$ coincides with the set $E_{n,\lambda}$ in Lemma 5.2.1 corresponding to the function $F_{n,\lambda}$ on I_n. Moreover, by that lemma, we know that $E_{n,\lambda}$ is either empty or the countable union of mutually disjoint intervals (α,β) with the property that $\lambda(\beta-\alpha) \leq \int_\alpha^\beta f(\xi)\,d\xi$. Hence,

$$\lambda|E_{n,\lambda}| \leq \int_{E_{n,\lambda}} f(\xi)\,d\xi.$$

After letting $n \nearrow \infty$, one quickly concludes from the above that

$$\lambda\big|\{s : \tilde{f}(s) > \lambda\}\big| \leq \int_{\{s:\tilde{f}(s)>\lambda\}} f(\xi)\,d\xi, \quad \lambda \in (0,\infty);$$

and so (5.2.4) results from taking left limits in the preceding.

Once one knows (5.2.4), one can get (5.2.5) for $p \in (1,\infty)$ and bounded, non-negative $f \in L^1(\mathbb{R})$ by simply noting that

$$\begin{aligned}
\|\tilde{f}\|_{L^p(\mathbb{R})}^p &= p\int_{(0,\infty)} \lambda^{p-1}\big|\{s : \tilde{f}(s) \geq \lambda\}\big|\,d\lambda \\
&\leq p\int_{(0,\infty)} \lambda^{p-2}\left(\int_{\{\xi:\tilde{f}(\xi)\geq\lambda\}} f(\xi)\,d\xi\right)d\lambda \\
&= \frac{p}{p-1}\int_{\mathbb{R}} \tilde{f}(\xi)^{p-1} f(\xi)\,d\xi \leq \frac{p}{p-1}\|\tilde{f}\|_{L^p(\mathbb{R})}^{p-1}\|f\|_{L^p(\mathbb{R})},
\end{aligned}$$

where we have used HÖLDER's inequality in the last step. The derivation of the general result is now an easy limit argument. Since (5.2.5) is obvious when $p = \infty$, the proof is now complete. ∎

We are now ready to start doing ergodic theory. Let $(\Omega,\mathcal{B})$ be a measurable space. The family $\Theta = \{\theta_t : t \in [0,\infty)\}$ is called a **measurable, one-parameter semigroup of transformations on** $(\Omega,\mathcal{B})$ if $(t,\omega) \in [0,\infty)\times\Omega \longmapsto \theta_t(\omega) \in \Omega$ is $\mathcal{B}_{[0,\infty)}\times\mathcal{B}$-measurable function from $[0,\infty)\times\Omega$ into $(\Omega,\mathcal{B})$ and $\theta_{s+t} = \theta_s\circ\theta_t$ for all $s,\, t \in [0,\infty)$. A set $A \subseteq \Omega$ is said to be Θ-**invariant** if $A = \theta_t^{-1}A$, $t \in [0,\infty)$; and a measure $Q \in \mathbf{M}_1\big((\Omega,\mathcal{B})\big)$ is said to be Θ-**invariant** if $Q = Q\circ\theta_t^{-1}$, $t \in [0,\infty)$. We will use $\mathfrak{I}_\Theta$ and $\mathbf{M}_1^\Theta\big((\Omega,\mathcal{B})\big)$, respectively, to denote the Θ-invariant subsets $A \in \mathcal{B}$ and Θ-invariant measures $Q \in \mathbf{M}_1\big((\Omega,\mathcal{B})\big)$.

5.2.6 Theorem. (MAXIMAL ERGODIC INEQUALITY) *Let $(\Omega, \mathcal{B})$ be a measurable space and $\Theta = \{\theta_t : t \in [0,\infty)\}$ a measurable, one-parameter semigroup of transformations on $(\Omega, \mathcal{B})$. Then the set $\mathfrak{I}_\Theta$ is a sub-σ-algebra of $\mathcal{B}$. Next, given a measurable $f : \Omega \longrightarrow \mathbb{R}$, let Ω_f be the set of $\omega \in \Omega$ with the property that $\int_0^T |f(\theta_t\omega)|\,dt < \infty$ for every $T \in [0,\infty)$. Then $\Omega_f \in \mathcal{B}$, and $Q(\Omega_f) = 1$ for all $Q \in \mathbf{M}_1^\Theta((\Omega, \mathcal{B}))$ and $f \in L^1(Q)$. Finally, given a measurable $f : \Omega \longrightarrow \mathbb{R}$, define $f_T : \Omega \longrightarrow \mathbb{R}$ for $T \in (0,\infty)$ by*

$$f_T(\omega) = \begin{cases} \frac{1}{T}\int_0^T f(\theta_t\omega)\,dt & \text{if } \omega \in \Omega_f \\ 0 & \text{otherwise}. \end{cases}$$

Then $(T,\omega) \in (0,\infty) \times \Omega \longmapsto f_T(\omega) \in \mathbb{R}$ is measurable, $T \in (0,\infty) \longmapsto f_T(\omega) \in \mathbb{R}$ is continuous for each $\omega \in \Omega$, and, for every $Q \in \mathbf{M}_1^\Theta((\Omega, \mathcal{B}))$, one has that

$$Q\left(\{\omega : Mf(\omega) \geq \lambda\}\right) \leq \frac{1}{\lambda}\|f\|_{L^1(Q)}, \quad \lambda \in (0,\infty), \tag{5.2.7}$$

and

$$\|Mf\|_{L^p(Q)} \leq \frac{p}{p-1}\|f\|_{L^p(Q)}, \quad p \in (1,\infty], \tag{5.2.8}$$

where

$$Mf(\omega) \equiv \sup_{T\in(0,\infty)} |f_T(\omega)|, \quad \omega \in \Omega.$$

PROOF: The only thing that we need to do is check that (5.2.7) and (5.2.8) hold for bounded measurable $f : \Omega \longrightarrow [0,\infty)$. Let such an f be given; and, for $m \in \mathbb{Z}^+$ and $\omega \in \Omega$, define

$$M_m f(\omega) = \sup_{T\in(0,m]} f_T(\omega)$$

and

$$f_{m,\omega}(t) = \begin{cases} f(\theta_t\omega) & \text{if } t \in [0,m] \\ 0 & \text{if } t \in \mathbb{R}\setminus[0,m]. \end{cases}$$

It is then an easy matter to see that for $1 \leq m < n$ and $t \in [0, n-m]$,

$$M_m f(\theta_t\omega) \leq \tilde{f}_{n,\omega}(t).$$

Hence, by (5.2.4), for all $\lambda \in (0,\infty)$ and $1 \leq m < n$

$$\begin{aligned} &\lambda Q\left(\{\omega : M_m f(\omega) \geq \lambda\}\right) \\ &\quad = \frac{\lambda}{n-m}\int_0^{n-m} Q\left(\{\omega : M_m f(\theta_t\omega) \geq \lambda\}\right)dt \\ &\quad \leq \frac{\lambda}{n-m}\int_\Omega \left|\{t \in [0,\infty) : \tilde{f}_{n,\omega}(t) \geq \lambda\}\right| Q(d\omega) \\ &\quad \leq \frac{1}{n-m}\int_\Omega \left(\int_0^n f(\theta_t\omega)\,dt\right) Q(d\omega) = \frac{n}{n-m}\|f\|_{L^1(Q)}. \end{aligned}$$

By first letting $n \nearrow \infty$ and then $m \nearrow \infty$, one easily gets (5.2.7) from this. Similarly, if $p \in (1, \infty)$ and one uses (5.2.5), then one has that

$$\begin{aligned}
&\int_\Omega (M_m f(\omega))^p \, Q(d\omega) \\
&\quad \le \frac{1}{n-m} \int_\Omega \left(\int_0^{n-m} \left(\tilde{f}_{n,\omega}(t)\right)^p dt \right) Q(d\omega) \\
&\quad \le \frac{1}{n-m} \left(\frac{p}{p-1}\right)^p \int_\Omega \left(\int_0^{n} f(\theta_t \omega)^p \, dt \right) Q(d\omega) \\
&\quad = \frac{n}{n-m} \left(\frac{p}{p-1}\right)^p \|f\|_{L^p(Q)}^p;
\end{aligned}$$

from which (5.2.8) follows when $p \in (1, \infty)$. Since (5.2.8) is trivial for $p = \infty$, we are now done. ∎

The most familiar application of the Maximal Ergodic Inequality is the renowned Individual Ergodic Theorem.

5.2.9 Theorem. (ERGODIC THEOREM) *Referring to* Theorem *5.2.6, let* $Q \in \mathbf{M}_1^\Theta((\Omega, \mathcal{B}))$ *and* $f \in L^1(Q)$ *be given. Then* $f_T \longrightarrow E^Q[f|\mathfrak{I}_\Theta]$ *Q-almost surely and in* $L^1(Q)$*. Moreover, for each* $p \in (1, \infty)$ *and* $f \in L^p(Q)$,

$$\lim_{t\to\infty} \left\| \sup_{T>t} \left| f_T - E^Q\left[f|\mathfrak{I}_\Theta\right] \right| \right\|_{L^p(Q)} = 0. \tag{5.2.10}$$

PROOF: Because $\|f_T\|_{L^1(Q)} = \|f\|_{L^1(Q)}$ for $f \in L^1(Q)^+$, the first assertion reduces to checking that the convergence takes place Q-almost surely. Moreover, because of (5.2.8), (5.2.10) will follow by LEBESGUE's Dominated Convergence Theorem as soon as we know the Q-almost sure convergence result. Thus, all that we will do is show that $f_T \longrightarrow E^Q[f|\mathfrak{I}_\Theta]$ Q-almost surely for $f \in L^1(Q)$.

Let $\mathfrak{F}$ denote the space of $f \in L^1(Q)$ such that

$$\lim_{T\to\infty} f_T = E^Q\left[f|\mathfrak{I}_\Theta\right] \quad \text{(a.s., } Q\text{)}$$

Clearly $\mathfrak{F}$ is a linear subspace of $L^1(Q)$, and, as a consequence of (5.2.7), we know that it is also closed in $L^1(Q)$. Thus, we will know that $\mathfrak{F} = L^1(Q)$ once we show that $\mathfrak{F}$ contains a dense subspace of $L^2(Q)$. To this end, let L be linear span of functions $f = g - g \circ \theta_t$ as g runs over bounded measurable functions and t runs over $[0, \infty)$. Noting that $E^Q[g|\mathfrak{I}_\Theta] = E^Q[g \circ \theta_t|\mathfrak{I}_\Theta]$

(a.s., Q), we see that $E^Q\big[f|\mathfrak{I}_\Theta\big]$ can be taken to be 0 for such an f. At the same time,

$$f_T(\omega) = \frac{1}{T}\left[\int_0^t g\big(\theta_s\omega\big)\,ds - \int_T^{T+t} g\big(\theta_s\omega\big)\,ds\right] \longrightarrow 0 \quad \text{as } T \nearrow \infty.$$

Thus, $L \subseteq \mathfrak{F}$, and so it remains only to check that the perpendicular complement $L^\perp$ of L in $L^2(Q)$ is also contained in $\mathfrak{F}$. But, if $h \in L^\perp$, then $h = h \circ \theta_t$ (a.s., Q) for every $t \in [0,\infty)$. Hence, by FUBINI's Theorem, for each $T \in (0,\infty)$, $h_T = h$ (a.s., Q); and, because $T \in (0,\infty) \longrightarrow h_T(\omega)$ is continuous for Q-almost every $\omega \in \Omega$, this means that there is one Q-null set $\Lambda \in \mathcal{B}$ such that $h_T(\omega) = h(\omega)$ for all $T \in (0,\infty)$ and $\omega \notin \Lambda$. In particular, if $G \equiv \{\omega : \lim_{T\to\infty} h_T(\omega) \text{ exists}\}$, then $G \in \mathfrak{I}_\Theta$ and $\Lambda^c \subseteq G$. Thus, if

$$\overline{h}(\omega) = \begin{cases} \lim_{T\to\infty} h_T(\omega) & \text{for } \omega \in G \\ 0 & \text{for } \omega \notin G, \end{cases}$$

then $\overline{h}(\omega) = \overline{h}\big(\theta_t\omega\big)$ for all $(t,\omega) \in [0,\infty) \times \Omega$ and $\overline{h} = h$ (a.s., Q). In particular, $\overline{h}$ is itself $\mathfrak{I}_\Theta$-measurable, and so we conclude that

$$E^Q\big[h|\mathfrak{I}_\Theta\big] = E^Q\big[\overline{h}|\mathfrak{I}_\Theta\big] = \overline{h} = \lim_{T\to\infty} h_T \quad \text{(a.s., } Q). \quad \blacksquare$$

We now want to study the structure of the set $\mathbf{M}_1^\Theta\big((\Omega,\mathcal{B})\big)$. Clearly it is a convex subset of $\mathbf{M}_1\big((\Omega,\mathcal{B})\big)$. In the following lemma, we show that the extreme elements of $\mathbf{M}_1^\Theta\big((\Omega,\mathcal{B})\big)$ coincide with the **ergodic** elements of $\mathbf{M}_1^\Theta\big((\Omega,\mathcal{B})\big)$ (i.e., those $Q \in \mathbf{M}_1^\Theta\big((\Omega,\mathcal{B})\big)$ with the property that $Q(I) \in \{0,1\}$, $I \in \mathfrak{I}_\Theta$). We will use $\mathbf{EM}_1^\Theta\big((\Omega,\mathcal{B})\big)$ to denote the set of ergodic elements of $\mathbf{M}_1^\Theta\big((\Omega,\mathcal{B})\big)$.

5.2.11 Lemma. *If Q_1 and Q_2 are elements of $\mathbf{M}_1^\Theta\big((\Omega,\mathcal{B})\big)$, then either $Q_1 = Q_2$ on $\mathcal{B}$ or there is an $I \in \mathfrak{I}_\Theta$ for which $Q_2(I) \neq Q_1(I)$. In particular, distinct elements of $\mathbf{EM}_1^\Theta\big((\Omega,\mathcal{B})\big)$ are singular, and $\mathbf{EM}_1^\Theta\big((\Omega,\mathcal{B})\big)$ is precisely the set of extreme elements of $\mathbf{M}_1^\Theta\big((\Omega,\mathcal{B})\big)$.*

PROOF: To prove the first statement, suppose that Q_1, $Q_2 \in \mathbf{M}_1^\Theta\big((\Omega,\mathcal{B})\big)$ and that $Q_1(\Gamma) \neq Q_2(\Gamma)$ for some $\Gamma \in \mathcal{B}$. Set $f = \chi_\Gamma$ and define $g = \overline{\lim}_{T\to\infty} f_T$. Then g is a bounded $\mathfrak{I}_\Theta$-measurable function and

$$\int_\Omega g\,dQ_1 = Q_1(\Gamma) \neq Q_2(\Gamma) = \int_\Omega g\,dQ_2.$$

Thus, Q_1 differs from Q_2 on $\mathfrak{I}_\Theta$.

The singularity of distinct ergodic measures is an immediate consequence of the preceding. Furthermore, if Q is ergodic and

$$Q = \alpha Q_1 + (1-\alpha)Q_2$$

for some $\alpha \in (0,1)$ and $Q_1, Q_2 \in \mathbf{M}_1^\Theta\big((\Omega,\mathcal{B})\big)$, then it is clear that Q_1 coincides with Q_2 on $\mathfrak{I}_\Theta$ and therefore, by the preceding, on $\mathcal{B}$. Thus, Q is extreme in $\mathbf{M}_1^\Theta\big((\Omega,\mathcal{B})\big)$ if it is ergodic. Finally, if $Q \in \mathbf{M}_1^\Theta\big((\Omega,\mathcal{B})\big)$ is not ergodic, choose $I \in \mathfrak{I}_\Theta$ so that $\alpha = Q(I) \in (0,1)$ and set

$$Q_1(d\omega) = \frac{\chi_I(\omega)}{\alpha}\,Q(d\omega) \quad \text{and} \quad Q_2(d\omega) = \frac{\chi_{I^c}(\omega)}{1-\alpha}\,Q(d\omega).$$

It is obvious that Q_1 and Q_2 are distinct elements of $\mathbf{M}_1^\Theta\big((\Omega,\mathcal{B})\big)$, and therefore we see that Q cannot be an extreme element of $\mathbf{M}_1^\Theta\big((\Omega,\mathcal{B})\big)$. ∎

By general principles of convex analysis, the preceding leads one to suspect that all the elements of $\mathbf{M}_1^\Theta\big((\Omega,\mathcal{B})\big)$ ought to be recoverable from elements of $\mathbf{EM}_1^\Theta\big((\Omega,\mathcal{B})\big)$. Indeed, the recovery process ought to consist of taking (finite) convex combinations followed by a "limit procedure." The rest of this section is devoted to making these ideas precise, at least when additional structure is imposed. In particular, from now on we will be assuming that Ω is a Polish space and that $\mathcal{B} = \mathcal{B}_\Omega$. In addition, we will assume that Θ is a measurable semigroup of continuous transformations $\theta_t : \Omega \longrightarrow \Omega$. What we are going to show, under these assumptions, is the existence for every $Q \in \mathbf{M}_1^\Theta(\Omega) \equiv \mathbf{M}_1^\Theta\big((\Omega,\mathcal{B}_\Omega)\big)$ of a $\rho_Q \in \mathbf{M}_1\big(\mathbf{M}_1(\Omega)\big)$ with the properties that ρ_Q is concentrated on $\mathbf{EM}_1^\Theta(\Omega) \equiv \mathbf{EM}_1^\Theta\big((\Omega,\mathcal{B})\big)$ and

$$Q = \int_{\mathbf{M}_1^\Theta(\Omega)} R\,\rho_Q(dR). \tag{5.2.12}$$

5.2.13 Remark.

Notice that the existence of ρ_Q ought to be a "trivial" consequence of the following line of reasoning. Namely, since Ω is Polish, there is a **conditional probability distribution** $\omega \in \Omega \longmapsto Q_\omega \in \mathbf{M}_1(\Omega)$ of Q given $\mathfrak{I}_\Theta$ (abbreviated by c.p.d. of $Q|\mathfrak{I}_\Theta$). That is, $\omega \longmapsto Q_\omega$ is $\mathfrak{I}_\Theta$-measurable and

$$Q(I \cap \Gamma) = \int_I Q_\omega(\Gamma)\,Q(d\omega), \quad I \in \mathfrak{I}_\Theta \text{ and } \Gamma \in \mathcal{B}_\Omega.$$

(See Theorem 1.1.6 in [**104**] for the proof that $\omega \longmapsto Q_\omega$ exists.) Moreover, it is not hard to believe that when $Q \in \mathbf{M}_1^\Theta(\Omega)$ then $Q_\omega \in \mathbf{M}_1^\Theta(\Omega)$ for Q-almost every $\omega \in \Omega$. Thus, if we knew that $\omega \longmapsto Q_\omega$ were regular (i.e.,

$Q_\omega(I) = \chi_I(\omega)$, $(\omega, I) \in \Omega \times \mathfrak{I}_\Theta)$, then we would be done. Indeed, we would then know that $Q_\omega \in \mathbf{EM}_1^\Theta(\Omega)$ for Q-almost every $\omega \in \Omega$, and we could therefore take ρ_Q to be the distribution of $\omega \longmapsto Q_\omega$ under Q. Unfortunately (cf. Exercise 5.2.21 below) the σ-algebra $\mathfrak{I}_\Theta$ will nearly never be countably generated, and therefore it is not clear that $\omega \longmapsto Q_\omega$ can be chosen to be regular. Thus, some more work is required.

In order to handle the problem raised above, it will be useful to have specified a countable class $\mathfrak{F} \subseteq C_b(\Omega; \mathbb{R})$ which has the properties that $\mathfrak{F}$ determines weak convergence (i.e., $Q_n \Longrightarrow Q$ if and only if $\int_\Omega f\, dQ_n \longrightarrow \int_\Omega f\, dQ$ for every $f \in \mathfrak{F}$) and that it generates, under bounded pointwise convergence, all of the bounded measurable functions on Ω to $\mathbb{R}$. For example, one can take $\mathfrak{F}$ to be a countable dense subset of the space of bounded, $\hat{\rho}$-uniformly continuous functions from Ω to $\mathbb{R}$, where $\hat{\rho}$ is some totally bounded metric for Ω.

Next, for $f \in C_b(\Omega; \mathbb{R})$, let $\Omega(f)$ be the set of $\omega \in \Omega$ at which

$$f^*(\omega) \equiv \lim_{T\to\infty} f_T(\omega)$$

exists. Then it is clear that

$$\Omega_0 \equiv \bigcap_{f\in\mathfrak{F}} \Omega(f) \in \mathfrak{I}_\Theta.$$

Also, define Ω_{00} to be the set of $\omega \in \Omega$ for which

$$\left\{\frac{1}{T}\int_0^T \delta_{\theta_s\omega}\, ds : T \in (0,\infty)\right\}$$

converges in $\mathbf{M}_1(\Omega)$ to some limit δ^*_ω as $T \nearrow \infty$. Although it is not clear whether $\Omega_{00} \in \mathcal{B}_\Omega$, it is obvious that Ω_{00} is $\{\theta_t : t > 0\}$-invariant and that $\delta^*_\omega \in \mathbf{M}_1^\Theta(\Omega)$ for every $\omega \in \Omega_{00}$. In addition,

$$\Omega_{00} \subseteq \bigcap_{f\in C_b(\Omega;\mathbb{R})} \Omega(f) \subseteq \Omega_0,$$

and $f^*(\omega) = \int_\Omega f\, d\delta^*_\omega$ for all $\omega \in \Omega_{00}$ and $f \in C_b(\Omega; \mathbb{R})$. Finally, given $Q \in \mathbf{M}_1^\Theta(\Omega)$ and a c.p.d. $\omega \longrightarrow Q_\omega$ of $Q|\mathfrak{I}_\Theta$, set

$$\Omega_Q = \left\{\omega \in \Omega_0 : f^*(\omega) = \int_\Omega f\, dQ_\omega \text{ for all } f \in \mathfrak{F}\right\}$$

and note both that $\Omega_Q \in \mathfrak{I}_\Theta$ and that

$$\Omega_Q = \{\omega \in \Omega_{00} : \delta^*_\omega = Q_\omega\}.$$

5.2.14 Lemma. *The set* $\mathbf{M}_1^\Theta(\Omega)$ *is closed in* $\mathbf{M}_1(\Omega)$, *and* $Q \in \mathbf{M}_1^\Theta(\Omega)$ *is an element of* $\mathbf{EM}_1^\Theta(\Omega)$ *if and only if*

$$\int_{\Omega_0} \left(f^*(\omega) - \int_\Omega f\, dQ \right)^2 Q(d\omega) = 0$$

for every $f \in \mathfrak{F}$. *In particular,* $\mathbf{EM}_1^\Theta(\Omega) \in \mathcal{B}_{\mathbf{M}_1(\Omega)}$. *Moreover,* $Q(\Omega_Q) = 1$ *for each* $Q \in \mathbf{M}_1^\Theta(\Omega)$; *and therefore, for each* $Q \in \mathbf{M}_1^\Theta(\Omega)$, $Q_\omega \in \mathbf{M}_1^\Theta(\Omega)$ *for* Q*-almost every* $\omega \in \Omega$.

PROOF: Since $Q \in \mathbf{M}_1^\Theta(\Omega)$ if and only if

$$\int_\Omega f(\theta_t\omega)\, Q(d\omega) = \int_\Omega f(\omega)\, Q(d\omega) \quad \text{for all } t \in (0,\infty) \text{ and } f \in C_b(\Omega;\mathbb{R});$$

and because $f \circ \theta_t \in C_b(\Omega;\mathbb{R})$, $t \in (0,\infty)$, whenever $f \in C_b(\Omega;\mathbb{R})$, it is clear how to write $\mathbf{M}_1^\Theta(\Omega)$ as the intersection of closed sets $C(t,f)$, $(t,f) \in (0,\infty) \times C_b(\Omega;\mathbb{R})$.

To prove the characterization of $\mathbf{EM}_1^\Theta(\Omega)$, it is enough to show that the stated condition is sufficient. But, if $f^* = \int_\Omega f\, dQ$ (a.s., Q) for every $f \in \mathfrak{F}$, then $E^Q[f|\mathfrak{I}_\Theta]$ is Q-almost surely constant for every $f \in \mathfrak{F}$. Since the class of $f \in B(\Omega;\mathbb{R})$ which have this property is closed under bounded point-wise convergence, we see that $E^Q[f|\mathfrak{I}_\Theta]$ is Q-almost surely constant for every $f \in B(\Omega;\mathbb{R})$; and obviously, this is tantamount to the assertion that Q is ergodic.

Finally, if $Q \in \mathbf{M}_1^\Theta(\Omega)$, then the equality $Q(\Omega_Q) = 1$ is an immediate consequence of the Individual Ergodic Theorem together with the fact that, for each $f \in B(\Omega;\mathbb{R})$, $\omega \in \Omega \longmapsto \int_\Omega f\, dQ_\omega$ is a version of $E^Q[f|\mathfrak{I}_\Theta]$. ∎

5.2.15 Lemma. *For every* $Q \in \mathbf{M}_1^\Theta(\Omega)$, $Q_\omega = \delta^*_\omega \in \mathbf{EM}_1^\Theta(\Omega)$ *for* Q*-almost every* $\omega \in \Omega$. *In particular, if* $\Omega'_Q = \{\omega \in \Omega_Q : Q_\omega \in \mathbf{EM}_1^\Theta(\Omega)\}$, *then*

$$\Omega'_Q \in \mathfrak{I}_\Theta, \ \Omega'_Q \subseteq \Omega_{00}, \ \text{ and } Q = \int_{\Omega'_Q} \delta^*_\omega\, Q(d\omega).$$

PROOF: Note that

$$Q\Big(\{\omega : Q_\omega \notin \mathbf{EM}_1^\Theta(\Omega)\}\Big) \\ \le \sum_{f \in \mathfrak{F}} Q\left(\left\{\omega \in \Omega_Q : \int_{\Omega_0} \left(f^* - f^*(\omega)\right)^2 d\delta^*_\omega > 0\right\}\right).$$

At the same time, for each $f \in \mathfrak{F}$ and $\epsilon > 0$,

$$Q\left(\left\{\omega \in \Omega_Q : \int_{\Omega_0} \left(f^* - f^*(\omega)\right)^2 d\delta^*_\omega \geq \epsilon\right\}\right) \leq \frac{1}{\epsilon}\int_{\Omega_Q}\left(\int_{\Omega_0}\left(f^*(\omega') - f^*(\omega)\right)^2 \delta^*_\omega(d\omega')\right) Q(d\omega)$$

and

$$\begin{aligned}
&\int_{\Omega_Q}\left(\int_{\Omega_0}\left(f^*(\omega') - f^*(\omega)\right)^2 \delta^*_\omega(d\omega')\right) Q(d\omega) \\
&= \lim_{T\to\infty}\int_{\Omega_Q}\left(\int_{\Omega_0}\left(f_T(\omega') - f^*(\omega)\right)^2 \delta^*_\omega(d\omega')\right) Q(d\omega) \\
&= \lim_{T\to\infty}\lim_{S\to\infty}\int_{\Omega_Q}\left(\frac{1}{S}\int_0^S \left(f_T(\theta_s\omega) - f^*(\omega)\right)^2 ds\right) Q(d\omega) \\
&= \lim_{T\to\infty}\lim_{S\to\infty}\frac{1}{S}\int_0^S \left(\int_{\Omega_0}\left(f_T(\theta_s\omega) - f^*(\omega)\right)^2 Q(d\omega)\right) ds \\
&= \lim_{T\to\infty}\int_{\Omega_0}\left(f_T(\omega) - f^*(\omega)\right)^2 Q(d\omega) = 0.
\end{aligned}$$

In the preceding, we have used the fact that $f_T \in C_b(\Omega;\mathbb{R})$ in order to pass from the second to the third lines, and we have used $\left(\chi_{\Omega_0} f^*\right) \circ \theta_s = \chi_{\Omega_0} f^*$, $s \in [0,\infty)$, in the passage to the last line. ∎

Clearly, the preceding shows that $\omega \longmapsto Q_\omega$ admits a regular version; and therefore, by the reasoning at the end of Remark 5.2.13, we have the following result as an immediate consequence of Lemma 5.2.15.

5.2.16 Theorem. (Ergodic Decomposition Theorem) *Let Ω be a Polish space and $\Theta = \{\theta_t : t \in [0,\infty)\}$ a measurable semigroup of continuous transformations on Ω. Then, for each $Q \in \mathbf{M}_1^\Theta(\Omega)$, there is a $\rho_Q \in \mathbf{M}_1\big(\mathbf{M}_1(\Omega)\big)$ with the properties that $\rho_Q\big(\mathbf{EM}_1^\Theta(\Omega)\big) = 1$ and (5.2.12) holds.*

Before closing this section, we record what our results look like in the case when $\Theta = \{\theta_t : t \in \mathbb{R}\}$ is a **measurable group of transformations** (i.e., $\theta_{s+t} = \theta_s \circ \theta_t$ for all s, $t \in \mathbb{R}$) on Ω. Note that invariance of measures or functions under Θ is equivalent to invariance under either of the semigroups $\Theta^+ \equiv \{\theta_t : t \in [0,\infty)\}$ or $\Theta^- \equiv \{\theta_{-t} : t \in [0,\infty)\}$. Thus, by treating Θ^+ and Θ^- separately, one sees that for every $Q \in \mathbf{M}_1^\Theta\big((\Omega,\mathcal{B})\big)$ and $f \in L^1(Q)$,

$$Q\left(\left\{\omega : \sup_{T\in(0,\infty)} \frac{1}{2T}\int_{-T}^{T} \left|f(\theta_t\omega)\right| dt \geq \lambda\right\}\right) \leq \frac{2}{\lambda}\|f\|_{L^1(Q)} \tag{5.2.17}$$

for $\lambda \in (0,\infty)$,

$$\left[\int_\Omega \sup_{T\in(0,\infty)} \left(\frac{1}{2T}\int_{-T}^{T} \left|f(\theta_t\omega)\right| dt\right)^p Q(d\omega)\right]^{1/p} \leq \frac{2p}{p-1}\|f\|_{L^p(Q)} \tag{5.2.18}$$

for $p \in (1,\infty)$,

$$\frac{1}{2T}\int_{-T}^{T} f(\theta_t\omega)\,dt \longrightarrow E^Q\left[f\middle|\mathfrak{I}_\Theta\right](\omega) \tag{5.1.19}$$

both Q-almost surely and in $L^1(Q)$, and

$$\lim_{t\to\infty}\left[\int_\Omega \sup_{T\geq t}\left|\frac{1}{2T}\int_{-T}^{T} f(\theta_s\omega)\,ds - E^Q\left[f\middle|\mathfrak{I}_\Theta\right](\omega)\right|^p Q(d\omega)\right]^{1/p} = 0 \tag{5.2.20}$$

if $p \in (1,\infty)$ and $f \in L^p(Q)$. Finally, when Ω is a Polish space and the θ_t 's are continuous, then the Ergodic Decomposition Theorem again applies and yields (5.2.12) with a ρ_Q which is concentrated on the ergodic elements of $\mathbf{M}_1^\Theta(\Omega)$.

5.2.21 Exercise.

As was mentioned in Remark 5.2.13, the σ-algebra $\mathfrak{I}_\Theta$ is hardly ever countably generated. To see why this is the case, assume that $\Theta = \{\theta_t : t \in \mathbb{R}\}$ is a measurable group of transformations on $(\Omega,\mathcal{B})$ with the property that every **orbit** $[\omega]_\Theta \equiv \{\theta_t\omega : t \in \mathbb{R}\}$, $\omega \in \Omega$, is an element of $\mathcal{B}$ and that there exists a $Q \in \mathbf{EM}_1^\Theta\big((\Omega,\mathcal{B})\big)$ such that $Q\big([\omega]_\Theta\big) = 0$ for every $\omega \in \Omega$. Under these circumstances, it is impossible for $\mathfrak{I}_\Theta$ to be countably generated. Indeed, suppose that $\mathfrak{I}_\Theta = \sigma\big(\{A_\ell\}_1^\infty\big)$. Choose $\{B_\ell\}_1^\infty$ so that

$$B_\ell = \begin{cases} A_\ell & \text{if } Q(A_\ell) = 1 \\ A_\ell^c & \text{if } Q(A_\ell) = 0. \end{cases}$$

Show that $C \equiv \bigcap_{\ell=1}^\infty B_\ell = [\omega]_\Theta$ for some $\omega \in \Omega$, and conclude that $1 = Q(C) = Q\big([\omega]_\Theta\big) = 0$. In particular, this rules out the possibility that $\mathfrak{I}_\Theta$ is countably generated. For a simple example of such a situation, take Ω to be the 2-torus $\mathbf{S}^1 \times \mathbf{S}^1$ and $\{\theta_t : t \in [0,\infty)\}$ to be the flow generated by the vector $\frac{\partial}{\partial x} + \gamma\frac{\partial}{\partial y}$ where γ is an irrational number. Check that all the orbits are then $\mathfrak{F}_\sigma$ subsets of Ω and that the normalized LEBESGUE measure on Ω is an ergodic, invariant measure which assigns measure 0 to each of these orbits.

5.2.22 Exercise.

For the sake of completeness, work out the theory developed in this section for the case of a discrete 1-parameter semigroup $\{\theta_n : n \in \mathbb{Z}^+\}$. Of course, since $\theta_n = \theta^n$ where $\theta \equiv \theta_1$, the appropriate notions of invariance are simply that $Q = Q \circ \theta^{-1}$ and $f = f \circ \theta^{-1}$.

(i) From the HARDY–LITTLEWOOD inequality, derive

$$\left|\left\{m \in \mathbb{Z}^+ : \sup_{n \in \mathbb{Z}^+} \frac{1}{n} \sum_{\ell=m}^{m+n-1} |a_\ell| \geq \lambda\right\}\right| \leq \frac{1}{\lambda} \sum_1^\infty |a_n| \tag{5.2.23}$$

for all $\lambda \in (0, \infty)$ and any sequence $\{a_n\}_1^\infty$. (Here we use $|\Gamma|$ to denote the LEBESGUE measure of $\Gamma \subseteq \mathbb{Z}^+$; in other words, the cardinality of Γ.)

(ii) Knowing **(i)**, prove that for any θ-invariant $Q \in \mathbf{M}_1\big((\Omega, \mathcal{B})\big)$ and any $f \in L^1(Q)$,

$$Q\left(\left\{\omega : \sup_{n \in \mathbb{Z}^+} \frac{1}{n} \sum_{m=1}^{n} \big|f(\theta^m \omega)\big| \geq \lambda\right\}\right) \leq \frac{1}{\lambda} \|f\|_{L^1(Q)} \tag{5.2.24}$$

for $\lambda \in (0, \infty)$,

$$\left[\int_\Omega \sup_{n \in \mathbb{Z}^+} \left(\frac{1}{n} \sum_{m=1}^{n} \big|f(\theta^m \omega)\big|\right)^p Q(d\omega)\right]^{1/p} \leq \frac{p}{p-1} \|f\|_{L^p(Q)} \tag{5.2.25}$$

for $p \in (1, \infty)$,

$$\frac{1}{n} \sum_{m=1}^{n} f(\theta^m \omega) \longrightarrow E^Q\big[f \big| \mathfrak{I}_\theta\big](\omega) \quad (\text{a.s.}, Q) \text{ and in } L^1(Q), \tag{5.2.26}$$

and

$$\lim_{m \to \infty} \left[\int_\Omega \sup_{n \geq m} \left|\frac{1}{n} \sum_{k=1}^{n} f(\theta^k \omega) - E^Q\big[f \big| \mathfrak{I}_\theta\big](\omega)\right|^p Q(d\omega)\right]^{1/p} = 0 \tag{5.2.27}$$

if $p \in (1, \infty)$ and $f \in L^p(Q)$.

(iii) Assuming that Ω is a Polish space and that θ is continuous, state and prove the appropriate version of the Ergodic Decomposition Theorem (i.e., Theorem 5.2.16).

5.2.28 Exercise.

Let $\Pi(\sigma,\cdot)$ be a transition probability function on the measurable space $(\Sigma,\mathcal{F})$ and define the operator $[\Pi\phi](\sigma) = \int_\Sigma \phi(\tau)\,\Pi(\sigma,d\tau)$, $\sigma\in\Sigma$, for $\phi\in B((\Sigma,\mathcal{F});\mathbb{R})$. Denote by $B_\Pi((\Sigma,\mathcal{F});\mathbb{R})$ the space of $\phi\in B((\Sigma,\mathcal{F});\mathbb{R})$ which are Π-invariant (i.e., $\phi=\Pi\phi$), and let $\mathbf{M}_1^\Pi((\Sigma,\mathcal{F}))$ be the space of Π-invariant $\mu\in\mathbf{M}_1((\Sigma,\mathcal{F}))$ (i.e., $\mu=\mu\Pi\equiv\int_\Sigma\Pi(\sigma,\cdot)\,\mu(d\sigma)$).

(**i**) Prove that, for any $\mu\in\mathbf{M}_1^\Pi((\Sigma,\mathcal{F}))$ and $\phi\in L^1(\mu)$,

$$\mu\left(\left\{\sigma:\ \sup_{n\in\mathbb{Z}^+}\frac{1}{n}\sum_{m=1}^n[\Pi^m\phi](\sigma)\geq\lambda\right\}\right)\leq\frac{1}{\lambda}\|\phi\|_{L^1(\mu)} \tag{5.2.29}$$

for $\lambda\in(0,\infty)$ and

$$\left[\int_\Sigma\sup_{n\in\mathbb{Z}^+}\left|\frac{1}{n}\sum_{m=1}^n[\Pi^m\phi](\sigma)\right|^p\mu(d\sigma)\right]^{1/p}\leq\frac{p}{p-1}\|\phi\|_{L^p(\mu)} \tag{5.2.30}$$

for $p\in(1,\infty]$.

(**ii**) Next, show that for each $\mu\in\mathbf{M}_1^\Pi((\Sigma,\mathcal{F}))$ there is a unique bounded linear operator $E_\mu: L^1(\mu)\longrightarrow L^1(\mu)$ with the property

$$\frac{1}{n}\sum_{m=1}^n[\Pi^m\phi](\sigma)\longrightarrow[E_\mu\phi](\sigma)\quad\mu\text{-almost surely and in } L^1(\mu).$$

Show that $E_\mu^2=E_\mu$, $E_\mu\phi\geq0$ if $\phi\geq0$, and $E_\mu\phi=\phi$ (a.s., μ) if $\phi\in B_\Pi((\Sigma,\mathcal{F});\mathbb{R})$. In particular, conclude that E_μ is a contraction on $L^p(\mu)$ for every $p\in[1,\infty]$. Finally, show that

$$\lim_{m\to\infty}\int_\Sigma\sup_{n\geq m}\left|\frac{1}{n}\sum_{\ell=1}^n[\Pi^\ell\phi](\sigma)-[E_\mu\phi](\sigma)\right|^p\mu(d\sigma)=0$$

for $p\in(1,\infty)$ and $\phi\in L^p(\mu)$.

(**iii**) Call an element μ of $\mathbf{M}_1^\Pi((\Sigma,\mathcal{F}))$ Π-**ergodic** if

$$\phi=\int_\Sigma\phi\,d\mu\quad\text{(a.s., }\mu)\quad\text{for each }\phi\in B_\Pi((\Sigma,\mathcal{F});\mathbb{R}).$$

Show that two Π-ergodic elements of $\mathbf{M}_1^\Pi((\Sigma,\mathcal{F}))$ are either equal or singular.

(iv) Set $\Omega = \Sigma^{\mathbb{N}}$, $\mathcal{B} = \mathcal{F}^{\mathbb{N}}$, and let $\{P_\sigma : \sigma \in \Sigma\}$ be the MARKOV family of probability measures on $(\Omega, \mathcal{B})$ whose transition function is $\Pi(\sigma, \cdot)$. Given $\mu \in \mathbf{M}_1\big((\Sigma, \mathcal{F})\big)$, set $P_\mu = \int_\Sigma P_\sigma \, \mu(d\sigma)$, and check that P_μ is invariant under the shift $\theta : \Omega \longrightarrow \Omega$ given by $(\theta\omega)_n = \omega_{n+1}$, $n \in \mathbb{Z}^+$, if and only if $\mu \in \mathbf{M}_1^\Pi\big((\Sigma, \mathcal{B})\big)$. Also, show that $\mu \in \mathbf{M}_1^\Pi\big((\Sigma, \mathcal{F})\big)$ is Π-ergodic if and only if $P_\mu \in \mathbf{M}_1^\Theta\big((\Omega, \mathcal{B})\big)$ is ergodic for θ.

5.2.31 Exercise.

Let $\Theta = \{\theta_t : t \in [0, \infty)\}$ be a measurable semigroup of transformations on the measurable space $(\Omega, \mathcal{F})$, and assume that there is a sub σ-algebra $\mathcal{F}_0 \subseteq \mathcal{F}$ with the property that $\bigcup_{t \in [0,\infty)} \theta_t^{-1} \mathcal{F}_0$ generates the whole of $\mathcal{F}$. Next, for each $T \in [0, \infty)$, let $\mathcal{F}_T$ and $\mathcal{F}^T$ be the σ-algebras generated by $\bigcup_{t \in [0,T]} \theta_t^{-1} \mathcal{F}_0$ and $\bigcup_{t \in [T,\infty)} \theta_t^{-1} \mathcal{F}_0$, respectively. Finally, define the **tail** σ-algebra $\mathcal{T} = \bigcap_{T \in [0,\infty)} \mathcal{F}^T$.

(i) Given any $f \in B\big((\Omega, \mathcal{F}); \mathbb{R}\big)$, set

$$f^*(\omega) \equiv \overline{\lim_{t \to \infty}} \, f_t(\omega) = \overline{\lim_{t \to \infty}} \, \frac{1}{t} \int_0^t f(\theta_s \omega) \, ds.$$

When f is $\mathcal{F}_T$-measurable for some $T \in [0, \infty)$, show that the function f^* is $\mathcal{T}$-measurable. Next, assuming that $Q \in \mathbf{M}_1^\Theta\big((\Omega, \mathcal{F})\big)$ and using $\overline{\mathcal{T}}^Q$ to denote the Q-completion of $\mathcal{T}$, show that f^* is $\overline{\mathcal{T}}^Q$-measurable for every $f \in B\big((\Omega, \mathcal{F}); \mathbb{R}\big)$.

(ii) Using (i), show that if $Q \in \mathbf{M}_1^\Theta\big((\Omega, \mathcal{F})\big)$, then $\mathfrak{I}_\Theta \subseteq \overline{\mathcal{T}}^Q$; and conclude that Q is ergodic if $Q(A) \in \{0, 1\}$ for every $A \in \mathcal{T}$.

5.3 The General Symmetric Markov Case

Our first application of the results obtained in Section 5.1 will be to the large deviation theory for the empirical distribution of the position of a symmetric MARKOV process. More precisely, let Σ, $P(t,\sigma,\cdot)$, and the associated MARKOV family $\{P_\sigma : \sigma \in \Sigma\} \subseteq \mathbf{M}_1\big((\Omega,\mathcal{B})\big)$ be as they were in Section 4.4; and define

$$\text{(5.3.1)} \qquad \mathbf{L}_t(\omega) = \overline{\lambda}_{[0,t]} \circ \Big(\Sigma.(\omega)|_{[0,t]}\Big)^{-1}, \quad (t,\omega) \in (0,\infty) \times \Omega,$$

as in Remark 4.2.2. Next, assume that there is a $P(t,\sigma,\cdot)$-reversing measure $m \in \mathbf{M}_1(\Sigma)$, and define the DIRICHLET form $\mathcal{E}$ and the associated functions $\Lambda_{\mathcal{E}} : B(\Sigma;\mathbb{R}) \longrightarrow \mathbb{R}$ and $J_{\mathcal{E}} : \mathbf{M}_1(\Sigma) \longrightarrow [0,\infty]$ as we did in the final part of Section 4.2 (cf. especially (4.2.47), (4.2.51), and (4.2.49)). Finally, set $P_m = \int_\Sigma P_\sigma \, m(d\sigma)$.

5.3.2 Lemma. *If $J_{\mathcal{E}}$ is lower semi-continuous, then*

$$\text{(5.3.3)} \quad J_{\mathcal{E}}(\mu) = \sup\left\{ \int_\Sigma V\,d\mu - \Lambda_{\mathcal{E}}(V) : V \in C_{\mathrm{b}}(\Sigma;\mathbb{R}) \right\}, \quad \mu \in \mathbf{M}_1(\Sigma),$$

and

$$\text{(5.3.4)} \qquad \varlimsup_{t\to\infty} \frac{1}{t} \log\Big[P_m\big(\mathbf{L}_t \in C\big)\Big] \le -\inf_C J_{\mathcal{E}}$$

for all $C \subset\subset \mathbf{M}_1(\Sigma)$. Moreover, if, in addition, $J_{\mathcal{E}}$ is good (or, equivalently, $\Lambda_{\mathcal{E}}$ is tight), then (5.3.4) *holds for every closed $C \subseteq \mathbf{M}_1(\Sigma)$.*

PROOF: In (**ii**) of Exercise 4.2.63, we saw that $J_{\mathcal{E}}$ is convex. Thus, by Theorem 2.2.15 and (4.2.51), if $J_{\mathcal{E}}$ is lower semi-continuous then (5.3.3) follows; and so, by the results in Section 5.1 (in particular part (**i**) of Exercise 5.1.12), all that we have to do is check that

$$\Lambda_{\mathcal{E}}(V) \ge \varlimsup_{t\to\infty} \frac{1}{t} \log\left(\int_\Omega \exp\left[\int_0^t V\big(\Sigma_s(\omega)\big)\,ds\right] P_m(d\omega)\right)$$

for $V \in C_{\mathrm{b}}(\Sigma;\mathbb{R})$. But, because $m \in \mathbf{M}_1(\Sigma)$, it is easy to see that

$$\begin{aligned}\|\overline{P}_t^V\|_{L^2(m)\to L^2(m)} &\ge \|P_t^V 1\|_{L^2(m)} \\ &= \int_\Omega \exp\left[\int_0^t V\big(\Sigma_s(\omega)\big)\,ds\right] P_m(d\omega). \quad \blacksquare\end{aligned}$$

We now want to show that, under reasonable conditions, one can prove the complementary lower bound. The approach which we are going to adopt is very reminiscent of the one which we used in the original proof that we gave in Section 1.2 of the classical CRAMÈR Theorem for real-valued random variables. That is, we will force certain ergodic behavior by the introduction of an appropriate RADON–NIKODYM factor and will get our lower bound by estimating the size of the factor which we have introduced. However, in order to carry out this program, we need to make the following mild assumption.

(E) If $\{Q_T : T > 0\} \subseteq \mathbf{M}_1((\Omega, \mathcal{B}))$ is consistent in the sense that $Q_{T_2} = Q_{T_1}$ on $\mathcal{B}_{T_1}$ for all $0 \le T_1 < T_2 < \infty$, then there exists a unique $Q \in \mathbf{M}_1((\Omega, \mathcal{B}))$ such that $Q = Q_T$ on $\mathcal{B}_T$ for each $T \in [0, \infty)$.

Note that (**E**) holds if Ω is a Polish space, $\mathcal{B} = \mathcal{B}_\Omega$, each $\mathcal{B}_t$ is countably generated, and $\mathcal{B}$ is generated by $\bigcup_{t\ge 0} \mathcal{B}_t$. (Cf. Theorem 1.1.10 in [**104**].)

5.3.5 Lemma. *Let $u \in \mathbf{D} \cap B(\Sigma; [1, \infty))$, set $V_u = -\frac{Lu}{u}$, define*

$$P_u(t, \sigma, \Gamma) = \frac{1}{u(\sigma)} \int_\Gamma u(\tau)\, P^{V_u}(t, \sigma, d\tau)$$

for $(t, \sigma) \in (0, \infty) \times \Sigma$, and set $\Gamma \in \mathcal{B}_\Sigma$ and

$$X_u(t, \omega) = \frac{u(\Sigma_t(\omega))}{u(\Sigma_0(\omega))} \exp\left[\int_0^t V_u(\Sigma_s(\omega))\, ds\right]$$

for $(t, \omega) \in [0, \infty) \times \Omega$. (See Lemma 4.2.23 and Theorem 4.2.25 for the notation here.) Then $P_u(t, \sigma, \cdot)$ is a transition probability function; and, for every $\sigma \in \Sigma$, $(X_u(t), \mathcal{B}_t, P_\sigma)$ is a non-negative martingale with mean-value 1. Moreover, for each $\sigma \in \Sigma$, there is a unique $P^u_\sigma \in \mathbf{M}_1((\Omega, \mathcal{B}))$ satisfying

$$P^u_\sigma(A) = \int_A X_u(t, \omega)\, P_\sigma(d\omega), \quad t \in [0, \infty) \text{ and } A \in \mathcal{B}_t.$$

In fact, the family $\{P^u_\sigma : \sigma \in \Sigma\}$ is measurable and, for each $\sigma \in \Sigma$,

$$P^u_\sigma(\{\omega : \Sigma_{s+t}(\omega) \in \Gamma\} | \mathcal{B}_s) = P_u(t, \Sigma_s, \Gamma) \quad (a.s., P^u_\sigma) \tag{5.3.6}$$

for all $s, t \in (0, \infty)$ and $A \in \mathcal{B}_s$. Finally, if

$$m_u(d\sigma) \equiv \left(\frac{u(\sigma)}{\|u\|_{L^2(m)}}\right)^2 m(d\sigma), \tag{5.3.7}$$

then m_u is a reversing measure for $P_u(t, \sigma, \cdot)$.

PROOF: We first check that $P_u(t, \sigma, \cdot)$ is a transition probability function. To this end, note that

$$P_u(t, \sigma, \Gamma) = \frac{1}{u(\sigma)} \left[P_t^{V_u}(u\chi_\Gamma) \right](\sigma).$$

Thus, the measurability of

$$(t, \sigma) \in (0, \infty) \times \Sigma \longmapsto P_u(t, \sigma, \cdot)$$

as well as the CHAPMAN–KOLMOGOROV equation are immediate. In addition, since $u = P_t^{V_u} u$ (cf. the proof of Lemma 4.2.35), it is clear that $P_u(t, \sigma, \Sigma) = 1$.

We next show that, for each $\sigma \in \Sigma$ and $\Gamma \in \mathcal{B}_\Sigma$,

$$\tag{5.3.8} \begin{aligned} &\int_A X_u(s+t, \omega) \chi_\Gamma\left(\Sigma_{s+t}(\omega)\right) P_\sigma(d\omega) \\ &\qquad = \int_A X_u(s, \omega) P_u\left(t, \Sigma_s(\omega), \Gamma\right) P_\sigma(d\omega) \end{aligned}$$

for $s,\, t \in (0, \infty)$ and $A \in \mathcal{B}_s$. Indeed, by the MARKOV property combined with (4.2.26),

$$\begin{aligned} &\int_A X_u(s+t, \omega) \chi_\Gamma\left(\Sigma_{s+t}(\omega)\right) P_\sigma(d\omega) \\ &\qquad = \int_A \frac{X_u(s, \omega)}{u(\Sigma_s(\omega))} \left[P_t^{V_u}(u\chi_\Gamma) \right] \left(\Sigma_s(\omega)\right) P_\sigma(d\omega), \end{aligned}$$

which is equivalent to (5.3.8). By taking $\Gamma = \Sigma$ in (5.3.8), we get the asserted martingale property; and therefore, by (**E**), the existence and uniqueness of P_σ^u have also been established. Moreover, the measurability of $\sigma \in \Sigma \longmapsto P_\sigma^u$ is a trivial consequence of the expression for P_σ^u on each of the $\mathcal{B}_t$'s, and (5.3.6) follows easily from (5.3.8).

Finally, to see that m_u is reversing for $P_u(t, \sigma, \cdot)$, note that for $\phi,\, \psi \in B(\Sigma; \mathbb{R})$

$$\begin{aligned} &\|u\|_{L^2(m)}^2 \int_{\Sigma^2} \phi(\sigma)\psi(\tau)\, P_u(t, \sigma, d\tau) m_u(d\sigma) \\ &= \int_{\Sigma^2} (u\phi)(\sigma)(u\psi)(\tau)\, P^{V_u}(t, \sigma, d\tau) m(d\sigma) = \left(u\phi, P_t^{V_u}(u\psi)\right)_{L^2(m)}. \end{aligned}$$

Since, by Lemma 4.2.50, $\overline{P}_t^{V_u}$ is self-adjoint on $L^2(m)$, it follows that the first expression in the above is symmetric in ϕ and ψ. ∎

5.3.9 Lemma. *Assume that $J_{\mathcal{E}}(\mu) = 0$ only if $\mu = m$. Then for every $u \in \mathbf{D} \cap B(\Sigma; [1,\infty))$ and every τ-open neighborhood (cf. the discussion preceding* Lemma *3.2.19) $G \in \mathcal{B}_{\mathbf{M}_1(\Sigma)}$ of m_u*

$$\lim_{t\to\infty} \int_{\{\omega:\, \mathbf{L}_t(\omega)\in G\}} X_u(t,\omega)\, P_\sigma(d\omega) = 1 \quad \textit{in } m\textit{-measure.}$$

PROOF: Note that it suffices to check that if $P^u \equiv \int_\Sigma P^u_\sigma\, m_u(d\sigma)$, then $P^u(\{\omega : \mathbf{L}_t(\omega) \in G\}) \longrightarrow 1$ as $t \to \infty$. Furthermore, since P^u is time-shift invariant, this latter statement will follow from the Individual Ergodic Theorem once we show that P^u is ergodic relative to time-shift. Thus, all that we have to do is show that if $\{t_n\}_1^\infty \subseteq [0,\infty)$, $F \in B(\Sigma^{\mathbb{Z}^+}; \mathbb{R})$, and

$$\Phi_t(\omega) \equiv F\left(\Sigma_{t_1+t}(\omega), \dots, \Sigma_{t_n+t}(\omega), \dots\right), \quad t \in [0,\infty),$$

then Φ_0 is P^u-almost surely constant if, for each $t \in (0,\infty)$, $\Phi_t = \Phi_0$ P^u-almost surely.

We begin by showing that if $\phi \in B(\Sigma; \mathbb{R})$ satisfies

$$\phi = \int_\Sigma \phi(\tau)\, P_u(t,\cdot,d\tau) \quad (\text{a.s., } m_u)$$

for each $t \in (0,\infty)$, then ϕ is m_u-almost surely constant. In fact, given such a ϕ, we can use symmetry to check that

$$\iint\limits_{\Sigma^2} \left(\phi(\tau) - \phi(\sigma)\right)^2 P_u(t,\sigma,d\tau) m_u(d\sigma) = 0, \quad t \in (0,\infty).$$

Since, for each $t \in (0,\infty)$, $P_u(t,\sigma,d\tau)m_u(d\sigma)$ is bounded above and below by constant positive multiples of $P(t,\sigma,d\tau)m(d\sigma)$, it follows from (4.2.54) that $\mathcal{E}(\phi,\phi) = 0$. But, this means that $J_{\mathcal{E}}(\mu) = 0$, where

$$\mu(d\sigma) = \left(\frac{\phi(\sigma)}{\|\phi\|_{L^2(m)}}\right)^2 m(d\sigma),$$

and therefore, by hypothesis, ϕ is m-almost surely constant.

Returning to the ergodicity question about P^u, suppose that $\Phi_t = \Phi_0$ P^u-almost surely for each $t \in (0,\infty)$. Set $\phi(\sigma) = \int_\Omega \Phi_0(\omega)\, P^u_\sigma(d\omega)$, and observe that for all $t \in (0,\infty)$ and m_u-almost every $\sigma \in \Sigma$

$$\phi(\sigma) = \int_\Omega \Phi_t(\omega)\, P^u_\sigma(d\omega) = \int_\Sigma \phi(\tau)\, P_u(t,\sigma,d\tau).$$

Thus, by the preceding, ϕ is m_u-almost surely constant. But this means that, for any $t \in (0,\infty)$ and $A \in \mathcal{B}_t$,

$$\begin{aligned}\int_A \Phi_0(\omega)\,P^u(d\omega) &= \int_A \Phi_t(\omega)\,P^u(d\omega)\\ &= \int_A \phi\big(\Sigma_t(\omega)\big)\,P^u(d\omega) = P^u(A)\int_\Omega \Phi_0(\omega)\,P^u(d\omega);\end{aligned}$$

and clearly this leads to the conclusion that

$$\int_\Omega \Phi_0(\omega)^2\,P^u(d\omega) = \left(\int_\Omega \Phi_0(\omega)\,P^u(d\omega)\right)^2.$$

In other words, Φ_0 must be P^u-almost surely constant. ∎

5.3.10 Theorem. *Assume that $J_{\mathcal{E}}(\mu) = 0$ only if $\mu = m$, and let $\nu \in \mathbf{M}_1(\Sigma)$ have the property that, for some $T \in [0,\infty)$, νP_T is not singular to m. Then for every τ-open set $G \in \mathcal{B}_{\mathbf{M}_1(\Sigma)}$*

$$\varliminf_{t\to\infty} \frac{1}{t}\log\Big(P_\nu\big(\{\omega : \mathbf{L}_t(\omega) \in G\}\big)\Big) \geq -\inf_G J_{\mathcal{E}}. \tag{5.3.11}$$

Hence, if, in addition, $J_{\mathcal{E}}$ is a good rate function, then

$$\begin{aligned}-\inf_{\Gamma^\circ} J_{\mathcal{E}} &\leq \varliminf_{t\to\infty} \frac{1}{t}\log\Big(P_m\big(\{\omega : \mathbf{L}_t(\omega) \in \Gamma\}\big)\Big)\\ &\leq \varlimsup_{t\to\infty} \frac{1}{t}\log\Big(P_m\big(\{\omega : \mathbf{L}_t(\omega) \in \Gamma\}\big)\Big) \leq -\inf_{\overline{\Gamma}} J_{\mathcal{E}}\end{aligned} \tag{5.3.12}$$

for every $\Gamma \in \mathcal{B}_{\mathbf{M}_1(\Sigma)}$.

PROOF: In view of Lemma 5.3.2, all that we have to do is check (5.3.11). Also, since, for any $T \in (0,\infty)$ and $\delta > 0$,

$$P_{\nu P_T}\big(\{\omega : \mathbf{L}_t(\omega) \in G\}\big) \leq P_\nu\Big(\big\{\omega : \big\|\mathbf{L}_t(\omega) - G\big\|_{\text{var}} < \delta\big\}\Big)$$

as soon as t is sufficiently large, we will assume, without loss of generality, that ν itself is not singular to m. In particular, this means, by Lemma 5.3.9, that

$$\varliminf_{t\to\infty} \int_{\{\omega:\mathbf{L}_t(\omega)\in G\}} X_u(t,\omega)\,P_\nu(d\omega) > 0.$$

We begin by showing that if $u \in \mathbf{D} \cap B(\Sigma;[1,\infty))$, then

$$\varliminf_{t\to\infty} \frac{1}{t}\log\Big(P_\nu\big(\{\omega : \mathbf{L}_t(\omega) \in G\}\big)\Big) \geq -J_{\mathcal{E}}(m_u) \tag{5.3.13}$$

for every τ-open $G \in \mathcal{B}_{\mathbf{M}_1(\Sigma)}$ containing m_u. To this end, set

$$G(r) = \left\{\mu \in G : \left|\int_\Sigma V_u\, d\mu - \int_\Sigma V_u\, dm_u\right| < r\right\}, \quad r \in (0,\infty).$$

Then

$$\begin{aligned} &P_\nu\big(\{\omega : \mathbf{L}_t(\omega) \in G\}\big) \geq P_\nu\big(\{\omega : \mathbf{L}_t(\omega) \in G(r)\}\big) \\ &\geq \frac{1}{\|u\|_B} \inf_{\mu \in G(r)} \exp\left[-t\int_\Sigma V_u\, d\mu\right] \int_{\{\omega:\mathbf{L}_t(\omega)\in G(r)\}} X_u(t,\omega)\, P_\nu(d\omega) \end{aligned}$$

for all $r \in (0,\infty)$; and therefore, by the remarks made above,

$$\begin{aligned} &\varliminf_{t\to\infty} \frac{1}{t} \log\Big(P_\nu\big(\{\omega : \mathbf{L}_t(\omega) \in G\}\big)\Big) \\ &\qquad \geq -\lim_{r\searrow 0} \sup_{\mu\in G(r)} \int_\Sigma V_u\, d\mu = \int_\Sigma \frac{Lu}{u}\, dm_u. \end{aligned}$$

Since

$$\int_\Sigma \frac{Lu}{u}\, dm_u = \frac{\big(u, Lu\big)_{L^2(m)}}{\|u\|^2_{L^2(m)}} = -J_\mathcal{E}\big(m_u\big),$$

(5.1.13) is now proved.

Finally, we will show that if $J_\mathcal{E}(\mu) < \infty$, then there exists a sequence $\{u_n\}_1^\infty \subseteq \mathbf{D} \cap B\big(\Sigma;[1,\infty)\big)$ such that $m_{u_n} \longrightarrow \mu$ in the strong topology on $\mathbf{M}_1(\Sigma)$ and $J_\mathcal{E}\big(m_{u_n}\big) \longrightarrow J_\mathcal{E}(\mu)$. Clearly, when combined with (5.3.13), this will complete the proof of (5.3.11).

Let $\mu \in \mathbf{M}_1(\Sigma)$ with $J_\mathcal{E}(\mu) < \infty$ be given. Then $\mu(d\sigma) = f(\sigma)\, m(d\sigma)$ where $f \in L^1(m)^+$ and $\mathcal{E}\big(f^{1/2}, f^{1/2}\big) = J_\mathcal{E}(\mu)$. For $n \in \mathbb{Z}^+$, set

$$f_n = \left(\frac{\big(f^{1/2} \wedge n\big) + \frac{1}{n}}{\big\|\big(f^{1/2}\wedge n\big) + \frac{1}{n}\big\|_{L^2(m)}}\right)^2.$$

Clearly $f_n^{1/2} \longrightarrow f^{1/2}$ in $L^2(m)$, and therefore, by (**i**) of Exercise 4.2.63,

$$\mathcal{E}\big(f^{1/2}, f^{1/2}\big) \leq \varliminf_{n\to\infty} \mathcal{E}\big(f_n^{1/2}, f_n^{1/2}\big).$$

At the same time, one can use (4.2.54) to check that

$$\mathcal{E}\big(f^{1/2}, f^{1/2}\big) \geq \varlimsup_{n\to\infty} \mathcal{E}\big(f_n^{1/2}, f_n^{1/2}\big).$$

Thus, our problem reduces to that of finding $\{u_n\}_1^\infty \subseteq \mathbf{D} \cap B(\Sigma;[1,\infty))$ for the case when $f \equiv \frac{d\mu}{dm}$ satisfies $\alpha^2 \le f \le 1/\alpha^2$ for some $\alpha \in (0,1]$. But in this case, set

$$u_n = \frac{n^2\left[\left(R_n^0\right)^2 f^{1/2}\right]}{\alpha},$$

where R_λ^0, $\lambda \in (0,\infty)$, is the operator defined in (4.2.32) with $V = 0$. Then, not only is $u_n \in B(\Sigma;[1,\infty))$ but also, by Lemma 4.2.31, $u_n \in \mathbf{D}$. At the same time, by the Spectral Theorem,

$$u_n = \frac{1}{\alpha}\int_{[0,\infty)}\left(\frac{n}{n+\lambda}\right)^2 dE_\lambda f^{1/2};$$

and so $\alpha u_n \longrightarrow f^{1/2}$ in $L^2(m)$ while $\alpha^2\mathcal{E}(u_n,u_n) \longrightarrow \mathcal{E}(f^{1/2},f^{1/2})$. From these, it is easy to see that $m_{u_n} \longrightarrow \mu$ strongly and that $J_{\mathcal{E}}(m_{u_n}) \longrightarrow J_{\mathcal{E}}(\mu)$. ∎

5.3.14 Exercise.

Assume that $P(t,\sigma,\cdot)$ is FELLER-continuous and that $P(t,\sigma,\cdot) << m$ for every $(t,\sigma) \in (0,\infty)\times\Sigma$.

(i) Show that m itself is the only $\{P_t : t > 0\}$-invariant probability measure, and conclude not only that $J_{\mathcal{E}}(\mu) = 0$ only if $\mu = m$ but also that

$$\inf_{\sigma\in\Sigma}\varliminf_{t\to\infty}\frac{1}{t}\log\left(P_\sigma\left(\{\omega : \mathbf{L}_t(\omega)\}\right)\right) \ge -\inf_G J_{\mathcal{E}}$$

for every τ-open $G \in \mathcal{B}_{\mathbf{M}_1(\Sigma)}$.

(ii) Next, using (4.2.42), (4.1.46), and Theorem 4.2.58, show that $J_{\mathcal{E}} = J_P$. In particular, this means that $J_{\mathcal{E}}$ is lower semi-continuous.

(iii) Finally, under the additional assumption that $\Lambda_{\mathcal{E}}$ is tight (equivalently, $J_{\mathcal{E}}$ is good), use **(i)** and **(ii)** above together with the considerations in Exercise 5.1.17 to conclude that

$$\begin{aligned}
-\inf_{\Gamma^\circ} J_{\mathcal{E}} &\le \inf_{\sigma\in\Sigma}\varliminf_{t\to\infty}\frac{1}{t}\log\left(P_\sigma\left(\{\omega : \mathbf{L}_t(\omega)\in\Gamma\}\right)\right)\\
&\le \sup_{\sigma\in\Sigma}\varlimsup_{t\to\infty}\frac{1}{t}\log\left(P_\sigma\left(\{\omega : \mathbf{L}_t(\omega)\in\Gamma\}\right)\right) \le -\inf_{\overline{\Gamma}} J_{\mathcal{E}}, \quad \Gamma \in \mathcal{B}_{\mathbf{M}_1(\Sigma)}.
\end{aligned}$$

5.3.15 Exercise.

(i) Suppose that there exist α, $\beta \in [0,\infty)$ such that

$$\mathbf{H}(\mu|m) \le \alpha J_{\mathcal{E}}(\mu) + \beta, \quad \mu \in \mathbf{M}_1(\Sigma). \tag{5.3.16}$$

Show that $J_{\mathcal{E}}$ is then a good rate function on $\mathbf{M}_1(\Sigma)$. In fact, show that $\{\mu : J_{\mathcal{E}}(\mu) \le L\}$ is a strongly compact subset of $\mathbf{M}_1(\Sigma)$ for every $L \in (0,\infty)$.

(ii) Next, suppose that (5.3.16) holds with $\beta = 0$, and show that, in this case, $J_{\mathcal{E}}(\mu) = 0$ if and only if $\mu = m$. (**Hint:** use (3.2.25).) In particular, (5.3.16) with $\beta = 0$ means that all the hypotheses of Theorem 5.3.10 are satisfied.

(iii) Define

$$\Lambda_{\tilde{m}}(V) = \log\left(\int_\Sigma \exp[V]\,dm\right), \quad V \in B(\Sigma;\mathbb{R})$$

as in Section 3.2. Using the preceding and Lemma 3.2.13, show that (5.3.16) holds for a given $\alpha > 0$ and $\beta = 0$ if and only if

$$\text{(5.3.17)} \qquad \Lambda_{\mathcal{E}}(V) \le \frac{1}{\alpha}\Lambda_{\tilde{m}}(\alpha V), \quad V \in B(\Sigma;\mathbb{R}).$$

Also, check that (5.3.17) holds as soon as the indicated domination holds for all $V \in C_b(\Sigma;\mathbb{R})$.

5.4 Large Deviations for Hypermixing Processes

In this section, we will present a very general, process-level large deviation principle. In order not to overburden the discussion with inessential technicalities, we will deal initially with continuous path processes. Afterwards, we will say what has to be done to extend the results to the SKOROKHOD setting; and we will leave the discrete time case as an exercise.

Let Σ be a Polish space and denote by Ω the Polish space $C(\mathbb{R};\Sigma)$ with the topology of uniform convergence on compacts. For $t \in \mathbb{R}$, define $\Sigma_t : \Omega \longrightarrow \Sigma$ so that $\Sigma_t(\omega)$ is the position of the path $\omega \in \Omega$ at time t. Next, given a closed interval $I \subseteq \mathbb{R}$, let Ω_I stand for $C(I;\Sigma)$, and for $t \in I$ define $\Sigma_t : \Omega_I \longrightarrow \Sigma$ accordingly. Also, define $\pi_I : \Omega \longrightarrow \Omega_I$ to be the natural projection map obtained by restriction to I; and set $\mathcal{B}_I = \pi_I^{-1}(\mathcal{B}_{\Omega_I})$. It is clear that $\mathcal{B}_I$ coincides with the σ-algebra over Ω generated by the maps Σ_t, $t \in I$. Given $\ell > 0$, $n \ge 2$, and real-valued functions $f_1, \dots, f_n$ on Ω, we will say that $f_1, \dots, f_n$ are **ℓ-measurably separated** if there exist intervals $I_1, \dots, I_n$ with the properties that $\operatorname{dist}(I_m, I_{m'}) \ge \ell$ for $1 \le m < m' \le n$ and f_m is $\mathcal{B}_{I_m}$-measurable for each $1 \le m \le n$.

With the preceding notation, we can now describe an important mixing property. Namely, we will say that $P \in \mathbf{M}_1(\Omega)$ is **hypermixing** if there exist a number $\ell_0 \ge 0$ and non-increasing functions $\alpha, \beta : (\ell_0, \infty) \longrightarrow [1, \infty)$ and $\gamma : (\ell_0, \infty) \longrightarrow [0, 1]$ which satisfy

$$\text{(5.4.1)} \qquad \lim_{\ell\to\infty} \alpha(\ell) = 1, \quad \overline{\lim_{\ell\to\infty}}\, \ell\big(\beta(\ell) - 1\big) < \infty, \quad \lim_{\ell\to\infty} \gamma(\ell) = 0,$$

and for which:

$$\text{(H-1)} \qquad \left\| f_1 \cdots f_n \right\|_{L^1(P)} \leq \prod_{m=1}^{n} \left\| f_m \right\|_{L^{\alpha(\ell)}(P)},$$

whenever $n \geq 2$, $\ell > \ell_0$, and $f_1, \dots, f_n$ are ℓ-measurably separated functions; and

$$\text{(H-2)} \quad \left| \int_\Omega \left(f(\omega) - \int_\Omega f\, dP \right) g(\omega)\, P(d\omega) \right| \leq \gamma(\ell) \left\| f \right\|_{L^{\beta(\ell)}(P)} \left\| g \right\|_{L^{\beta(\ell)}(P)}$$

whenever $\ell > \ell_0$ and $f,\ g \in L^1(P)$ are ℓ-measurably separated.

Finally, define the **time-shift** transformation group $\{\theta_t : t \in \mathbb{R}\}$ on Ω by $\Sigma_s(\theta_t \omega) = \Sigma_{s+t}(\omega)$ for $s,\ t \in \mathbb{R}$ and $\omega \in \Omega$. Clearly, $\theta_s \circ \theta_t = \theta_{s+t}$ for all $s,\ t \in \mathbb{R}$, $\pi_I \circ \theta_t = \pi_{t+I}$ for all $t \in \mathbb{R}$ and intervals I, and $(t, \omega) \in \mathbb{R} \times \Omega \longmapsto \theta_t \omega \in \Omega$ is continuous. We will use $\mathbf{M}_1^{\mathrm{S}}(\Omega)$, $\mathbf{EM}_1^{\mathrm{S}}(\Omega)$, and $\mathfrak{I}_{\mathbf{S}}$, to denote, respectively, the $\{\theta_t : t \in \mathbb{R}\}$-invariant $Q \in \mathbf{M}_1(\Omega)$, the ergodic $Q \in \mathbf{M}_1^{\mathrm{S}}(\Omega)$, and the $\{\theta_t : t \in \mathbb{R}\}$-invariant elements of $\mathcal{B}_\Omega$. For $\omega \in \Omega$ and $T \in (0, \infty)$, let $\mathbf{R}_T(\omega) \in \mathbf{M}_1(\Omega)$ be the **empirical process measure** given by

$$\text{(5.4.2)} \qquad \mathbf{R}_T(\omega) = \frac{1}{2T} \int_{-T}^{T} \delta_{\theta_t \omega}\, dt.$$

Obviously, $(T, \omega) \in (0, \infty) \times \Omega \longmapsto \mathbf{R}_T(\omega) \in \mathbf{M}_1(\Omega)$ is continuous.

Throughout the rest of this section, P will denote a fixed hypermixing element of $\mathbf{M}_1^{\mathrm{S}}(\Omega)$. As a consequence of (**H-2**) and part (**ii**) of Exercise 5.2.31, it is clear that P is ergodic and therefore that $\mathbf{R}_T(\omega) \Longrightarrow P$ for P-almost every $\omega \in \Omega$. What we want to do is derive an associated large deviation principle. The procedure which we will use is based on the following outline.

STEP 1: For any compact interval I and $V \in B(\Omega_I; \mathbb{R})$, we will use (**H-1**) to show that

$$\text{(5.4.3)} \quad \Lambda_I(V) \equiv \lim_{T \to \infty} \frac{1}{2T} \log \left(\int_\Omega \exp \left[\int_{-T}^{T} V \circ \pi_I(\theta_t \omega)\, dt \right] P(d\omega) \right)$$

exists. At the same time, we will show that $V \in C_{\mathrm{b}}(\Omega_I; \mathbb{R}) \longmapsto \Lambda_I(V)$ is tight and thereby derive the upper bound

$$\text{(5.4.4)} \qquad \varlimsup_{T \to \infty} \frac{1}{2T} \log \Big[P\big(\{\omega : \mathbf{R}_T(\omega) \in F\}\big) \Big] \leq - \inf_F \Lambda^*$$

for closed $F \subseteq \mathbf{M}_1(\Omega)$, where $\Lambda^* : \mathbf{M}_1(\Omega) \longrightarrow [0, \infty]$ is the good rate function defined by

$$(5.4.5) \qquad \Lambda^*(Q) = \sup\left\{\Lambda_I^*(Q \circ \pi_I^{-1}) : I \text{ is a compact interval}\right\},$$

and, for each I, $\Lambda_I^* : \mathbf{M}_1(\Omega_I) \longrightarrow [0, \infty]$ given by

$$(5.4.6) \qquad \Lambda_I^*(Q_I) \equiv \sup\left\{\int_{\Omega_I} V\, dQ_I - \Lambda_I(V) : V \in C_{\mathrm{b}}(\Omega_I; \mathbb{R})\right\}$$

is the LEGENDRE transform of Λ_I.

STEP 2: Given a compact interval I, define $\mathbf{H}_I(Q|P)$ for $Q \in \mathbf{M}_1(\Sigma)$ to be the relative entropy $\mathbf{H}(Q \circ \pi_I^{-1} | P \circ \pi_I^{-1})$ of $Q \circ \pi_I^{-1}$ given $P \circ \pi_I^{-1}$. Again using (**H-1**), we will show that

$$(5.4.7) \qquad \Lambda^*(Q) = \begin{cases} \lim_{I \nearrow \mathbb{R}} \frac{1}{|I|} \mathbf{H}_I(Q|P) & \text{for } Q \in \mathbf{M}_1^{\mathrm{S}}(\Omega) \\ \infty & \text{otherwise.} \end{cases}$$

Thus, if we define the **specific entropy** function

$$(5.4.8) \qquad \mathbf{H}(Q) = \begin{cases} \lim_{I \nearrow \mathbb{R}} \frac{1}{|I|} \mathbf{H}_I(Q|P) & \text{for } Q \in \mathbf{M}_1^{\mathrm{S}}(\Omega) \\ \infty & \text{for } Q \in \mathbf{M}_1(\Omega) \setminus \mathbf{M}_1^{\mathrm{S}}(\Omega), \end{cases}$$

then $\mathbf{H}$ is a good rate function and

$$(5.4.9) \qquad \overline{\lim_{T \to \infty}} \frac{1}{2T} \log\left[P(\{\omega : \mathbf{R}_T(\omega) \in F\})\right] \le -\inf_F \mathbf{H}$$

for closed $F \subseteq \mathbf{M}_1(\Omega)$.

STEP 3: Having established the upper bound (5.4.9), we will turn to complementary lower bound. We will use the Ergodic Theorem to check that, for $Q \in \mathbf{EM}_1^{\mathrm{S}}(\Omega)$,

$$(5.4.10) \qquad \underline{\lim_{T \to \infty}} \frac{1}{2T} \log\Big(P(\{\omega : \mathbf{R}_T(\omega) \in G\})\Big) \ge -\mathbf{H}(Q) \quad \text{for open } G \ni Q.$$

In order to remove the restriction that Q be ergodic, we will introduce the lower semi-continuous function $J : \mathbf{M}_1(\Omega) \longrightarrow [0, \infty]$ given by

$$(5.4.11) \qquad J(Q) = -\inf\left\{ \underline{\lim_{T \to \infty}} \frac{1}{2T} \log\left[P(\{\omega : \mathbf{R}_T(\omega) \in G\})\right] : G \text{ is an open set containing } Q\right\}$$

for $Q \in \mathbf{M}_1(\Omega)$. Clearly $J \le \mathbf{H}$ on $\mathbf{M}_1(\Omega) \setminus \mathbf{M}_1^{\mathrm{S}}(\Omega)$; and, by (5.4.10), we also know that $J \le \mathbf{H}$ on $\mathbf{EM}_1^{\mathrm{S}}(\Omega)$. Thus, all that remains is to check that the domination of J by $\mathbf{H}$ on $\mathbf{EM}_1^{\mathrm{S}}(\Omega)$ extends to the whole of $\mathbf{M}_1^{\mathrm{S}}(\Omega)$; and our proof of this fact will turn on The Ergodic Decomposition Theorem. Namely, at the one place where we use (**H-2**), we will show that J is a convex function. At the same time, we already know (cf. Exercise 4.4.41) that $\mathbf{H}$ is affine; and these two observations will be used to conclude that

$$\begin{aligned} J(Q) &\le \int_{\mathbf{EM}_1^{\mathrm{S}}(\Omega)} J(R)\,\rho_Q(dR) \\ \text{and}\quad \mathbf{H}(Q) &= \int_{\mathbf{EM}_1^{\mathrm{S}}(\Omega)} \mathbf{H}(R)\,\rho_Q(dR), \end{aligned} \tag{5.4.12}$$

where $\rho_Q \in \mathbf{M}_1\big(\mathbf{EM}_1^{\mathrm{S}}(\Omega)\big)$ is the measure described in the Ergodic Decomposition Theorem (cf. Theorem 5.2.16). Obviously, in conjunction with the preceding, (5.4.12) is more than enough to complete the proof of the lower bound.

Keeping the preceding outline in mind, we now get down to business.

5.4.13 Lemma. *For every compact interval I and every $V \in B\big(\Omega_I;\mathbb{R}\big)$, the limit $\Lambda_I(V)$ in (5.4.3) exists. In addition, for $\ell > \ell_0$,*

$$\Lambda_I(V) \le \frac{\log\Big(\int_\Omega \exp\Big[(\ell+|I|)\alpha(\ell) V\circ\pi_I(\omega)\Big]\,P(d\omega)\Big)}{(\ell+|I|)\alpha(\ell)}. \tag{5.4.14}$$

In particular, the map $V \in C_{\mathrm{b}}(\Omega_I;\mathbb{R}) \longmapsto \Lambda_I(V)$ is a tight, convex function which satisfies $\Lambda_I(c\mathbf{1}) = c$ for $c \in \mathbb{R}$.

PROOF: Without loss of generality, we will assume throughout that V is non-negative, and we will use M to denote $\|V\|_B$.

To prove the existence of $\Lambda_I(V)$, set

$$f(T) = \log\left(\int_\Omega \exp\left[\int_0^T V\circ\pi_I(\theta_t\omega)\,dt\right]\,P(d\omega)\right), \quad T \in (0,\infty).$$

Because of shift-invariance, all that we have to do is check that the limit $\lim_{T\to\infty} \frac{f(T)}{T}$ exists. To this end, let $S \in (0,\infty)$ be given and write $T = n_T S + r_T$, where $n_T \in \mathbb{Z}^+$ and $r_T \in [0,S)$, for $T > S$. Then, by (**H-1**) and shift-invariance, for every $\ell > \ell_0$,

$$
\begin{aligned}
f(T) &\le r_T M + \log\left(\int_\Omega \exp\left[\int_0^{n_T(S+\ell+|I|)} V\circ\pi_I(\theta_t\omega)\,dt\right] P(d\omega)\right)\\
&\le \big(r_T + n_T(\ell+|I|)\big)M\\
&\qquad + \log\left(\int_\Omega \exp\left[\sum_{m=0}^{n_T-1}\int_0^S V\circ\pi_I(\theta_{t+m(S+\ell+|I|)}\omega)\,dt\right] P(d\omega)\right)\\
&\le \big(r_T + n_T(\ell+|I|)\big)M\\
&\qquad + \frac{n_T}{\alpha(\ell)}\log\left(\int_\Omega \exp\left[\alpha(\ell)\int_0^S V\circ\pi_I(\theta_t\omega)\,dt\right] P(d\omega)\right)\\
&\le \Big(r_T + n_T(\ell+|I|) + \big(\alpha(\ell)-1\big)T\Big)M\\
&\qquad + n_T\log\left(\int_\Omega \exp\left[\int_0^S V\circ\pi_I(\theta_t\omega)\,dt\right] P(d\omega)\right).
\end{aligned}
$$

Hence,

$$
\varlimsup_{T\to\infty}\frac{f(T)}{T} \le \frac{f(S)}{S} + \left[\frac{\ell+|I|}{S} + \big(\alpha(\ell)-1\big)\right]M
$$

for $S \in (0,\infty)$ and $\ell > \ell_0$; and, since $\alpha(\ell) \searrow 1$ as $\ell \nearrow \infty$, this clearly this implies that

$$
\varlimsup_{T\to\infty}\frac{f(T)}{T} \le \varliminf_{S\to\infty}\frac{f(S)}{S}.
$$

In order to prove (5.4.14), let $\ell > \ell_0$ be given and set $T = \ell + |I|$. Then, again by shift-invariance and (**H-1**),

$$
\begin{aligned}
&\int_\Omega \exp\left[\int_0^{nT} V\circ\pi_I(\theta_t\omega)\,dt\right] P(d\omega)\\
&\quad = \int_\Omega \exp\left[\sum_{m=0}^{n-1}\frac{1}{T}\int_0^T TV\circ\pi_I(\theta_{t+mT}\omega)\,dt\right] P(d\omega)\\
&\quad \le \frac{1}{T}\int_0^T\left(\int_\Omega \exp\left[\sum_{m=0}^{n-1} TV\circ\pi_I(\theta_{t+mT}\omega)\right] P(d\omega)\right) dt\\
&\quad \le \left(\int_\Omega \exp\left[T\alpha(\ell)V\circ\pi_I(\omega)\right] P(d\omega)\right)^{n/\alpha(\ell)},
\end{aligned}
$$

where we have used JENSEN's inequality in the passage from the second to the third line. After dividing through by nT and then letting $n \to \infty$, we arrive at (5.4.14).

Finally, the convexity of Λ_I as well as the equality $\Lambda_I(c\mathbf{1}) = c$, $c \in \mathbb{R}$, are both immediate consequences of the definition of Λ_I. Moreover, given (5.4.14), it is clear how to choose the sets $K(M) \subset\subset \Omega_I$ to check tightness. Namely, let $\ell > \ell_0 \vee 1$ be given and choose $K(M) \subset\subset \Omega_I$ so that

$$P\big(\{\omega : \pi_I(\omega) \notin K(M)\}\big) \le \exp\big[-(\ell + |I|)\alpha(\ell)M\big]. \quad \blacksquare$$

Now let $\Lambda_I^* : \mathbf{M}_1(\Omega_I) \longrightarrow [0,\infty]$ be the function defined in (5.4.6). Then, by Lemma 5.4.13 and Corollary 5.1.11, Λ_I^* is a good rate function on $\mathbf{M}_1(\Omega_I)$ and

$$\overline{\lim_{T\to\infty}} \frac{1}{2T} \log\Big(P\big(\{\omega : \mathbf{R}_T(\omega)\circ\pi_I^{-1} \in F\}\big)\Big) \le -\inf_F \Lambda_I^*$$

for closed $F \subseteq \mathbf{M}_1\big(\Omega_I\big)$. Thus, by **(ii)** of Exercise 2.1.21 (cf. **(ii)** of Exercise 3.2.22 as well), the function $\Lambda^* : \mathbf{M}_1(\Omega) \longrightarrow [0,\infty]$ in (5.4.5) is also a good rate function; and, just as in **(iii)** of Exercise 2.1.21, we now have (5.4.4).

Having completed STEP 1, we now begin STEP 2 by checking that

$$\Lambda^*(Q) = \infty \quad \text{when } Q \in \mathbf{M}_1(\Omega) \setminus \mathbf{M}_1^{\mathrm{S}}(\Omega). \tag{5.4.15}$$

To this end, suppose that $Q \notin \mathbf{M}_1^{\mathrm{S}}(\Omega)$ is given. One can then choose a compact interval I and a $V \in C_{\mathrm{b}}\big(\Omega_I;\mathbb{R}\big)$ so that

$$\int_\Omega V\circ\pi_I\big(\theta_\ell\omega\big)\,Q(d\omega) \ge \int_\Omega V\circ\pi_I(\omega)\,Q(d\omega) + 1 \quad \text{for some } \ell \in \mathbb{R}. \tag{5.4.16}$$

In particular, if the compact interval J is chosen so that $\big(\ell + I\big) \cup I \subseteq J$ and $W \in C_{\mathrm{b}}\big(\Omega_J;\mathbb{R}\big)$ is defined by $W\circ\pi_J \equiv V\circ\pi_I\circ\theta_\ell - V\circ\pi_I$, then (5.4.16) leads to

$$\Lambda^*(Q) \ge \sup\Big\{M - \Lambda_J\big(MW\big) : M \in (0,\infty)\Big\}.$$

Thus, we will have completed the proof of (5.4.15) once we show that $\Lambda_J(MW) \le 0$ for every $M \in (0,\infty)$. But it is clear that, for any $T > \ell$,

$$\begin{aligned}
&\left|\int_{-T}^{T} W\circ\pi_J\big(\theta_t\omega\big)\,dt\right| \\
&= \left|\int_{-T}^{T} V\circ\pi_I\big(\theta_{\ell+t}\omega\big)\,dt - \int_{-T}^{T} V\circ\pi_I\big(\theta_t\omega\big)\,dt\right| \le 2|\ell|\|V\|_B;
\end{aligned}$$

and, therefore,

$$\frac{1}{2T}\log\left(\int_\Omega \exp\left[\int_{-T}^{T} MW\circ\pi_J\big(\theta_t\omega\big)\,dt\right] P(d\omega)\right) \le \frac{\ell M\|V\|_B}{T} \longrightarrow 0$$

as $T \longrightarrow \infty$.

To complete the proof of (5.4.7), we will use the following lemma.

5.4.17 Lemma. *Let I be a compact interval. Then*

$$\frac{1}{(\ell+|I|)\alpha(\ell)}\mathbf{H}_I(Q|P) \leq \Lambda_I^*(Q\circ\pi_I^{-1}) \tag{5.4.18}$$

for all $Q \in \mathbf{M}_1(\Omega)$ and $\ell > \ell_0$; and, for every $Q \in \mathbf{M}_1^{\mathrm{S}}(\Omega)$,

$$\begin{aligned}\int_\Omega V\circ\pi_I\,dQ - \frac{1}{2T}\log\left(\int_\Omega \exp\left[\int_{-T}^{T} V\circ\pi_I(\theta_t\omega)\,dt\right] P(d\omega)\right)\\ \leq \frac{1}{2T}\mathbf{H}_{I(T)}(Q|P)\end{aligned} \tag{5.4.19}$$

for $T \in (0,\infty)$ and $V \in B(\Omega_I;\mathbb{R})$, where $I(T) \equiv \{t:\ |t-I| \leq T\}$.

PROOF: Recall (cf. Lemma 3.2.13) that $\mathbf{H}_I(Q|P)$ is given by

$$\sup\left\{\int_\Omega V\circ\pi_I\,dQ - \log\left(\int_\Omega \exp\left[V\circ\pi_I\right]dP\right) : V \in C_{\mathrm{b}}(\Omega_I;\mathbb{R})\right\}. \tag{5.4.20}$$

Thus, (5.4.18) is an immediate consequence of (5.4.14).

To prove (5.4.19), let $Q \in \mathbf{M}_1^{\mathrm{S}}(\Omega)$ and $V \in B(\Omega_I;\mathbb{R})$ be given. For $T \in (0,\infty)$, define $V_T \in B(\Omega_{I(T)};\mathbb{R})$ so that

$$V_T\circ\pi_{I(T)}(\omega) = \int_{-T}^{T} V\circ\pi_I(\theta_t\omega)\,dt, \quad \omega \in \Omega.$$

Because Q is shift-invariant, one then has that

$$\begin{aligned}\int_\Omega V\circ\pi_I\,dQ - \frac{1}{2T}\log\left(\int_\Omega \exp\left[\int_{-T}^{T} V\circ\pi_I(\theta_t\omega)\,dt\right] P(d\omega)\right)\\ = \frac{1}{2T}\left[\int_\Omega V_T\circ\pi_{I(T)}\,dQ - \log\left(\int_\Omega \exp\left[V_T\circ\pi_{I(T)}\right]dP\right)\right].\end{aligned}$$

Finally, by (5.4.20), the right hand side of the preceding is dominated by $\frac{1}{2T}\mathbf{H}_{I(T)}(Q|P)$ when $V \in C_{\mathrm{b}}(\Omega_I;\mathbb{R})$ and therefore for general $V \in B(\Omega_I;\mathbb{R})$. ∎

From here, it is an easy matter to complete STEP 2. Indeed, by (5.4.18), for any $Q \in \mathbf{M}_1(\Omega)$, we have that

$$\varlimsup_{I\nearrow\mathbb{R}} \frac{1}{|I|}\mathbf{H}_I(Q|P) \leq \Lambda^*(Q).$$

On the other hand, if $Q \in \mathbf{M}_1^{\mathrm{S}}(\Omega)$, then both $\mathbf{H}_I(Q|P)$ and $\Lambda_I^*(Q)$ depend on I only through $|I|$, and, by (5.4.19),

$$\int_\Omega V \circ \pi_{[-S,S]}\, dQ - \Lambda_{[-S,S]}(V) \le \varliminf_{T\to\infty} \frac{1}{2T}\mathbf{H}_{[-S-T,S+T]}(Q|P)$$

for any $S \in (0,\infty)$ and $V \in C_{\mathrm{b}}(\Omega_{[-S,S]};\mathbb{R})$. Clearly this leads immediately to $\Lambda^*_{[-S,S]}(Q \circ \pi^{-1}_{[-S,S]}) \le \varliminf_{T\to\infty} \frac{1}{2T}\mathbf{H}_{[-T,T]}(Q|P)$; and the rest of STEP 2 is now simply a matter of notation.

We next turn to STEP 3 and verify that (5.4.10) holds for ergodic $Q \in \mathbf{M}_1^{\mathrm{S}}(\Omega)$.

5.4.21 Lemma. *If $Q \in \mathbf{EM}_1^{\mathrm{S}}(\Omega)$ and I is a compact interval, then for any $G_I \in \mathcal{B}_{\mathbf{M}_1(\Omega_I)}$ which is a τ-open neighborhood of $Q \circ \pi_I^{-1}$*

$$\varliminf_{T\to\infty} \frac{1}{2T}\log\Big(P\big(\{\omega : \mathbf{R}_T(\omega)\circ\pi_I^{-1} \in G_I\}\big)\Big) \ge -\mathbf{H}(Q).$$

PROOF: The argument is very much like the one used in (**ii**) of Exercise 3.2.23, only here the Ergodic Theorem plays the role that the Law of Large Numbers did there.

Set $I(T) = \{t : |t - I| \le T\}$ and

$$F_T = \frac{dQ \circ \pi^{-1}_{I(T)}}{dP \circ \pi^{-1}_{I(T)}} \circ \pi_{I(T)},$$

and let $A_T = \big\{\omega : \mathbf{R}_T(\omega)\circ\pi_I^{-1} \in G_I \text{ and } F_T(\omega) > 0\big\}$. Then, by the Ergodic Theorem, $Q(A_T) \longrightarrow 1$ as $T \to \infty$. Thus, by JENSEN's inequality,

$$\begin{aligned}
&\varliminf_{T\to\infty} \frac{1}{2T}\log\Big(P\big(\{\omega : \mathbf{R}_T(\omega) \in G\}\big)\Big)\\
&\qquad \ge \varliminf_{T\to\infty} \frac{1}{2T}\log\left(\frac{1}{Q(A_T)}\int_{A_T}\frac{1}{F_T(\omega)}\,Q(d\omega)\right)\\
&\qquad \ge \varliminf_{T\to\infty} \frac{-1}{2TQ(A_T)}\int_{A_T}\log\big(F_T(\omega)\big)\,Q(d\omega) \ge -\mathbf{H}(Q),
\end{aligned}$$

since

$$\begin{aligned}
-&\int_{A_T}\log\big(F_T(\omega)\big)\,Q(d\omega)\\
&= \int_{A_T^{\mathrm{c}}} F_T(\omega)\log\big(F_T(\omega)\big)\,P(d\omega) - \mathbf{H}_{I(T)}(Q|P)\\
&\ge -e^{-1} - \mathbf{H}_{I(T)}(Q|P). \quad \blacksquare
\end{aligned}$$

As an essentially immediate consequence of Lemma 5.4.21, we see that

$$\varliminf_{T\to\infty} \log\big(P\big(\{\omega : \mathbf{R}_T(\omega) \in G\}\big)\big) \ge -\mathbf{H}(Q)$$

for any open $G \subseteq \mathbf{M}_1(\Omega)$ and any ergodic $Q \in G$.

Continuing with STEP 3, we next define the lower semi-continuous function $J : \mathbf{M}_1(\Omega) \longrightarrow [0,\infty]$ as in (5.4.11). Our goal is to prove that $J \le \mathbf{H}$. At the moment (cf. the preceding paragraph), we know that $J \le \mathbf{H}$ on $\mathbf{EM}_1^{\mathrm{S}}(\Omega) \cup \big(\mathbf{M}_1(\Omega) \setminus \mathbf{M}_1^{\mathrm{S}}(\Omega)\big)$.

5.4.22 Lemma. *The function J in (5.4.11) is convex.*

PROOF: Since J is lower semi-continuous, it suffices to check that

$$J\left(\frac{Q_1+Q_2}{2}\right) \le \frac{J(Q_1)+J(Q_2)}{2}$$

for Q_1, $Q_2 \in \mathbf{M}_1(\Omega)$ satisfying $J(Q_1) \vee J(Q_2) < \infty$. To this end, let G be an open set containing $Q \equiv \big(Q_1+Q_2\big)/2$. Choose $S \in (0,\infty)$ and $r > 0$ so that

$$G \supseteq \pi_I^{-1}\left(\frac{B_I(Q_1,2r)+B_I(Q_2,2r)}{2}\right),$$

where $I = [-S,S]$ and the balls B_I are defined relative to the LÉVY metric on $\mathbf{M}_1(\Omega_I)$. Set

$$u_i(T) = P\big(\{\omega : \mathbf{R}_T(\omega)\circ\pi_I^{-1} \in B_I(Q_i,r)\}\big), \quad i \in \{1,2\}$$

and

$$w(T) = P\big(\{\omega : \mathbf{R}_T(\omega) \in G\}\big).$$

Then, by (**H-2**):

$$\begin{aligned}
w(2T) &= P\big(\{\omega : \mathbf{R}_{2T}(\omega) \in G\}\big) \\
&\ge P\big(\{\omega : \mathbf{R}_T(\theta_{-T}\omega)\circ\pi_I^{-1} \in B_I(Q_1,2r) \\
&\qquad\qquad \text{and } \mathbf{R}_T(\theta_T\omega)\circ\pi_I^{-1} \in B_I(Q_2,2r)\}\big) \\
&\ge P\big(\{\omega : \mathbf{R}_T(\theta_{-T-S-\ell/2}\omega)\circ\pi_I^{-1} \in B_I(Q_1,r) \\
&\qquad\qquad \text{and } \mathbf{R}_T(\theta_{T+S+\ell/2}\omega)\circ\pi_I^{-1} \in B_I(Q_2,r)\}\big) \\
&= u_1(T)u_2(T) + \int_\Omega \Big(\chi_{B_I(Q_1,r)}\big(\mathbf{R}_T(\theta_{-T-S-\ell/2}\omega)\big) - u_1(T)\Big) \\
&\qquad\qquad \times \chi_{B_I(Q_2,r)}\big(\mathbf{R}_T(\theta_{T+S+\ell/2}\omega)\big)\,P(d\omega) \\
&\ge u_1(T)u_2(T)\left[1 - \gamma(\ell)\big(u_1(T)u_2(T)\big)^{-1/\beta(\ell)'}\right]
\end{aligned}$$

as long as $\ell > \ell_0$ and $T > (2S+\ell)/2r$. (The number $\beta(\ell)'$ is the HÖLDER conjugate of $\beta(\ell)$.) Since $J(Q_1) \vee J(Q_2) < \infty$ means that $u_1(T)u_2(T) \ge$

$\exp\left[-MT\right]$ for some $M < \infty$ and all sufficiently large T's, we now see that

$$\begin{aligned} w(2T) &\geq u_1(T)u_2(T)\left(1 - \gamma(2rT - 2S)\exp\left[-\frac{MT}{\beta(2rT - 2S)'}\right]\right) \\ &\geq \frac{1}{2}u_1(T)u_2(T) \end{aligned}$$

for all sufficiently large T's; and clearly this leads to

$$\begin{aligned} &\varliminf_{T\to\infty} \frac{1}{4T}\log\Big(P\big(\{\omega : \mathbf{R}_{2T}(\omega) \in G\}\big)\Big) \\ &\quad \geq \frac{1}{2}\varliminf_{T\to\infty}\frac{1}{2T}\log\Big(P\big(\{\omega : \mathbf{R}_T(\omega)\circ\pi_I^{-1} \in B_I(Q_1, r)\}\big)\Big) \\ &\qquad\qquad + \frac{1}{2}\varliminf_{T\to\infty}\frac{1}{2T}\log\Big(P\big(\{\omega : \mathbf{R}_T(\omega)\circ\pi_I^{-1} \in B_I(Q_2, r)\}\big)\Big) \\ &\quad \geq -\frac{J(Q_1) + J(Q_2)}{2}. \quad \blacksquare \end{aligned}$$

We are now in the following situation. Both of the functions J and $\mathbf{H}$ are lower semi-continuous and convex; and we know that $\mathbf{H}(Q) \geq J(Q)$ for all $Q \in \big(\mathbf{M}_1(\Omega) \setminus \mathbf{M}_1^{\mathrm{S}}(\Omega)\big) \cup \mathbf{EM}_1^{\mathrm{S}}(\Omega)$. Furthermore, the function $\mathbf{H}$ is **affine** on $\mathbf{M}_1^{\mathrm{S}}(\Omega)$ in the sense that

$$\mathbf{H}\big(\alpha Q_1 + (1-\alpha)Q_2\big) = \alpha\mathbf{H}(Q_1) + (1-\alpha)\mathbf{H}(Q_2) \tag{5.4.23}$$

for $\alpha \in [0,1]$ and Q_1, $Q_2 \in \mathbf{M}_1^{\mathrm{S}}(\Omega)$. To see this, simply observe that (cf. **(ii)** of Exercise 4.4.41)

$$\begin{aligned} \alpha\mathbf{H}_I(Q_1|P) + (1-\alpha)\mathbf{H}_I(Q_2|P) &\geq \mathbf{H}_I\big(\alpha Q_1 + (1-\alpha)Q_2\big|P\big) \\ &\geq \alpha\mathbf{H}_I(Q_1|P) + (1-\alpha)\mathbf{H}_I(Q_2|P) - \frac{2}{e}. \end{aligned}$$

From these remarks, it should be clear that the following lemma is all that we need in order to complete STEP 3.

5.4.24 Lemma. *Let $\Phi : \mathbf{M}_1(\Omega) \longrightarrow [0,\infty]$ be a lower semi-continuous function. If Φ is convex, then for every $\rho \in \mathbf{M}_1(\Omega)$*

$$\Phi\left(\int_{\mathbf{M}_1(\Omega)} R\,\rho(dR)\right) \leq \int_{\mathbf{M}_1(\Omega)} \Phi(R)\,\rho(dR). \tag{5.4.25}$$

On the other hand, if Φ is affine on $\mathbf{M}_1^{\mathrm{S}}(\Omega)$ and $\rho \in \mathbf{M}_1\big(\mathbf{M}_1^{\mathrm{S}}(\Omega)\big)$, then

$$\Phi\left(\int_{\mathbf{M}_1^{\mathrm{S}}(\Omega)} R\,\rho(dR)\right) = \int_{\mathbf{M}_1^{\mathrm{S}}(\Omega)} \Phi(R)\,\rho(dR). \tag{5.4.26}$$

PROOF: We begin with the case in which $\rho(K) = 1$ for some compact subset K of $\mathbf{M}_1(\Omega)$. Throughout, $B(Q, r)$ denotes the LÉVY-metric ball in $\mathbf{M}_1(\Omega)$ of radius r around Q.

For $m \in \mathbb{Z}^+$, choose a finite set $\{R_{m,\ell}\}_{\ell=1}^{L_m} \subseteq K$ so that the balls $B_{m,\ell} = B(R_{m,\ell}, 1/m)$, $1 \le \ell \le L_m$, cover K; set $A_{m,1} = K \cap B_{m,1}$ and

$$A_{m,\ell} = K \cap \left(B_{m,\ell} \setminus \bigcup_{j=1}^{\ell-1} B_{m,j} \right)$$

for $2 \le \ell \le L_m$; and take $\alpha_{m,\ell} = \rho(A_{m,\ell})$. Next, for $m \in \mathbb{Z}^+$ and $1 \le \ell \le L_m$, choose $P_{m,\ell} \in K \cap B_{m,\ell}$ so that

$$\Phi(P_{m,\ell}) \le \inf\Big\{ \Phi(R) : R \in K \cap B_{m,\ell} \Big\} + \frac{1}{m};$$

and define $\overline{P}_{m,\ell}$ by

$$\overline{P}_{m,\ell} = \begin{cases} \frac{1}{\alpha_{m,\ell}} \int_{A_{m,\ell}} R\, \rho(dR) & \text{if } \alpha_{m,\ell} \ne 0 \\ R_{m,\ell} & \text{if } \alpha_{m,\ell} = 0. \end{cases}$$

Assuming that Φ is convex, we have that

$$\Phi\left(\sum_{\ell=1}^{L_m} \alpha_{m,\ell} P_{m,\ell} \right) \le \sum_{\ell=1}^{L_m} \alpha_{m,\ell} \Phi\big(P_{m,\ell}\big) = \int_K \Phi_m(R)\, \rho(dR),$$

where $\Phi_m(R) = \Phi(P_{m,\ell})$ for $R \in A_{m,\ell}$. Since Φ is lower semi-continuous, $\Phi_m(R) \longrightarrow \Phi(R)$ for each $R \in K$. Thus, when Φ is bounded, LEBESGUE's Dominated Convergence Theorem shows that

$$\int_K \Phi_m(R)\, \rho(dR) \longrightarrow \int_{\mathbf{M}_1(\Omega)} \Phi(R)\, \rho(dR)$$

as $m \longrightarrow \infty$. At the same time,

$$\sum_{\ell=1}^{L_m} \alpha_{m,\ell} P_{m,\ell} \Longrightarrow \int_{\mathbf{M}_1(\Omega)} R\, \rho(dR);$$

and so, again by lower semi-continuity,

$$\Phi\left(\int_{\mathbf{M}_1(\Omega)} R\, \rho(dR) \right) \le \varliminf_{m \to \infty} \Phi\left(\sum_{\ell=1}^{L_m} \alpha_{m,\ell} P_{m,\ell} \right);$$

and together, these imply the desired result when Φ is bounded. Thus, even if Φ is not bounded, we have that (5.4.25) holds for $\Phi \wedge n$; and, therefore, a passage to the limit as $n \longrightarrow \infty$ yields the result for Φ's which are not necessarily bounded.

Next, assume that Φ is affine on $\mathbf{M}_1^{\mathrm{S}}(\Omega)$. Because $\mathbf{M}_1^{\mathrm{S}}(\Omega)$ is closed, we may and will assume that the K for which $\rho(K) = 1$ is contained in $\mathbf{M}_1^{\mathrm{S}}(\Omega)$; and therefore that each of the measures $\overline{P}_{m,\ell}$ is an element of $\mathbf{M}_1^{\mathrm{S}}(\Omega)$. Thus, since $\int_{\mathbf{M}_1^{\mathrm{S}}(\Omega)} R\,\rho(dR) = \sum_{\ell=1}^{L_m} \alpha_{m,\ell}\overline{P}_{m,\ell}$,

$$\Phi\left(\int_{\mathbf{M}_1^{\mathrm{S}}(\Omega)} R\,\rho(dR)\right) = \sum_{\ell=1}^{L_m} \alpha_{m,\ell}\Phi\left(\overline{P}_{m,\ell}\right) = \int_K \overline{\Phi}_m(R)\,\rho(dR),$$

where $\overline{\Phi}_m(R) = \Phi(\overline{P}_{m,\ell})$ for $R \in A_{m,\ell}$. Noting that, by lower semicontinuity,

$$\Phi(R) \leq \varliminf_{m\to\infty} \overline{\Phi}_m(R)$$

for each $R \in K$, we can now use FATOU's Lemma to conclude that the left hand side of (5.4.26) dominates the right hand side. At the same time, by the result in the preceding paragraph, the opposite inequality also holds.

We have now completed the proof in the case when ρ is compactly supported. To handle the case when ρ is not compactly supported, choose a non-decreasing sequence of compact sets K_n so that $\rho(K_n) \geq (n-1)/n$; set $\alpha_n = \rho(K_n)$; and define $\sigma_n(\Gamma) = \frac{1}{\alpha_n}\rho(\Gamma \cap K_n)$ and $\tau_n(\Gamma) = \frac{1}{1-\alpha_n}\rho(\Gamma \cap K_n^{\mathrm{c}})$ for $\Gamma \in \mathcal{B}_{\mathbf{M}_1(\Omega)}$. Since each σ_n is compactly supported and $\int R\,\sigma_n(dR) \Longrightarrow \int R\,\rho(dR)$, we see from the above that

$$\Phi\left(\int_{\mathbf{M}_1(\Omega)} R\,\rho(dR)\right) \leq \varliminf_{n\to\infty} \Phi\left(\int_{\mathbf{M}_1(\Omega)} R\,\sigma_n(dR)\right) \leq \int_{\mathbf{M}_1(\Omega)} \Phi(R)\,\rho(dR)$$

when Φ is convex. On the other hand, if Φ is affine, then

$$\begin{aligned}
&\Phi\left(\int_{\mathbf{M}_1^{\mathrm{S}}(\Omega)} R\,\rho(dR)\right)\\
&= \Phi\left(\alpha_n \int_{\mathbf{M}_1^{\mathrm{S}}(\Omega)} R\,\sigma_n(dR) + (1-\alpha_n)\int_{\mathbf{M}_1^{\mathrm{S}}(\Omega)} R\,\tau_n(dR)\right)\\
&= \alpha_n\Phi\left(\int_{\mathbf{M}_1(\Omega)} R\,\sigma_n(dR)\right) + (1-\alpha_n)\Phi\left(\int_{\mathbf{M}_1(\Omega)} R\,\tau_n(dR)\right)\\
&\geq \alpha_n\Phi\left(\int_{\mathbf{M}_1(\Omega)} R\,\sigma_n(dR)\right) = \alpha_n\int_{\mathbf{M}_1(\Omega)} \Phi(R)\,\sigma_n(dR).
\end{aligned}$$

Since it is clear that

$$\alpha_n \int_{\mathbf{M}_1(\Omega)} \Phi(R)\,\sigma_n(dR) \nearrow \int_{\mathbf{M}_1(\Omega)} \Phi(R)\,\rho(dR),$$

we are done. ∎

Applying Lemma 5.4.24 to J and $\mathbf{H}$, we now see that

$$\begin{aligned} J(Q) &\le \int_{\mathbf{EM}_1^{\mathrm{S}}(\Omega)} J(R)\,\rho_Q(dR) \\ &\le \int_{\mathbf{EM}_1^{\mathrm{S}}(\Omega)} \mathbf{H}(R)\,\rho_Q(dR) = \mathbf{H}(Q), \quad Q \in \mathbf{M}_1^{\mathrm{S}}(\Omega), \end{aligned}$$

where $\rho_Q \in \mathbf{M}_1\big(\mathbf{EM}_1^{\mathrm{S}}(\Omega)\big)$ is the measure described in the Ergodic Decomposition Theorem. Hence, we have now completed STEP 3; and therefore we have derived the following version of a theorem proved originally by T. CHIYONOBU and S. KUSUOKA in [**18**].

5.4.27 Theorem. *Assume that $P \in \mathbf{M}_1^{\mathrm{S}}(\Omega)$ is hypermixing. Then the specific entropy function $\mathbf{H} : \mathbf{M}_1(\Omega) \longrightarrow [0,\infty]$ in (5.4.8) exists (i.e., the indicated limit exists) and defines a good rate function which governs the large deviations of $\{P \circ \mathbf{R}_T^{-1} : T \in (0,\infty)\}$ as $T \longrightarrow \infty$.*

At the beginning of this section we mentioned that there are certain technical difficulties associated with taking Ω to be the SKOROKHOD space $D(\mathbb{R};\Sigma)$ of right-continuous paths $\omega : \mathbb{R} \longrightarrow \Sigma$ which have a left-limit at each $t \in \mathbb{R}$. The difficulties alluded to stem from the problem of putting a Polish topology on Ω which is the projective limit of Polish topologies on the SKOROKHOD spaces of paths on finite time intervals. To be precise, let I be a compact interval and denote by $D(I;\Sigma)$ the space of right-continuous paths $\omega_I : I \longrightarrow \Sigma$ which have a left-limit at each $t \in I$ and are left-continuous at the right hand end of I. Using SKOROKHOD's prescription, one can then put a metric ρ_I on $D(I;\Sigma)$ in such a way that $\big(D(I;\Sigma),\rho_I\big)$ is a complete, separable metric space and ρ_I-convergence of $\{\omega_{I,\ell}\}_{\ell=1}^{\infty}$ to ω_I is equivalent to

$$\lim_{\ell\to\infty} \inf \left\{ \sup_{t\in I} \mathrm{dist}_\Sigma\big(\Sigma_t(\omega_{I,\ell}), \Sigma_{\lambda(t)}(\omega_I)\big) + \sup_{t\in I} |\lambda(t) - t| : \lambda \in L_I \right\} = 0,$$

where dist_Σ denotes the distance on Σ determined by the Σ's metric and L_I stands for the group of increasing homeomorphisms of I onto itself. Furthermore, the ρ_I's can be chosen so that if $I = [a,b]$ and $J = [c,d]$, where $c \le a$ and $b \le d$, and if ω_J, ω'_J are elements of $D(J;\Sigma)$ which are left-continuous at b, then

$$\rho_I\left(\omega_J\big|_I, \omega'_J\big|_I\right) \le \rho_J\left(\omega_J, \omega'_J\right). \tag{5.4.28}$$

The problem comes from the fact that $\omega_J \in D(J;\Sigma)$ and $I \subseteq J$ do not guarantee that $\omega_J\big|_I$ is an element of $D(I;\Sigma)$, since ω_J need not be left-continuous at the right end of I. Worse, even if one replaces the restriction map by $\pi_I : D(J;\Sigma) \longrightarrow D(I;\Sigma)$ given by

$$\Sigma_s(\pi_I\omega_J) = \begin{cases} \Sigma_s(\omega_J) & \text{if } s \in [a,b) \\ \Sigma_{b-}(\omega_J) \equiv \lim_{t \nearrow b} \Sigma_t(\omega_J) & \text{if } s = b, \end{cases}$$

the situation in (5.4.28) does not improve substantially (i.e., the topologies still do not mesh correctly). For this reason, we will adopt a scheme for introducing a topology on $D(\mathbb{R};\Sigma)$ which is slightly different from the one which we used for $C(\mathbb{R};\Sigma)$.

From now on, Ω will denote $D(\mathbb{R};\Sigma)$; and, for compact intervals I, ρ_I will be the metric introduced by SKOROKHOD on $D(I;\Sigma)$. Given $T \in (0,\infty)$, we will use Ω_T to denotes the space $D\big((-T,T);\Sigma\big)$ of paths $\omega_T : (-T,T) \longrightarrow \Sigma$ which are right-continuous and have a left limit at each $t \in (-T,T)$. Next, we define the metric d_T on Ω_T by

$$d_T(\omega_T, \omega'_T) = \int_{(0,T)} e^{-t} \frac{\rho_{[-t,t]}\left(\pi_{[-t,t]}\omega_T, \pi_{[-t,t]}\omega'_T\right)}{1 + \rho_{[-t,t]}\left(\pi_{[-t,t]}\omega_T, \pi_{[-t,t]}\omega'_T\right)}\, dt, \quad \omega_T, \omega'_T \in \Omega_T;$$

and we take

$$d(\omega, \omega') = \int_{(0,\infty)} e^{-t} \frac{\rho_{[-t,t]}\left(\pi_{[-t,t]}\omega, \pi_{[-t,t]}\omega'\right)}{1 + \rho_{[-t,t]}\left(\pi_{[-t,t]}\omega, \pi_{[-t,t]}\omega'\right)}\, dt, \quad \omega, \omega' \in \Omega.$$

Finally, we define $\pi_{T,S} : \Omega_T \longrightarrow \Omega_S$, $0 < S < T$, and $\pi_S : \Omega \longrightarrow \Omega_S$, $S \in (0,\infty)$, to be the natural restriction mappings.

As a relatively straight-forward application of the fact that each $\omega \in \Omega$ can have at most countably many points of discontinuity, one can use (5.4.28) to check all but the final assertion in the following lemma. The final assertion is a consequence of the well-known facts that, for each compact interval I, the SKOROKHOD topology on $D(I;\Sigma)$ restricts to the uniform topology on $C(I;\Sigma)$ and that the Borel field of the SKOROKHOD topology is the σ-algebra generated by the evaluation maps Σ_t, $t \in I$.

5.4.29 Lemma. *Each of the spaces (Ω_T, d_T), $T \in (0,\infty)$ is a complete separable metric space; and, for all $0 < S < T$,*

$$d_S\big(\pi_{T,S}\omega, \pi_{T,S}\omega'_T\big) \le d_T(\omega_T, \omega'_T), \quad \omega_T,\, \omega'_T \in \Omega_T.$$

Moreover, (Ω, d) is a complete, separable metric space which is homeomorphic to the projective limit of the sequence $\{(\Omega_n, \pi_{n+1,n}, d_n) : n \in \mathbb{Z}^+\}$; and $(t,\omega) \in (0,\infty)\times\Omega \longmapsto \theta_t\omega$ is continuous. Finally, the relative topology which $C(\mathbb{R};\Sigma)$ inherits as a subset of (Ω, d) coincides with the topology of uniform convergence on compacts, and $\mathcal{B}_\Omega$ is the σ-algebra over Ω generated by the maps $\omega \in \Omega \longmapsto \Sigma_t(\omega) \in \Sigma$, $t \in \mathbb{R}$.

Once one has the facts contained in Lemma 5.4.29, the argument used to prove Theorem 5.4.27 with $\Omega = C(\mathbb{R};\Sigma)$ applies without change to the case when $\Omega = D(\mathbb{R};\Sigma)$.

5.4.30 Exercise.

Formulate and prove the analogue of Theorem 5.4.27 for the discrete-parameter setting.

5.4.31 Exercise.

Let Ω be either $C(\mathbb{R};\Sigma)$ or $D(\mathbb{R};\Sigma)$ and let Σ' be a second Polish space. Suppose that $F : \Omega \longrightarrow \Sigma'$ is a $\mathcal{B}_{[-T,T]}$-measurable map for some $T \in [0,\infty)$, and assume that $t \in \mathbb{R} \longmapsto F(\theta_t\omega) \in \Sigma'$ is an element of $\Omega' \equiv D(\mathbb{R};\Sigma')$ for each $\omega \in \Omega$. Finally, define $\Phi : \Omega \longrightarrow \Omega'$ so that $\Sigma'_t\big(\Phi(\omega)\big) = F\big(\theta_t\omega\big)$ for $t \in \mathbb{R}$. Given a $P \in \mathbf{M}_1^{\mathrm{S}}(\Omega)$ which is hypermixing, show that $P' \equiv P \circ \Phi^{-1}$ is a hypermixing element of $\mathbf{M}_1^{\mathrm{S}}(\Omega')$.

5.4.32 Exercise.

Let $\Omega = C(\mathbb{R};\Sigma)$, and suppose that $P \in \mathbf{M}_1(\Omega)$ admits a good rate function $J : \mathbf{M}_1(\Omega) \longrightarrow [0,\infty]$ which governs the large deviations of $\{P \circ \mathbf{R}_T^{-1} : T > 0\}$. Next, define the empirical position measure

$$\mathbf{L}_T(\omega) = \frac{1}{2T}\int_{-T}^{T} \delta_{\Sigma_t(\omega)}\, dt,$$

and observe that $\mathbf{L}_T(\omega) = \mathbf{R}_T(\omega) \circ \Sigma_0^{-1}$. Thus, since $\omega \in \Omega \longmapsto \Sigma_0(\omega) \in \Sigma$ is continuous, and, therefore, so is $R \in \mathbf{M}_1(\Omega) \longmapsto R \circ \Sigma_0^{-1} \in \mathbf{M}_1(\Sigma)$, the final part of Lemma 2.1.4 says that

$$I(\mu) = \inf\big\{J(R) :\ R \in \mathbf{M}_1(\Omega) \text{ and } \mu = R \circ \Sigma_0^{-1}\big\}, \quad \mu \in \mathbf{M}_1(\Sigma),$$

is a good rate function which governs the large deviations of $\{P \circ \mathbf{L}_T^{-1} : T > 0\}$.

Now let $\Omega = D(\mathbb{R};\Sigma)$ and suppose that there exist $P \in \mathbf{M}_1(\Omega)$ and a good rate function $J : \mathbf{M}_1(\Omega) \longrightarrow [0,\infty]$ which is related to P as in the preceding paragraph. What one would like is to repeat the argument just given and thereby show that the large deviations of $\{P \circ \mathbf{L}_T : T > 0\}$ are governed by a rate function of the sort described above. The problem is, of course, that $\omega \in \Omega \longmapsto \Sigma_0(\omega) \in \Sigma$ is no longer a continuous mapping. In order to circumvent this problem, one can take the following sequence of easy steps.

(i) Set $\Omega^0 = \{\omega : \Sigma_0(\omega) = \Sigma_{0-}(\omega)\}$ and show that Ω^0 is a $\mathfrak{G}_\delta$-subset of Ω and that $\omega \in \Omega^0 \longmapsto \Sigma_0(\omega) \in \Sigma$ is continuous. Conclude that $\mathbf{M}_1^0(\Omega) \equiv \{Q \in \mathbf{M}_1(\Omega) : Q(\Omega^0) = 1\}$ is a $\mathfrak{G}_\delta$ subset of $\mathbf{M}_1(\Omega)$ and that $Q \in \mathbf{M}_1^0(\Omega) \longmapsto Q \circ \Sigma_0^{-1}$ is continuous. Finally, check that $\mathbf{M}_1^{\mathrm{S}}(\Omega) \subseteq \mathbf{M}_1^0(\Omega)$.

(ii) For $(T,\omega) \in (0,\infty) \times \Omega$, define $\tilde{\omega}_T \in \Omega$ so that $\tilde{\omega}_T\big|_{[-T,T)} = \omega_{[-T,T)}$ and $\theta_{2T}\tilde{\omega}_T = \tilde{\omega}_T$. Show that $(T,\omega) \in (0,\infty) \times \Omega \longmapsto \tilde{\omega}_T \in \Omega$ is measurable and therefore so is $(T,\omega) \in (0,\infty) \times \Omega \longmapsto \tilde{\mathbf{R}}_T(\omega) \equiv \mathbf{R}_T(\tilde{\omega}_T) \in \mathbf{M}_1^{\mathrm{S}}(\Omega)$. In addition, check that, for each $S \in [0,\infty)$,

$$\lim_{T\to\infty} \sup_{\omega\in\Omega} \left\| \mathbf{R}_T(\omega) \circ \pi_{[-S,S]}^{-1} - \tilde{\mathbf{R}}_T(\omega) \circ \pi_{[-S,S]}^{-1} \right\|_{\mathrm{var}} = 0.$$

(iii) Suppose that $P \in \mathbf{M}_1(\Omega)$ and that $J : \mathbf{M}_1(\Omega) \longrightarrow [0,\infty]$ is a good rate function which governs the large deviations of $\{P \circ \mathbf{R}_T^{-1} : T \in (0,\infty)\}$ as $T \longrightarrow \infty$. Show that $J\big|_{\mathbf{M}_1^{\mathrm{S}}(\Omega)}$ is a good rate function which governs the large deviations of $\{P \circ \tilde{\mathbf{R}}_T^{-1} : T \in (0,\infty)\}$ as $T \longrightarrow \infty$. Next, define

$$\mathbf{L}_T(\omega) = \frac{1}{T}\int_0^T \delta_{\Sigma_t(\omega)}\,dt$$

and show that $\{P \circ \mathbf{L}_T^{-1} : T \in (0,\infty)\}$ satisfies the full large deviation principle with respect to the good rate function

$$\mu \in \mathbf{M}_1(\Sigma) \longmapsto I(\mu) \equiv \inf\Big\{J(Q) : Q \in \mathbf{M}_1^{\mathrm{S}}(\Omega) \text{ and } Q \circ \Sigma_0^{-1} = \mu\Big\}.$$

In particular, when $P \in \mathbf{M}_1^{\mathrm{S}}(\Omega)$ is hypermixing, conclude that $\{P \circ \mathbf{L}_T^{-1} : T \in (0,\infty)\}$ satisfies the full large deviation principle with the good rate function $I : \mathbf{M}_1(\Sigma) \longrightarrow [0,\infty]$ given by

$$I(\mu) = \inf\Big\{\mathbf{H}(Q) : Q \circ \Sigma_0^{-1} = \mu\Big\}. \tag{5.4.33}$$

5.4.34 Exercise.

Let $P \in \mathbf{M}_1^{\mathrm{S}}(\Omega)$ be hypermixing. Starting from (5.4.14), show that, for each compact interval I, $V \in B(\Omega_I;\mathbb{R}) \longmapsto \Lambda_I(V) \in \mathbb{R}$ is a continuous function of bounded, point-wise convergence. (**Hint:** See the proof of Lemma 4.1.40.) Conclude that

$$\Lambda_I^*(Q \circ \pi_I^{-1}) = \overline{\Lambda}_I{}^*(Q \circ \pi_I^{-1}) \tag{5.4.35}$$
$$\equiv \sup\left\{\int_\Omega V \circ \pi_I \, dQ - \Lambda_I(V) : V \in B(\Omega_I; \mathbb{R})\right\}$$

for $Q \in \mathbf{M}_1(\Omega)$.

5.4.36 Exercise.

Let $P(t,\sigma,\cdot)$ be a transition probability function on Σ and assume that the corresponding MARKOV family $\{P_\sigma : \sigma \in \Sigma\}$ can be realized on $D([0,\infty);\Sigma)$. Also, suppose that there is precisely one $P(t,\sigma,\cdot)$-invariant $\mu \in \mathbf{M}_1(\Sigma)$; and denote by P the unique element of $\mathbf{M}_1^{\mathrm{S}}(\Omega)$ with the property that

$$P\big(\{\omega : \Sigma_t(\omega) \in \Gamma\}\big|\mathcal{B}_{(-\infty,s]}\big) = P(t-s, \Sigma_s, \Gamma) \quad (\text{a.s.}, P) \tag{5.3.37}$$

for $-\infty < s < t < \infty$ and $\Gamma \in \mathcal{B}_\Sigma$. (Obviously, $P \circ \Sigma_t^{-1} = \mu$ for all $t \in \mathbb{R}$.) Finally, assume that P is hypermixing. The purpose of this exercise is to see when the rate function I in (5.4.33) can be identified with one of the rate functions which we produced in Section 4.2.

(**i**) Show that if $\mu = m$ is $P(t,\sigma,\cdot)$-reversing, then $I = J_{\mathcal{E}}$, where $J_{\mathcal{E}}$ is defined from the associated DIRICHLET form $\mathcal{E}$ (cf. (4.2.47)) as in (4.2.49). (**Hint:** Use (5.4.18) with $I = \{0\}$ and Exercise 5.3.15.)

(**ii**) The non-reversible case is not so satisfactory. To see what sort of thing can be said, define $\tilde{\Lambda}_P$, $\tilde{\Lambda}_P^*$, and $\overline{\tilde{\Lambda}}_P^*$ as in Exercise 5.1.17, and J_P and $\overline{J}_P$ as in (4.2.38) and (4.2.36). Noting that $\Lambda_{\{0\}} \leq \Lambda_P$ (cf. (4.2.21)), show that $I \geq \overline{J}_P$. Next, if, for some $V \in B(\Sigma;\mathbb{R})$,

$$\sup_{\sigma\in\Sigma} \varlimsup_{T\to\infty} \frac{1}{T} \log\left(\int_{D([0,\infty);\Sigma)} \exp\left[\int_0^T V(\Sigma_t(\omega))\,dt\right] P_\sigma(d\omega)\right) \tag{5.4.38}$$
$$= \inf_{\sigma\in\Sigma} \varliminf_{T\to\infty} \frac{1}{T} \log\left(\int_{D([0,\infty);\Sigma)} \exp\left[\int_0^T V(\Sigma_t(\omega))\,dt\right] P_\sigma(d\omega)\right)$$

show that $\tilde{\Lambda}_P(V) \leq \Lambda_{\{0\}}(V)$. Conclude from this, Exercise 5.1.17, and Exercise 5.4.34 that $I = \overline{J}_P$ when (5.4.38) holds for every $V \in B(\Sigma;\mathbb{R})$. Similarly, when $P(t,\sigma,\cdot)$ is FELLER-continuous, show that, when (5.4.38) holds for every $V \in C_{\mathrm{b}}(\Sigma;\mathbb{R})$, I must equal J_P.

5.4.39 Exercise.

One of the more remarkable features of the hypermixing property is its behavior under products. To be precise, let $\mathfrak{I}$ be a countable index set and for each $i \in \mathfrak{I}$ let P_i be a hypermixing element of $\mathbf{M}_1^{\mathrm{S}}\big(D(\mathbb{R};\Sigma_i)\big)$, where each Σ_i is a Polish space. Further, assume that there are functions α, β, and γ satisfying (5.4.1) such that (**H-1**) and (**H-2**) hold with $P = P_i$ for all $i \in \mathfrak{I}$. After making the obvious identification of

$$\prod_{i\in\mathfrak{I}} D(\mathbb{R};\Sigma_i) \quad \text{with} \quad D\left(\mathbb{R};\prod_{i\in\mathfrak{I}}\Sigma_i\right),$$

show that $\prod_{i\in\mathfrak{I}} P_i$ determines an element of

$$\mathbf{M}_1^{\mathrm{S}}\left(D\left(\mathbb{R};\prod_{i\in\mathfrak{I}}\Sigma_i\right)\right)$$

which is hypermixing with the same choice of functions α, β, and γ.

5.4.40 Exercise.

Define the τ-topology on $\mathbf{M}_1(\Omega)$ to be the weakest topology with respect to which the mapping

$$Q \in \mathbf{M}_1(\Omega) \longmapsto \int_\Omega V \circ \pi_I \, dQ$$

is continuous for each compact interval I and $V \in B\big(\Omega_I;\mathbb{R}\big)$. Given a $\Gamma \subseteq \mathbf{M}_1(\Omega)$, let $\Gamma^{\circ\,\tau}$ and $\overline{\Gamma}^{\,\tau}$ denote, respectively, the interior and closure of Γ in the τ-topology. Assuming that $P \in \mathbf{M}_1^{\mathrm{S}}(\Omega)$ is hypermixing, show that, for every measurable $\Gamma \subseteq \mathbf{M}_1(\Omega)$,

$$\begin{aligned}
-\inf_{Q\in\Gamma^{\circ\,\tau}} \mathbf{H}(Q) &\le \varliminf_{t\to\infty} \frac{1}{t}\log\Big(P\big(\{\omega : \mathbf{R}_t(\omega) \in \Gamma\}\big)\Big) \\
&\le \varlimsup_{t\to\infty} \frac{1}{t}\log\Big(P\big(\{\omega : \mathbf{R}_t(\omega) \in \Gamma\}\big)\Big) \le -\inf_{Q\in\overline{\Gamma}^{\,\tau}} \mathbf{H}(Q).
\end{aligned}$$

(**Hint:** Use the estimate on which (5.4.14) is based and apply Theorem 3.2.21.)

With the preceding in hand, one sees that it would have been possible to avoid some of the difficulties associated with the Skorokhod topology by proceeding along a line of reasoning like the one which we used to complete the program in Section 4.4.

5.5 Hypermixing in the Epsilon Markov Case

In this section, we develop a sufficient condition for the hypermixing property to hold. Throughout, Ω will denote the space $D(\mathbb{R};\Sigma)$ (cf. the discussion following Theorem 5.4.27) and P will denote a fixed element of $\mathbf{M}_1^{\mathrm{S}}(\Omega)$.

Recall the σ-algebras $\mathcal{B}_I = \sigma\big(\{\Sigma_t : t \in I\}\big)$, where I runs over intervals in $\mathbb{R}$. We will use $B_I(\Omega;\mathbb{R})$ to denote the subset of $f \in B(\Omega;\mathbb{R})$ which are $\mathcal{B}_I$-measurable. Also, given I, choose $\omega \in \Omega \longmapsto P_\omega^I \in \mathbf{M}_1(\Omega)$ to be a regular conditional probability distribution of P given $\mathcal{B}_I$ and define $E_I : B(\Omega;\mathbb{R}) \longrightarrow B_I(\Omega;\mathbb{R})$ so that $E_I f(\omega) = \int_\Omega f(\omega')\, P_\omega^I(d\omega')$. Notice that, by JENSEN's inequality,

$$\|E_I\|_{L^p(P)\to L^p(P)} = 1, \quad p \in [1,\infty], \tag{5.5.1}$$

where

$$\|K\|_{L^p(\mu)\to L^q(\mu)} \equiv \sup\left\{ \frac{\|K\phi\|_{L^q(\mu)}}{\|\phi\|_{L^p(\mu)}} : \phi \in B\big((E,\mathcal{F});\mathbb{R}\big) \setminus \{0\} \right\} \tag{5.5.2}$$

for p, $q \in [1,\infty]$ and any operator K defined on the bounded measurable functions $B\big((E,\mathcal{F});\mathbb{R}\big)$ of a measure space $(E,\mathcal{F},\mu)$. In addition, by shift-invariance, one has that

$$E_{s+I} f = \big[E_I(f \circ \theta_{-s})\big] \circ \theta_s \quad \text{(a.s., } P\text{)} \tag{5.5.3}$$

for all $s \in \mathbb{R}$ and $f \in B(\Omega;\mathbb{R})$.

Using E_s^- and E_s^+ to denote $E_{(-\infty,s]}$ and $E_{[s,\infty)}$, respectively; we now define $\mathcal{P}_t : B(\Omega;\mathbb{R}) \longrightarrow B(\Omega;\mathbb{R})$ for $t \in (0,\infty)$ by

$$\mathcal{P}_t f = E_0^-\big[E_0^+(f \circ \theta_t)\big] = E_0^-\big[(E_{-t}^+ f) \circ \theta_t\big]. \tag{5.5.4}$$

Obviously,

$$\|\mathcal{P}_t\|_{L^p(P)\to L^p(P)} = 1, \quad p \in [1,\infty]. \tag{5.5.5}$$

In addition, if $f \in B_{-s}^+(\Omega;\mathbb{R}) \equiv B_{[-s,\infty)}(\Omega;\mathbb{R})$ and $0 < s < t < \infty$, then (cf. (5.5.3)) P-almost surely:

$$\begin{aligned} \mathcal{P}_t f &= E_0^-(f \circ \theta_t) = E_0^- E_{t-s}^-(f \circ \theta_t) \\ &= E_0^-\Big(\big[E_0^-(f \circ \theta_s)\big] \circ \theta_{t-s}\Big) = E_0^-\big[(\mathcal{P}_s f) \circ \theta_{t-s}\big]; \end{aligned}$$

and therefore, by (5.5.1), we see that

$$\|\mathcal{P}_t f\|_{L^p(P)} \le \|\mathcal{P}_s f\|_{L^p(P)}, \quad 0 < s < t < \infty \tag{5.5.6}$$

for $p \in [1,\infty]$ and $f \in B_{-s}^+(\Omega;\mathbb{R})$. Finally, another application of (5.5.3) yields

$$E_s^- E_s^+ f = \big[\mathcal{P}_t(f \circ \theta_{-s-t})\big] \circ \theta_s \quad \text{(a.s., } P\text{)} \tag{5.5.7}$$

for $s \in \mathbb{R}$, $t \in (0,\infty)$, and $f \in B(\Omega;\mathbb{R})$.

We are now ready to describe an extension of the usual MARKOV property. Given $\epsilon \in [0, \infty)$, we will say that P is ϵ-**Markov** if

(EM) $$E_0^- E_0^+ f = E_{[-\epsilon,0]} f \quad \text{(a.s., } P), \quad f \in B(\Omega; \mathbb{R}).$$

Notice that, with only a P-negligible alteration in the definition, we may and will assume that

$$\mathcal{P}_t : B(\Omega;\mathbb{R}) \longrightarrow B_{[-\epsilon,0]}(\Omega;\mathbb{R}), \quad t \in (0,\infty) \tag{5.5.8}$$

when P is ϵ-MARKOV. Thus, if P is ϵ-MARKOV and $s,\, t \in [\epsilon, \infty)$, then, by (5.5.7),

$$\begin{aligned}\mathcal{P}_{s+t} f &= E_0^-(f \circ \theta_{s+t}) = E_0^- E_s^-(f \circ \theta_{s+t}) \\ &= E_0^-\big[(\mathcal{P}_t f)\circ\theta_s\big] = \mathcal{P}_s\big(\mathcal{P}_t f\big) \quad \text{(a.s., } P)\end{aligned}$$

for $f \in B^+_{-\epsilon}(\Omega;\mathbb{R})$. Hence, we have the following semigroup property.

5.5.9 Lemma. *If P is ϵ-MARKOV, then, for $s \in [\epsilon,\infty)$, $\mathcal{P}_s$ maps $B^+_{-\epsilon}(\Omega;\mathbb{R})$ into itself and*

$$\mathcal{P}_{s+t} f = \mathcal{P}_s\big(\mathcal{P}_t f\big) \quad \text{(a.s., } P) \tag{5.5.10}$$

for $s,\, t \in [\epsilon,\infty)$ and $f \in B^+_{-\epsilon}(\Omega;\mathbb{R})$.

We next introduce a property which, in conjunction with the ϵ-MARKOV property, will guarantee hypermixing. Namely, if $P \in \mathbf{M}_1^{\mathrm{S}}(\Omega)$ is ϵ-MARKOV and $T_0 \in [\epsilon,\infty)$, we say that P is T_0-**hypercontractive** if

(HC) $$\big\|\mathcal{P}_{T_0} f\big\|_{L^4(P)} \le \|f\|_{L^2(P)} \quad \text{for } f \in B^+_{-\epsilon}(\Omega;\mathbb{R}).$$

5.5.11 Lemma. *If P is T_0-hypercontractive, then*

$$\left\|\mathcal{P}_{T_0} f - \int_\Omega f\,dP\right\|_{L^2(P)} \le 3^{-1/2}\|f\|_{L^2(P)}, \quad f \in B^+_{-\epsilon}(\Omega;\mathbb{R}).$$

PROOF: Let $f \in B^+_{-\epsilon}(\Omega;\mathbb{R})$ with $\int_\Omega f\,dP = 0$ be given. Then, by the T_0-hypercontraction property, for every $a \in \mathbb{R}$:

$$\begin{aligned}\big(a^2 + \|f\|^2_{L^2(P)}\big)^2 &= \|a1 + f\|^4_{L^2(P)} \ge \big\|\mathcal{P}_{T_0}(a1+f)\big\|^4_{L^4(P)} \\ &\ge a^4 + 6a^2\big\|\mathcal{P}_{T_0} f\big\|^2_{L^2(P)} - 4|a|\big\|\mathcal{P}_{T_0} f\big\|^3_{L^3(P)} + \big\|\mathcal{P}_{T_0} f\big\|^4_{L^4(P)};\end{aligned}$$

and, therefore,

$$2a^2\|f\|^2_{L^2(P)} \ge 6a^2\big\|\mathcal{P}_{T_0} f\big\|^2_{L^2(P)} + \mathcal{O}(a) \quad \text{as } a \longrightarrow \infty.$$

Clearly, this shows that $\big\|\mathcal{P}_{T_0} f\big\|_{L^2(P)} \le 3^{-1/2}\|f\|_{L^2(P)}$, which is all that we needed to know. ∎

5.5.12 Theorem. *Assume that P is ϵ-MARKOV and T_0-hypercontractive. Then,*

$$\left\| \mathcal{P}_t f - \int_\Omega f\,dP \right\|_{L^2(P)} \leq 3^{1/2} \exp\left[-\frac{t \log 3}{2T_0} \right] \|f\|_{L^2(P)} \tag{5.5.13}$$

for $t \in [T_0, \infty)$ and $f \in B^+_{-\epsilon}(\Omega; \mathbb{R})$. Moreover, if $\alpha \equiv \frac{\log(3/2)}{4T_0}$, then for $1 < p < q < \infty$ and $t \geq 4T_0$:

$$\left\| \mathcal{P}_t f \right\|_{L^q(P)} \leq \|f\|_{L^p(P)} \quad \text{for } f \in B^+_{-\epsilon}(\Omega; \mathbb{R}) \tag{5.5.14}$$

as long as $e^{\alpha t} \geq \frac{q-1}{p-1}$. In particular, if $p(t) = 1 + \exp(-\alpha t)$ and $q(t) = 1 + \exp(\alpha t)$, then, for $t \in [4T_0, \infty)$:

$$\left\| \mathcal{P}_t f \right\|_{L^2(P)} \leq \|f\|_{L^{p(t)}(P)} \quad \text{and} \quad \left\| \mathcal{P}_t f \right\|_{L^{q(t)}(P)} \leq \|f\|_{L^2(P)} \tag{5.5.15}$$

whenever $f \in B^+_{-\epsilon}(\Omega; \mathbb{R})$.

PROOF: The first assertion is an immediate consequence of Lemma 5.5.9 and Lemma 5.5.11.

In proving the rest of the theorem, we will use p' to denote the HÖLDER conjugate of $p \in [1, \infty]$. Set $\theta = 4'/2' = 2/3$, and define r_n for $n \in \mathbb{N}$ so that $1/r_n' = \theta^n/4'$. It is then clear that $r_n \searrow 1$ and that

$$\frac{1}{r_n} = \frac{1-\theta}{1} + \frac{\theta}{r_{n-1}} \quad \text{and} \quad \frac{1}{r_{n+1}} = \frac{1-\theta}{1} + \frac{\theta}{r_n}.$$

Hence, by the RIESZ–THORIN Interpolation Theorem applied to the LEBESGUE spaces

$$L^p_{-\epsilon}(P) \equiv \left\{ \phi \in L^p(P) : \phi \text{ is } \mathcal{B}_{[-\epsilon,\infty)}\text{-measurable} \right\},$$

we see that

$$\left\| \mathcal{P}_{T_0} \right\|_{L^{r_{n+1}}_{-\epsilon}(P) \to L^{r_n}_{-\epsilon}(P)} \leq \left\| \mathcal{P}_{T_0} \right\|^{1-\theta}_{L^1_{-\epsilon}(P) \to L^1_{-\epsilon}(P)} \left\| \mathcal{P}_{T_0} \right\|^{\theta}_{L^{r_n}_{-\epsilon}(P) \to L^{r_{n-1}}_{-\epsilon}(P)};$$

and so, by induction and Lemma 5.5.9, we get

$$\left\| \mathcal{P}_{(n-m)T_0} \right\|_{L^{r_n}_{-\epsilon}(P) \to L^{r_m}_{-\epsilon}(P)} \leq 1, \quad 0 \leq m \leq n.$$

Because $\|\mathcal{P}_t\|_{L^p(P) \to L^q(P)} = 1$ for all $t \in (0, \infty)$ and $q \leq p$, it is easy to deduce from the above that $\|\mathcal{P}_t\|_{L^p_{-\epsilon}(P) \to L^q_{-\epsilon}(P)} = 1$ whenever $1 < p < q \leq 2$ and $t \in [2T_0, \infty)$ satisfies $\exp\left[\frac{t}{2T_0} \log(3/2)\right] \geq \frac{q-1}{p-1}$.

Next, set $\overline{\theta} = 2/4 = 1/2$ and define s_n for $n \in \mathbb{N}$ so that $1/s_n = \overline{\theta}^n/2$. Proceeding as in the preceding, one then finds that $\|\mathcal{P}_t\|_{L^p_{-\epsilon}(P) \to L^q_{-\epsilon}(P)} = 1$ for $2 \leq p < q < \infty$ and $t \in [2T_0, \infty)$ satisfying $\exp\left[\frac{t}{2T_0} \log 2\right] \geq \frac{q-1}{p-1}$; and, after combining this with the result obtained before, one arrives at the desired conclusion. ∎

5.5.16 Corollary. *Let everything be as in* Theorem 5.5.12, *and define*

$$\rho(t) = \frac{(1 + \exp[\alpha t])(1 + \exp[-\alpha t])}{\exp[\alpha t] - \exp[-\alpha t]}, \quad t \in (0, \infty).$$

Then for $t \in [8T_0, \infty)$, $0 \le b < c < \infty$ *satisfying* $c \ge b + 2t$, $f \in B_{[-\epsilon,b]}(\Omega; \mathbb{R})$, *and* $g \in B_{[c,\infty)}(\Omega; \mathbb{R})$,

$$\left\|\mathcal{P}_t(fg)\right\|_{L^2(P)} \le \|f\|_{L^{\rho(t)}(P)} \left\|\mathcal{P}_t(g \circ \theta_{-c})\right\|_{L^2(P)}.$$

PROOF: Note that, by (5.5.7),

$$\begin{aligned}\mathcal{P}_t(fg) &= E_0^-\left[(f \circ \theta_t)\left(E_{b+t}^- E_{c+t}^+(g \circ \theta_t)\right)\right] \\ &= E_0^-\left[(f \circ \theta_t)\left(\left[\mathcal{P}_{c-b}(g \circ \theta_{-c})\right] \circ \theta_{b+t}\right)\right] \\ &= \mathcal{P}_t\left[f\left(\left[\mathcal{P}_{c-b}(g \circ \theta_{-c})\right] \circ \theta_{\mathrm{b}}\right)\right] \qquad \text{(a.s., } P).\end{aligned}$$

Thus, by (5.5.15), HÖLDER's inequality, with $r = \rho(t)/p(t)$, and (5.5.6):

$$\begin{aligned}\left\|\mathcal{P}_t(fg)\right\|_{L^2(P)} &\le \left\|f\left(\left[\mathcal{P}_{c-b}(g \circ \theta_{-c})\right] \circ \theta_{\mathrm{b}}\right)\right\|_{L^{p(t)}(P)} \\ &\le \|f\|_{L^{rp(t)}(P)} \left\|\mathcal{P}_{c-b}(g \circ \theta_{-c})\right\|_{L^{q(t)}(P)} \\ &\le \|f\|_{L^{\rho(t)}(P)} \left\|\mathcal{P}_{c-b-t}(g \circ \theta_{-c})\right\|_{L^2(P)} \\ &\le \|f\|_{L^{\rho(t)}(P)} \left\|\mathcal{P}_t(g \circ \theta_{-c})\right\|_{L^2(P)}. \quad \blacksquare\end{aligned}$$

5.5.17 Theorem. *Assume that* P *is* ϵ-MARKOV *and* T_0-*hypercontractive, and define*

$$p : (0, \infty) \longrightarrow (1, \infty) \quad \text{and } \rho : (0, \infty) \longrightarrow (1, \infty)$$

as in Theorem 5.5.12 *and* Corollary 5.5.16, *respectively. Then* P *is hypermixing with* $\ell_0 = 12T_0$, $\alpha(\ell) = \rho(\ell/2)$ *in* (**H-1**), *and* $\beta(\ell) = p(\ell/3)$ *and*

$$\gamma(\ell) = 3^{1/2} \exp\left[-\frac{\ell \log(3/2)}{6T_0}\right]$$

in (**H-2**).

Conversely, if $P \in \mathbf{M}_1^{\mathrm{S}}(\Omega)$ *satisfies* (**H-1**), *then it is* T_0-*hypercontractive for some* $T_0 \in (0, \infty)$; *and so, for* ϵ-MARKOV P*'s, hypercontractivity is equivalent to* (**H-1**), *and* (**H-1**) *implies hypermixing.*

PROOF: In proving (**H-1**), we assume, without loss of generality, that $f_m \in B_{[a_m,b_m]}(\Omega; [0, \infty))$ for $1 \le m \le n$, where $a_1 = 0$ and $a_{m+1} - b_m \ge \ell$ for $1 \le m < n$. Since, for any $t \in (0, \infty)$,

$$\left\|f_1 \cdots f_n\right\|_{L^1(P)} = \left\|\mathcal{P}_t(f_1 \cdots f_n)\right\|_{L^1(P)} \le \left\|\mathcal{P}_t(f_1 \cdots f_n)\right\|_{L^2(P)},$$

it suffices for us to show that

$$(5.5.18) \qquad A(f_1,\dots,f_n) \equiv \left\|\mathcal{P}_t(f_1\cdots f_n)\right\|_{L^2(P)} \le \prod_{m=1}^{n} \|f_m\|_{L^{\alpha(\ell)}(P)}$$

with $t = \ell/2$. To this end, note that, by Corollary 5.5.16,

$$\left\|\mathcal{P}_t(f_1\cdots f_n)\right\|_{L^2(P)} \le \|f_1\|_{L^{\alpha(\ell)}(P)}\left\|\mathcal{P}_t g\right\|_{L^2(P)},$$

where $g = (f_2\cdots f_n)\circ\theta_{-a_2}$. In particular, by (5.5.15),

$$\begin{aligned} A(f_1,f_2) &\le \|f_1\|_{L^{\alpha(\ell)}(P)}\left\|\mathcal{P}_t(f_2\circ\theta_{-a_2})\right\|_{L^2(P)} \\ &\le \|f_1\|_{L^{\alpha(\ell)}(P)}\|f_2\|_{L^{p(t)}(P)} \le \|f_1\|_{L^{\alpha(\ell)}(P)}\|f_2\|_{L^2(P)}; \end{aligned}$$

and, for $n \ge 3$,

$$A(f_1,\dots,f_n) \le \|f_1\|_{L^{\alpha(\ell)}(P)} A(f_2\circ\theta_{-a_2},\dots,f_n\circ\theta_{-a_2}).$$

Hence (5.5.18) follows easily by induction on n.

We now turn to the proof of (**H-2**). Since

$$\int_\Omega \left(f(\omega) - \int_\Omega f\,dP\right) g(\omega)\,P(d\omega) = \int_\Omega f(\omega)\left(g(\omega) - \int_\Omega g\,dP\right)\,P(d\omega),$$

we may and will assume that $f \in B_{[\ell,\infty)}(\Omega;\mathbb{R})$, $g \in B_{(-\infty,0]}(\Omega;\mathbb{R})$ and that $\int_\Omega f\,dP = 0$. But then, by (5.5.7) (with g in place of f),

$$E_0^-(fg) = g(E_0^- f) = g\Big(\big[\mathcal{P}_\ell(f\circ\theta_{-\ell})\big]\Big);$$

and so, by HÖLDER's inequality, (5.5.14), (5.5.13), and (5.5.15):

$$\begin{aligned} \left|\int_\Omega f(\omega)g(\omega)\,P(d\omega)\right| &= \left|\int_\Omega g(\omega)\big[\mathcal{P}_\ell(f\circ\theta_{-\ell})\big](\omega)\,P(d\omega)\right| \\ &\le \|g\|_{L^{p(t)}(P)}\left\|\mathcal{P}_{2t}(f\circ\theta_{-\ell})\right\|_{L^2(P)} \\ &\le 3^{1/2}\exp\left[-\frac{t\log\left(\frac{3}{2}\right)}{2T_0}\right]\|g\|_{L^{p(t)}(P)}\left\|\mathcal{P}_t(f\circ\theta_{-\ell})\right\|_{L^2(P)} \\ &\le 3^{1/2}\exp\left[-\frac{t\log\left(\frac{3}{2}\right)}{2T_0}\right]\|g\|_{L^{p(t)}(P)}\|f\|_{L^{p(t)}(P)}, \end{aligned}$$

where $t = \ell/3$.

To prove the converse assertion, note that if $P \in \mathbf{M}_1^{\mathrm{S}}(\Omega)$ satisfies **(H-1)** for some $\alpha(\ell) \searrow 1$, then there is an $\ell \in (0,\infty)$ with the property that

$$\|g(f \circ \theta_\ell)\|_{L^1(P)} \leq \|g\|_{L^{4/3}(P)}\|f\|_{L^{4/3}(P)} \leq \|g\|_{L^{4/3}(P)}\|f\|_{L^2(P)}$$

for $g \in B_{(-\infty,0]}(\Omega;\mathbb{R})$ and $f \in B_+(\Omega;\mathbb{R})$. But, for such f and g,

$$E_0^-\big[g(f\circ\theta_\ell)\big] = E_0^-\big[g(\mathcal{P}_\ell f)\big];$$

and so it is clear that P is ℓ-hypercontractive. ∎

5.5.19 Exercise.

The notion of ϵ-MARKOV has various manifestations. A more precise name for the one which we have adopted would be **backward** ϵ-MARKOV since it says that "the future given the past depends only the past back to time $-\epsilon$."

(i) Show that $E_0^- E_0^+ = E_{[-\epsilon,0]}$ if and only if

$$E_{[-\epsilon,0]}\big[fg\big] = \big(E_{[-\epsilon,0]}f\big)\big(E_{[-\epsilon,0]}g\big)$$

for $f \in B_{(-\infty,0]}(\Omega;\mathbb{R})$ and $g \in B_{[0,\infty)}(\Omega;\mathbb{R})$.

(ii) The **forward** ϵ-MARKOV property can be expressed as the equality $E_0^+E_0^- = E_{[0,\epsilon]}$. Show that an equivalent formulation is the statement that

$$E_{[0,\epsilon]}\big[fg\big] = \big(E_{[0,\epsilon]}f\big)\big(E_{[0,\epsilon]}g\big)$$

for $f \in B_{(-\infty,0]}(\Omega;\mathbb{R})$ and $g \in B_{[0,\infty)}(\Omega;\mathbb{R})$; and check that Theorem 5.5.17 continues to hold when one adopts this notion of ϵ-MARKOV.

(iii) A more symmetric notion of ϵ-MARKOV (and the one which was adopted in [**18**]) is contained in the equality

$$E_{(-\infty,\epsilon/2]}E_{[-\epsilon/2,\infty)} = E_{[-\epsilon/2,\epsilon/2]}.$$

Show that this definition is equivalent to

$$E_{[-\epsilon/2,\epsilon/2]}\big[fg\big] = \big(E_{[-\epsilon/2,\epsilon/2]}f\big)\big(E_{[-\epsilon/2,\epsilon/2]}g\big)$$

for $f \in B_{(-\infty,\epsilon/2]}(\Omega;\mathbb{R})$ and $g \in B_{[-\epsilon/2,\infty)}(\Omega;\mathbb{R})$, and conclude that when it holds then so do both of the one-sided versions of the ϵ-MARKOV property.

VI Analytic Considerations

6.1 When Is a Markov Process Hypermixing?

In this section, we pick up the project, initiated in Exercise 5.4.36, of connecting the results obtained in Chapter V to those in Chapter IV, especially those in Sections 4.2 and 4.4. Thus, $P(t, \sigma, \cdot)$ will be a transition probability function on the Polish space Σ, and we will be assuming that the associated measurable MARKOV family $\{P_\sigma : \sigma \in \Sigma\}$ can be realized on the SKOROKHOD space $D\big([0,\infty);\Sigma\big)$. Also, we will use $\{P_t : t > 0\}$ to denote the MARKOV semigroup on $B(\Sigma;\mathbb{R})$ which is determined by $P(t,\sigma,\cdot)$, and we will suppose that there is a $\{P_t : t > 0\}$-invariant $\mu \in \mathbf{M}_1(\Sigma)$. Finally, we will denote by P the unique element of $\mathbf{M}_1^{\mathrm{S}}(\Omega)$ ($\Omega \equiv D(\mathbb{R};\Sigma)$) with the properties that

$$P \circ \Sigma_0^{-1} = \mu \quad \text{and} \quad P(A \cap B) = \int_A P_{\Sigma_0(\omega)}(B)\, P(d\omega) \tag{6.1.1}$$

for $A \in \mathcal{B}_{(-\infty,0]}$ and $B \in \mathcal{B}[0,\infty)$.

It should be obvious that, in the terminology of Section 5.5, P is 0-MARKOV. In fact, the $\mathcal{P}_t$ in (5.5.4) is given by

$$[\mathcal{P}_t f](\omega) = \int_\Sigma \left(\int_\Omega f(\omega')\, P_\sigma(d\omega') \right) P(t, \Sigma_0(\omega), d\sigma) \tag{6.1.2}$$

for $f \in B_+(\Omega;\mathbb{R})$; and, as a consequence, Theorem 5.5.17 is easily seen to become the following statement.

6.1.3 Theorem. *The P in (6.1.1) is hypermixing if and only if*

$$\left\| P_T \right\|_{L^2(\mu)\to L^4(\mu)} = 1 \quad \textit{for some } T \in (0,\infty). \tag{6.1.4}$$

A MARKOV semigroup for which (6.1.4) holds is said to be μ-**hypercontractive**; and it is our goal to find conditions which guarantee this hypercontractive property. As a preliminary step in this direction, the following result is often useful.

6.1.5 Lemma. *If*

$$\|P_T\|_{L^2(\mu)\to L^4(\mu)} = 1$$

then

$$\left\|P_t\phi - \langle\phi\rangle_\mu\right\|_{L^2(\mu)} \leq 3^{-\frac{[t/T]}{2}}\|\phi\|_{L^2(\mu)} \tag{6.1.6}$$

for $t \in [T, \infty)$ and $\phi \in B(\Sigma; \mathbb{R})$; where we have introduced the notation

$$\langle\phi\rangle_\mu \equiv \int_\Sigma \phi\, d\mu. \tag{6.1.7}$$

Conversely, if, for some T_0, $T_1 \in (0, \infty)$,

$$\begin{aligned} &M_0 \equiv \|P_{T_0}\|_{L^2(\mu)\to L^4(\mu)} < \infty \\ &\rho \equiv \sup\left\{\frac{\left\|P_{T_1}\phi - \langle\phi\rangle_\mu\right\|_{L^2(\mu)}}{\|\phi\|_{L^2(\mu)}} : \phi \in B(\Sigma; \mathbb{R}) \setminus \{0\}\right\} < 1, \end{aligned} \tag{6.1.8}$$

then $\{P_t : t > 0\}$ is μ-hypercontractive.

PROOF: The first assertion is simply a translation of Lemma 5.5.11 into the present context.

To prove the second part, write $\phi = a + \psi$, where $a = \langle\phi\rangle_\mu$. Then, by HÖLDER's inequality, for $t > T_0 \vee T_1$:

$$\begin{aligned} \left\|P_t\phi\right\|^4_{L^4(\mu)} &\leq a^4 + 6a^2\left\|P_t\psi\right\|^2_{L^2(\mu)} + 4|a|\left\|P_t\psi\right\|^3_{L^3(\mu)} + \left\|P_t\psi\right\|^4_{L^4(\mu)} \\ &\leq a^4 + 8a^2\left\|P_t\psi\right\|^2_{L^2(\mu)} + 3\left\|P_t\psi\right\|^4_{L^4(\mu)} \\ &\leq a^4 + 8\rho^{2[t/T_1]}a^2\|\psi\|_{L^2(\mu)} + 3M_0^4\rho^{4[(t-T_0)/T_1]}\|\psi\|^4_{L^2(\mu)}, \end{aligned}$$

where we have used (6.1.8) in the passage to the last line. Finally, we choose $t > T_0 \vee T_1$ so that

$$8\rho^{2[t/T_1]} \leq 2 \quad \text{and} \quad 3M_0^4\rho^{4[(t-T_0)/T_1]} \leq 1,$$

and thereby obtain

$$\begin{aligned} \left\|P_t\phi\right\|^4_{L^4(\mu)} &\leq a^4 + 2a^2\|\psi\|^2_{L^2(\mu)} + \|\psi\|^4_{L^2(\mu)} \\ &= \left(a^2 + \|\psi\|^2_{L^2(\mu)}\right)^2 = \|\phi\|^4_{L^2(\mu)}. \quad \blacksquare \end{aligned}$$

The next result is a typical application of Lemma 6.1.5.

6.1.9 Theorem. *Suppose that there exist $T_0, T_1 \in (0,\infty)$ for which*

$$P(T_i,\sigma,d\tau) = p(T_i,\sigma,\tau)\,\mu(d\tau), \quad i \in \{0,1\} \text{ and } \sigma \in \Sigma,$$

where the maps $(\sigma,\tau) \in \Sigma^2 \longmapsto p(T_i,\sigma,\tau) \in [0,\infty)$ are measurable,

$$\int_\Sigma \left(\int_\Sigma p(T_0,\sigma,\tau)^2\,\mu(d\tau) \right)^2 \mu(d\sigma) < \infty,$$

and there is an $\epsilon > 0$ such that

$$p(T_1,\sigma,\tau) \ge \epsilon \quad \text{for } (\sigma,\tau) \in \Sigma^2.$$

Then $\{P_t : t > 0\}$ is μ-hypercontractive.

PROOF: Obviously, $\|P_{T_0}\|_{L^2(\mu)\to L^4(\mu)} < \infty$. Thus, all that we have to do is check that

$$\sup\left\{ \left\|P_t\phi - \langle\phi\rangle_\mu\right\|_{L^2(\mu)} : \phi \in B(\Sigma;\mathbb{R}) \text{ with } \|\phi\|_{L^2(\mu)} = 1 \right\} < 1$$

for some $t \in (0,\infty)$. But, by **(ii)** of Exercise 4.1.48 with $\overline{\Pi}(\sigma,\cdot) = P(T_1,\sigma,\cdot)$, we see that (4.1.50) says that

$$\sup_{\sigma\in\Sigma} \left\|P(nT_1,\sigma,\cdot) - \mu\right\|_{\text{var}} \le 2(1-\epsilon)^n, \quad n \in \mathbb{Z}^+.$$

Hence, if E_0 is the operator on $L^1(\mu)$ which takes $\phi \in L^1(\mu)$ into the constant function $\langle\phi\rangle_\mu$, then (because μ is $P(t,\sigma,\cdot)$-invariant)

$$\left\|P_t - E_0\right\|_{L^1(\mu)\to L^1(\mu)} \le 2 \quad \text{for every } t \in (0,\infty),$$

and, by the preceding,

$$\left\|P_{nT_1} - E_0\right\|_{L^\infty(\mu)\to L^\infty(\mu)} \le 2(1-\epsilon)^n \text{ for every } n \in \mathbb{Z}^+.$$

Hence, by the RIESZ–THORIN Interpolation Theorem,

$$\left\|P_{nT_1} - E_0\right\|_{L^2(\mu)\to L^2(\mu)} \le 2(1-\epsilon)^{n/2}, \quad n \in \mathbb{Z}^+;$$

and clearly this means that we need only take $t = nT_1$ for some sufficiently large $n \in \mathbb{Z}^+$. ∎

6.1.10 Remark.

Theorem 6.1.9 makes it reasonably clear where hypermixing stands in relation to the hypothesis $(\tilde{\mathbf{U}})$ under which we proved our large deviation principle in Section 4.2. Namely, hypermixing is implied by the following strong version of $(\tilde{\mathbf{U}})$:

$$(\mathbf{SU}) \qquad P(T_1,\sigma,\cdot) \le MP(T_2,\tau,\cdot), \quad (\sigma,\tau)\in\Sigma^2,$$

for some T_1, $T_2 \in (0,\infty)$ and $M \in [1,\infty)$. Indeed, there is then (cf. Exercise 4.2.59) precisely one $\{P_t : t>0\}$-invariant $\mu \in \mathbf{M}_1(\Sigma)$; and, by Theorem 6.1.9, the corresponding P is necessarily hypermixing. Even though (**SU**) implies hypermixing, it is easy to see that $(\tilde{\mathbf{U}})$ itself does not always lead to hypermixing processes. For example, uniform rotation on $\mathbf{S}^1$ certainly satisfies $(\tilde{\mathbf{U}})$ and is certainly not hypermixing. On the other hand, as the following example demonstrates, there are important hypermixing processes for which $(\tilde{\mathbf{U}})$ fails.

6.1.11 Example.

Define $\gamma_t : \mathbb{R} \longrightarrow (0,\infty)$ for $t \in (0,\infty)$ by

$$\gamma_t(x) = \frac{1}{(2\pi t)^{1/2}} \exp\left[-x^2/2t\right], \quad x\in\mathbb{R},$$

and let

$$(6.1.12) \quad P(t,x,dy) = \gamma_{1-\exp(-t)}\left(y - e^{-t/2}x\right) dy, \quad (t,x)\in(0,\infty)\times\mathbb{R}.$$

The corresponding MARKOV process is the famous **Ornstein–Uhlenbeck process**; and, as is well known, the associated measures $\{P_x : x\in\mathbb{R}\}$ live on $C\big([0,\infty);\mathbb{R}\big)$. In fact, P_x is the distribution under WIENER's measure $\mathcal{W}$ of the solution $X : [0,\infty)\times\Theta \longrightarrow \mathbb{R}$ to

$$X(t,\theta) = x + \theta(t) - \frac{1}{2}\int_0^t X(s,\theta)\,ds.$$

(See Section 1.3 for the notation here.) Furthermore, it is obvious that $m(dx) = \gamma_1(x)\,dx$ is the one and only $\{P_t : t>0\}$-invariant measure but that $(\tilde{\mathbf{U}})$ cannot be satisfied by $P(t,x,\cdot)$ for any choice of ρ_1 and ρ_2. Nonetheless, as we are about to see, the $\{P_t : t>0\}$ is m-hypercontractive, and therefore the corresponding P in (6.1.1) is hypermixing.

To verify the preceding assertion, first note that

$$P(t,x,dy) = p(t,x,y)\,m(dy),$$

where

$$p(t,x,y) = \left(1-e^{-t}\right)^{-1/2} \exp\left[-\frac{e^{-t}x^2 - 2e^{-t/2}xy + e^{-t}y^2}{2(1-e^{-t})}\right].$$

From this expression it is easily seen that

$$\int_{\Sigma^2} p(t,x,y)^4 m^2(dx \times dy) < \infty,$$

and therefore $\|P_t\|_{L^2(m)\to L^4(m)} < \infty$ for all sufficiently large $t \in (0,\infty)$. Thus, by Lemma 6.1.5, all that remains is to check that the second part of (6.1.8) holds. To this end, observe that (as the preceding expression makes explicit) m is $P(t,x,\cdot)$-reversing, and therefore, by (4.2.46) and (4.2.57),

$$\|\overline{P}_t\phi - \langle\phi\rangle_m\|_{L^2(m)} \le e^{-\lambda t}\|\phi\|_{L^2(m)}, \quad t \in (0,\infty) \text{ and } \phi \in L^2(m)$$

where

(6.1.13) $$\lambda \equiv \inf\left\{\mathcal{E}(\phi,\phi) : \phi \in L^2(m) \text{ and } \|\phi - \langle\phi\rangle_m\|_{L^2(m)} = 1\right\}.$$

(We are using primes here to denote derivatives with respect to x.) Since $C_b^2(\mathbb{R};\mathbb{R})$ is $\{P_t : t > 0\}$-invariant and

$$\begin{aligned}\mathcal{E}(\phi,\phi) &= \frac{d}{dt}\Big|_{t=0}\left(\phi - P_t\phi, \phi\right)_{L^2(m)} \\ &= -\frac{1}{2}\int_\Sigma \left(\phi''(x) - x\phi'(x)\right)\phi(x)\, m(dx) = \frac{1}{2}\|\phi'\|^2_{L^2(m)}\end{aligned}$$

and $\left(P_t\phi\right)' = e^{-t/2}P_t(\phi')$ for $\phi \in C_b^2(\mathbb{R};\mathbb{R})$, we know that

$$\begin{aligned}-\frac{d}{dt}\|P_t\phi\|^2_{L^2(m)} &= 2\mathcal{E}\left(P_t\phi, P_t\phi\right) = e^{-t}\|P_t(\phi')\|^2_{L^2(m)} \\ &\le e^{-t}\|\phi'\|^2_{L^2(m)} = 2e^{-t}\mathcal{E}(\phi,\phi)\end{aligned}$$

first for all $\phi \in C_b^2(\mathbb{R};\mathbb{R})$ and thence for all $\phi \in L^2(m)$. Finally, since $\overline{P}_t\phi \longrightarrow \langle\phi\rangle_m$ in $L^2(m)$ as $t \to \infty$, we now have:

$$\begin{aligned}\|\phi - \langle\phi\rangle_m\|^2_{L^2(m)} &= \|\phi\|^2_{L^2(m)} - \langle\phi\rangle^2_m \\ &= -\int_0^\infty \frac{d}{ds}\|P_s\phi\|^2_{L^2(m)}\, ds \le 2\mathcal{E}(\phi,\phi).\end{aligned}$$

Hence, $\lambda \ge \frac{1}{2}$, and therefore the second part of (6.1.8) holds for all $T_1 \in (0,\infty)$. Actually, $\lambda = \frac{1}{2}$, since

$$\|\phi - \langle\phi\rangle_m\|^2_{L^2(m)} = 2\mathcal{E}(\phi,\phi)$$

when $\phi(x) = x$, $x \in \mathbb{R}$.

At least when μ is $P(t,\sigma,\cdot)$-reversing, the preceding example indicates that the property of μ-hypercontractivity is closely related to properties of the associated DIRICHLET form. This connection is spelled out most precisely in the following version of a theorem due to L. GROSS [**56**].

6.1.14 Theorem. (GROSS) *Suppose that $m \in \mathbf{M}_1(\Sigma)$ is $P(t,\sigma,\cdot)$-reversing; and let $\mathcal{E}$ be the associated* DIRICHLET *form (cf. (4.2.47)). Given $\alpha \in (0,\infty)$ and $\beta \geq 0$,*

$$\textbf{(LS)} \qquad \int_\Sigma \phi^2 \log \frac{|\phi|^2}{\|\phi\|^2_{L^2(m)}}\, dm \leq \alpha\mathcal{E}(\phi,\phi) + \beta\|\phi\|^2_{L^2(m)}, \quad \phi \in L^2(m),$$

if and only if

$$\|P_t\|_{L^p(\mu)\to L^q(\mu)} \leq \exp\left[4\beta\left(\frac{1}{p}-\frac{1}{q}\right)\right] \tag{6.1.15}$$

for $1 < p \leq q < \infty$ and $t \in (0,\infty)$ with $e^{4t/\alpha} \geq (q-1)/(p-1)$. In fact, (6.1.15) *with $p = 2$ implies* (**LS**) *and therefore* (6.1.15) *for general $p \in (1,\infty)$.*

PROOF: Recall the operator $\overline{L}$ which generates the semigroup $\{\overline{P}_t : t > 0\}$ on $L^2(m)$ (cf. the discussion preceding (4.2.46)), let $\phi \in B(\Sigma;(0,\infty)) \cap \mathbf{Dom}(\overline{L})$ be given, and set $\phi_t = P_t\phi$. Then, for any $q \in [1,\infty)$,

$$\frac{d}{dt}\|\phi_t\|^q_{L^q(m)} = q\big(\phi_t^{q-1}, \overline{L}\phi_t\big)_{L^2(m)}$$

Note (cf. the argument leading to (4.2.54)) that, for any $\psi \in B(\Sigma;[0,\infty)) \cap \mathbf{Dom}(\overline{L})$,

$$\begin{aligned}
\big(\psi^{q-1}, \overline{L}\psi\big)_{L^2(m)} &= -\lim_{s\searrow 0}\frac{1}{s}\big(\psi - P_s\psi, \psi^{q-1}\big)_{L^2(m)} \\
&= -\lim_{s\searrow 0}\frac{1}{2s}\int_{\Sigma^2}\big(\psi^{q-1}(\tau)-\psi^{q-1}(\sigma)\big)\big(\psi(\tau)-\psi(\sigma)\big)\,P(s,\sigma,d\tau)m(d\sigma) \\
&\leq -\frac{4(q-1)}{q^2}\lim_{s\searrow 0}\frac{1}{2s}\int_{\Sigma^2}\big(\psi^{q/2}(\tau)-\psi^{q/2}(\sigma)\big)^2 P(s,\sigma,d\tau)m(d\sigma) \\
&= -\frac{4(q-1)}{q^2}\mathcal{E}\big(\psi^{q/2},\psi^{q/2}\big);
\end{aligned}$$

where we have used the fact that, for any a, $b \in (0,\infty)$ and $q \in [1,\infty)$,

$$\big(a^{q/2}-b^{q/2}\big)^2 \leq \frac{q^2}{4(q-1)}\big(a^{q-1}-b^{q-1}\big)(a-b),$$

which follows, in turn, from

$$\begin{aligned}\frac{\eta^{q-1}-1}{\eta-1} &= \frac{q-1}{\eta-1}\int_1^\eta \left(\xi^{q/2-1}\right)^2 d\xi \\ &\geq (q-1)\left(\frac{1}{\eta-1}\int_1^\eta \xi^{q/2-1}\,d\xi\right)^2 \\ &= \frac{4(q-1)\left(\eta^{q/2}-1\right)^2}{q^2(\eta-1)^2}\end{aligned}$$

for $\eta \in (1,\infty)$. Hence, we now see that

$$\|\phi_t\|_{L^q(m)}^{q-1}\frac{d}{dt}\|\phi_t\|_{L^q(m)} \leq -\frac{4(q-1)}{q^2}\mathcal{E}\left((\phi_t)^{q/2},(\phi_t)^{q/2}\right), \quad t \in (0,\infty).$$

At the same time, if $t \in (0,\infty) \longmapsto q(t) \in (1,\infty)$ is smooth and $\psi \in B\big(\Sigma;[0,\infty)\big)$, then

$$\|\psi\|_{L^{q(t)}(m)}^{q(t)-1}\frac{d}{dt}\|\psi\|_{L^{q(t)}(m)} = \frac{q'(t)}{q(t)}\int_\Sigma \psi^{q(t)}\log\frac{\psi}{\|\psi\|_{L^{q(t)}(m)}}\,dm.$$

Therefore, after combining this with the above, we have that for smooth $t \in (0,\infty) \longmapsto q(t) \in (1,\infty)$ and $\phi \in B\big(\Sigma;[0,\infty)\big) \cap \mathbf{Dom}(\overline{L})$:

$$\begin{aligned}&\|\phi_t\|_{L^{q(t)}(m)}^{q(t)-1}\frac{d}{dt}\|\phi_t\|_{L^{q(t)}(m)} \\ \text{(6.1.16)}\quad &\leq -\frac{4(q(t)-1)}{q(t)^2}\mathcal{E}\left((\phi_t)^{q(t)/2},(\phi_t)^{q(t)/2}\right) \\ &\qquad + \frac{q'(t)}{q(t)^2}\int_\Sigma \phi_t^{q(t)}\log\frac{\phi_t^{q(t)}}{\|\phi_t^{q(t)/2}\|_{L^2(m)}^2}\,dm.\end{aligned}$$

Now suppose that (**LS**) holds and, for given $p \in (1,\infty)$, set $q(t) = 1+(p-1)e^{4t/\alpha}$. Then $q'(t) = 4\big(q(t)-1\big)/\alpha$ and so (6.1.16) says that

$$\frac{d}{dt}\|\phi_t\|_{L^{q(t)}(m)} \leq \frac{4q'(t)\beta}{q(t)^2}\|\phi_t\|_{L^{q(t)}(m)};$$

and therefore that

$$\|P_t\phi\|_{L^{q(t)}(m)} \leq \exp\left[4\beta\left(\frac{1}{p}-\frac{1}{q(t)}\right)\right]\|\phi\|_{L^p(m)},$$

at least for $\phi \in B\big(\Sigma;[0,\infty)\big) \cap \mathbf{Dom}(\overline{L})$. Since the passage from this to general $\phi \in L^p(m)$ is trivial and $\|P_t\|_{L^r(m)\to L^r(m)} = 1$ for all $r \in [1,\infty]$, we have now proved that (**LS**) implies (6.1.15). On the other hand, if one takes $q(t) = 1 + e^{4t/\alpha}$, then one finds that (6.1.16) becomes an equality at $t = 0$. Hence, when (6.1.15) holds with $p = 2$ and therefore $\frac{d}{dt}\big\|\phi_t\big\|_{L^{q(t)}(m)} \leq 0$ at $t = 0$, (**LS**) follows for $\phi \in B\big(\Sigma;[0,\infty)\big) \cap \mathbf{Dom}(\overline{L})$. At this point it is an easy step to (**LS**) for all $\phi \in B\big(\Sigma;[0,\infty)\big)$ and thence, via (4.2.54), for all $\phi \in L^2(m)$. ∎

An estimate of the form in (**LS**) is called a **logarithmic Sobolev inequality**.

6.1.17 Corollary. *Assume that $m \in \mathbf{M}_1(\Sigma)$ is $P(t,\sigma,\cdot)$-reversing and define $\Lambda_{\mathcal{E}}$ and $J_{\mathcal{E}}$ accordingly (as in (4.2.62) and (4.2.60), respectively). Then the following three properties are equivalent:*

$$\mathbf{H}(\nu|m) \leq \alpha J_{\mathcal{E}}(\nu), \quad \nu \in \mathbf{M}_1(\Sigma), \tag{6.1.18}$$

$$\|P_t\|_{L^p(m)\to L^q(m)} = 1 \quad \text{for } 1 < p \leq q < \infty \text{ and } t \in (0,\infty) \tag{6.1.19}$$

with $e^{4t/\alpha} \geq (q-1)/(p-1)$, and

$$\Lambda_{\mathcal{E}}(V) \leq \frac{1}{\alpha}\log\left(\int_\Sigma \exp\big[\alpha V\big]\,dm\right), \quad V \in C_{\mathrm{b}}(\Sigma;\mathbb{R}). \tag{6.1.20}$$

Moreover, if any one of these holds, then

$$\big\|\overline{P}_t\phi - \langle\phi\rangle_m\big\|_{L^2(m)} \leq e^{-2t/\alpha}\|\phi\|_{L^2(m)} \tag{6.1.21}$$

for $t \in (0,\infty)$ and $\phi \in L^2(m)$.

PROOF: Note that (6.1.18) is equivalent to (**LS**), first for non-negative ϕ's and then (by (4.2.54)) for all ϕ's. Thus, by Theorem 6.1.14, (6.1.18) and (6.1.19) are equivalent. At the same time, the equivalence of (6.1.18) to (6.1.20) is the content of Exercise 5.3.15.

Finally, by (6.1.6), one knows that $\|P_t\|_{L^2(m)\to L^4(m)} = 1$ implies that $\big\|P_t\phi - \langle\phi\rangle_m\big\|_{L^2(m)} \leq 3^{-1/2}\|\phi\|_{L^2(m)}$. In particular, when (6.1.19) holds, then one can take $t = (\alpha\log 3)/4$. After combining this with the Spectral Theorem (cf. (4.2.57)), one concludes that $E_0\phi = \langle\phi\rangle_m$, that $E_\lambda - E_0 = 0$ for $\lambda \in [0, 2/\alpha)$, and therefore that (6.1.21) holds. ∎

We conclude this section with a result which sharpens for the reversible setting the sort of topics treated in Theorem 5.5.12 and Lemma 6.1.5.

6.1.22 Theorem. *Assume that m is $P(t,\sigma,\cdot)$-reversing.*

(i) *Suppose that*

$$\|P_T\|_{L^p(m)\to L^q(m)} = 1$$

for some $T \in (0,\infty)$ *and* $1 < p < q < \infty$. *Then* (6.1.18) *holds with*

$$\alpha(T,p,q) = \frac{pqT}{q-p}. \tag{6.1.23}$$

In particular, if $\{P_t : t > 0\}$ *is* m*-hypercontractive at time* $T \in (0,\infty)$, *then*

$$\|P_t\|_{L^p(m)\to L^q(m)} = 1 \tag{6.1.24}$$

for $1 < p < q < \infty$ *and* $t \in (0,\infty)$ *with* $e^{t/T} \geq \frac{q-1}{p-1}$.

(ii) *Assume that*

$$\|\phi - \langle\phi\rangle_m\|^2_{L^2(m)} \leq \gamma\mathcal{E}(\phi,\phi), \quad \phi \in L^2(m), \tag{6.1.25}$$

and that **(LS)** *holds for some* α, β, $\gamma \in (0,\infty)$. *Then*

$$\mathbf{H}(\nu|m) \leq \Big[\alpha + (\beta+2)\gamma\Big] J_{\mathcal{E}}(\nu), \quad \nu \in \mathbf{M}_1(\Sigma),$$

and so $\{P_t : t > 0\}$ *is* m*-hypercontractive.*

PROOF: To prove **(i)** we will use the criterion provided by the equivalence of (6.1.18) and (6.1.20). To this end, we first show that, for given $V \in B(\Sigma;\mathbb{R})$,

$$\varlimsup_{n\to\infty} \frac{1}{nT} \log\big\|P^V_{nT}\big\|_{L^p(m)\to L^p(m)} \leq \frac{1}{\alpha}\log\left(\int_\Sigma e^{\alpha V}\,dm\right),$$

where $\alpha = \alpha(T,p,q)$. Indeed, for $\phi \in B\big(\Sigma;[0,\infty)\big)$, set

$$\Phi_{n,t}(\sigma) = \int_\Omega \exp\left[\sum_{m=0}^{n-1} TV\big(\Sigma_{mT}(\omega)\big)\right]\phi\big(\Sigma_{nT-t}(\omega)\big)\,P_\sigma(d\omega), \quad \sigma \in \Sigma.$$

Then, by Theorem 4.2.25, JENSEN's inequality, and the MARKOV property:

$$\begin{aligned}
\big[P^V_{nT}\phi\big](\sigma) &= \int_\Omega \exp\left[\sum_{m=0}^{n-1}\int_0^T V\big(\Sigma_{mT+t}(\omega)\big)\,dt\right]\phi\big(\Sigma_{nT}(\omega)\big)\,P_\sigma(d\omega) \\
&\leq \frac{1}{T}\int_0^T\int_\Omega \exp\left[\sum_{m=0}^{n-1} TV\big(\Sigma_{mT+t}(\omega)\big)\right]\phi\big(\Sigma_{nT}(\omega)\big)\,P_\sigma(d\omega) \\
&= \frac{1}{T}\int_0^T \big[P_t\Phi_{n,t}\big](\sigma)\,dt;
\end{aligned}$$

and so

$$\left\|P_{nT}^V\phi\right\|_{L^p(m)} \le \sup_{t\in[0,1)} \left\|\Phi_{n,t}\right\|_{L^p(m)}.$$

But

$$\begin{aligned}\Phi_{n,t}(\sigma) &\le e^{T\|V\|_B} \int_\Omega \exp\left[\sum_{m=0}^{n-2} TV\left(\Sigma_{mT}(\omega)\right)\right] \left[P_{T-t}\phi\right]\left(\Sigma_{(n-1)T}\right) P_\sigma(d\omega) \\ &= e^{T\|V\|_B}\left[\left(e^{TV}P_T\right)^{n-1} \circ P_{T-t}\phi\right](\sigma);\end{aligned}$$

and therefore

$$\left\|\Phi_{n,t}\right\|_{L^p(m)} \le e^{T\|V\|_B}\left\|e^{TV}P_T\right\|_{L^p(m)\to L^p(m)}^{n-1} \|\phi\|_{L^p(m)}.$$

Since, by our hypothesis and HÖLDER's inequality, it is easy to see that

$$\left\|e^{TV}P_T\right\|_{L^p(m)\to L^p(m)} \le \left(\int_\Sigma \exp[\alpha V]\, dm\right)^{T/\alpha},$$

we now get the asserted estimate after letting $n \longrightarrow \infty$.

To complete the proof of (**i**), we reason as follows. If $p = 2$, there is nothing to do, since

$$\Lambda_{\mathcal{E}}(V) = \lim_{t\to\infty} \frac{1}{t} \log\left(\left\|P_t^V\right\|_{L^2(m)\to L^2(m)}\right).$$

On the other hand, if $1 < p < 2$, then (by precisely the same argument as we used to prove (5.5.14)) we can find a $T_1 \in (0,\infty)$ for which $\left\|P_{T_1}\right\|_{L^p(m)\to L^2(m)} = 1$; and therefore

$$\begin{aligned}\Lambda_{\mathcal{E}}(V) &= \lim_{n\to\infty} \frac{1}{nT} \log\left(\left\|P_{nT+T_1}^V\right\|_{L^2(m)\to L^2(m)}\right) \\ &\le \varliminf_{n\to\infty} \frac{1}{nT} \log\left(\left\|P_{nT}^V\right\|_{L^p(m)\to L^p(m)}\right).\end{aligned}$$

A similar argument applies when $2 < p < \infty$.

To prove (**ii**), we will show that

$$\begin{aligned}&\int_\Sigma \phi^2 \log\left(\frac{\phi^2}{\|\phi\|_{L^2(m)}^2}\right) dm \\ &\le \int_\Sigma \left(\phi - \langle\phi\rangle_m\right)^2 \log\left(\frac{\left(\phi^2 - \langle\phi\rangle_m\right)^2}{\left\|\phi - \langle\phi\rangle_m\right\|_{L^2(m)}^2}\right) dm \\ &\qquad + 2\left\|\phi - \langle\phi\rangle_m\right\|_{L^2(m)}^2;\end{aligned} \tag{6.1.26}$$

and clearly this will lead immediately to the desired result.

Note that in order to prove (6.1.26), it suffices to show that

$$\int_{\Sigma} (1+t\psi)^2 \log\left(\frac{(1+t\psi)^2}{1+t^2}\right) dm \le t^2 \int_{\Sigma} \psi^2 \log(\psi^2)\, dm + 2t^2, \quad t \in \mathbb{R}$$

for all $\psi \in L^2(m)$ with $\langle \psi \rangle_m = 0$ and $\|\psi\|_{L^2(m)} = 1$. To this end, let $\delta > 0$ be given and set

$$f_\delta(t) = \int_{\Sigma} (1+t\psi)^2 \log\left(\frac{(1+t\psi)^2+\delta}{1+t^2}\right) dm - t^2 \int_{\Sigma} \psi^2 \log(\psi^2)\, dm$$

for $t \in \mathbb{R}$. Then $f_\delta(0) = \log(1+\delta)$,

$$\begin{aligned} f_\delta'(t) = 2\int_{\Sigma} (1+t\psi)\psi \log((1+t\psi)^2+\delta)\, dm + 2\int_{\Sigma} \frac{(1+t\psi)^3\psi}{(1+t\psi)^2+\delta}\, dm \\ - 2t\left[1+\log(1+t^2) + \int_{\Sigma} \psi^2\log(\psi^2)\, dm\right], \end{aligned}$$

and

$$\begin{aligned} f_\delta''(t) &= 2\int_{\Sigma} \psi^2 \log\left(\frac{(1+t\psi)^2+\delta}{(1+t^2)\psi^2}\right) dm + 10\int_{\Sigma} \psi^2 \frac{(1+t\psi)^2}{(1+t\psi)^2+\delta}\, dm \\ &\quad - 4\int_{\Sigma} \psi^2 \frac{(1+t\psi)^4}{\left[(1+t\psi)^2+\delta\right]^2}\, dm - 2 - \frac{4t^2}{1+t^2} \\ &\le 2\log\left(1+\frac{\delta}{1+t^2}\right) - \left[4A(t,\delta)^2 - 10A(t,\delta)\right] - 2 \\ &\le 2\log(1+\delta) + 4 \end{aligned}$$

where

$$A(t,\delta) \equiv \int_{\Sigma} \psi^2 \frac{(1+t\psi)^2}{(1+t\psi)^2+\delta}\, dm \in [0,1]$$

and we have used JENSEN's inequality in the passage to the last line. From these and TAYLOR's Theorem, we conclude that

$$f_\delta(t) \le \log(1+\delta) + \big(2+\log(1+\delta)\big)t^2;$$

and therefore the required estimate follows once one lets $\delta \searrow 0$. ∎

6.1.27 Exercise.

Referring to Lemma 5.3.5, let $u \in \mathbf{D} \cap B(\Sigma;[1,\infty))$ be given and define $m_u \in \mathbf{M}_1(\Sigma)$ and the transition probability function $P_u(t,\sigma,\cdot)$ accordingly.

(**i**) Show that for any $\phi \in B(\Sigma;\mathbb{R})$ and $\mu \in \mathbf{M}_1(\Sigma)$

$$\begin{aligned}&\int_\Sigma \phi^2 \log\left(\frac{\phi^2}{\|\phi\|^2_{L^2(\mu)}}\right) d\mu \\ &= \inf\left\{\int_\Sigma \left[\phi^2\log\phi^2 - \phi^2\log t - \phi^2 + t\right] d\mu : t \in (0,\infty)\right\}.\end{aligned} \tag{6.1.28}$$

Next, check that $x\log x - x\log t - x + t \geq 0$ for all $(t,x) \in (0,\infty)\times[0,\infty)$; and use this in conjunction with (6.1.28) to show that

$$\mathbf{H}(\nu|m_u) \leq \|u\|_B^2 \mathbf{H}(\nu|m), \quad \nu \in \mathbf{M}_1(\Sigma). \tag{6.1.29}$$

(**ii**) Let $\mathcal{E}_u$ denote the DIRICHLET form associated with $P_u(t,\sigma,\cdot)$ and m_u. Using (4.2.54), show that

$$\mathcal{E}(\phi,\phi) \leq \|u\|^2_{L^1(m)}\mathcal{E}_u(\phi,\phi), \quad \phi \in B(\Sigma;\mathbb{R}). \tag{6.1.30}$$

(**iii**) By combining parts (**i**) and (**ii**), show that (6.1.18) implies that

$$\mathbf{H}(\nu|m_u) \leq \alpha\|u\|_B^4 J_{\mathcal{E}_u}(\nu), \quad \nu \in \mathbf{M}_1(\Sigma).$$

In particular, this means that the hypermixing property is preserved by the transformation described in Lemma 5.3.5.

6.1.31 Exercise.

Let $m \in \mathbf{M}_1(\Sigma)$ be a $P(t,\sigma,\cdot)$-reversing measure. More familiar than logarithmic SOBOLEV inequalities are **classical Sobolev inequalities** of the form

$$\|\phi\|^2_{L^p(m)} \leq A\big(\mathcal{E}(\phi,\phi) + B\|\phi\|^2_{L^2(m)}\big), \quad \phi \in B(\Sigma;\mathbb{R}), \tag{6.1.32}$$

for some $p \in (2,\infty)$ and A, $B \in [0,\infty)$. One naturally expects that a classical SOBOLEV inequality ought to be a stronger statement than a logarithmic one. To verify this, let $\phi \in B(\Sigma;\mathbb{R})$ with $\|\phi\|_{L^2(m)} = 1$ be given, and use JENSEN's inequality to check that

$$\begin{aligned}\int_\Sigma \phi^2\log\phi^2\,dm &= \frac{2}{p-2}\int_\Sigma \phi^2\log\phi^{p-2}\,dm \leq \frac{p}{p-2}\log\|\phi\|^2_{L^p(m)} \\ &\leq \frac{p}{p-2}\|\phi\|^2_{L^p(m)} \leq \frac{pA}{p-2}\left[\mathcal{E}(\phi,\phi)+B\right].\end{aligned}$$

Thus, (6.1.32) implies that

$$\int_\Sigma \phi^2 \log \frac{\phi^2}{\|\phi\|^2_{L^2(m)}}\, dm \le \frac{pA}{p-2}\left[\mathcal{E}(\phi,\phi) + B\|\phi\|^2_{L^2(m)}\right].$$

In particular, if one has, in addition to (6.1.32), that

$$\left\|\phi - \langle\phi\rangle_m\right\|^2_{L^2(m)} \le C\mathcal{E}(\phi,\phi), \quad \phi \in B(\Sigma;\mathbb{R}),$$

then, by part (**ii**) of Theorem 6.1.22,

$$\mathbf{H}(\nu|m) \le \left[\frac{pA(1+BC)}{p-2} + 2C\right] J_{\mathcal{E}}(\nu), \quad \nu \in \mathbf{M}_1(\Sigma).$$

6.1.33 Exercise.

In his article [**56**], GROSS considered the "two-point" space $\Sigma = \{-1, 1\}$ with the BERNOULLI measure $m = (\delta_{-1} + \delta_1)/2$ and the transition probability function

$$P(t,\sigma,\tau) = \begin{cases} \frac{1+e^{-t}}{2} & \text{if } \sigma = \tau \\ \frac{1-e^{-t}}{2} & \text{if } \sigma = -\tau. \end{cases}$$

Obviously, m is $P(t,\sigma,\cdot)$-reversing. Using $\mathcal{E}$ to denote the DIRICHLET form associated with $P(t,\sigma,\cdot)$ and m, show that

$$\int_\Sigma \phi^2 \log \frac{|\phi|^2}{\|\phi\|^2_{L^2(m)}}\, dm \le 2\mathcal{E}(\phi,\phi), \quad \phi \in B(\Sigma;\mathbb{R}), \tag{6.1.34}$$

and conclude from this that the associated semigroup $\{P_t : t > 0\}$ has the property that $\|P_t\|_{L^p(m)\to L^q(m)} = 1$ as long as $1 < p < q < \infty$ and $e^{2t} \ge (q-1)/(p-1)$. Finally, check that (6.1.34) is optimal.

Hint: First observe that it suffices to prove (6.1.34) for ϕ's of the form $\phi_b(\sigma) = 1 + b\sigma$, where $b \in [0,1]$; and then show that (6.1.34) for ϕ_b is equivalent to

$$h(b) \equiv (1+b)^2\log(1+b) + (1-b)^2\log(1-b) - \left(1+b^2\right)\log\left(1+b^2\right) \le 2b^2$$

for $b \in [0,1]$. Finally, prove the preceding by checking that $h(0) = h'(0) = 0$ and that $h''(b) \le 4$.

6.1.35 Exercise.

Referring to the situation in Corollary 6.1.17 and assuming (6.1.18) holds, show that

$$\text{(6.1.36)}\quad \mathbf{H}\big(\nu P_t\big|m\big) \le \exp\left[-\frac{4t}{\alpha}\right]\mathbf{H}(\nu|m), \quad \nu \in \mathbf{M}_1(\Sigma) \text{ and } t \in [0,\infty).$$

Hint: Assuming that f is a uniformly positive element of $\mathbf{Dom}(\overline{L})$ which is bounded, set $f_t = [P_t f]$ and check that

$$\frac{d}{dt}\int_\Sigma f_t \log f_t \, dm = -\mathcal{E}\big(\log f_t, f_t\big).$$

Next, using (4.2.54) in the same sort of way that we used it in the proof of Theorem 6.1.14, show that

$$\mathcal{E}\big(\log f_t, f_t\big) \ge 4\mathcal{E}\big(f_t^{1/2}, f_t^{1/2}\big).$$

6.2 Symmetric Diffusions on a Manifold

The purpose of this section is to provide a ready source of examples to which the results in Chapter V and Section 6.1 are applicable. The setting in which we will working is that of differentiable manifolds. Thus, we will assume that Σ is a separable, connected, N-dimensional C^∞-manifold on which there is given a complete RIEMANNian structure; and we will denote by λ the associated RIEMANNian measure on Σ. Given vector fields X, $Y \in \Gamma\big(\mathbf{T}(\Sigma)\big)$, $(X|Y) \in C^\infty(\Sigma;\mathbb{R})$ will be the RIEMANNian inner product of X and Y; and $|X| \equiv (X|X)^{1/2}$ is the length of X. (We use $\mathbf{T}(\Sigma)$ to denote the tangent bundle over Σ and $\Gamma\big(\mathbf{T}(\Sigma)\big)$ to denote the space of smooth sections.) Also, we use $\nabla_X Y \in \Gamma\big(\mathbf{T}(\Sigma)\big)$ to denote the associated (LEVI–CIVITA) RIEMANNian covariant derivative of Y with respect to X. That is, ∇ is defined to be the KOSUL connection which satisfies

$$\text{(6.2.1)}\qquad \nabla_X Y - \nabla_Y X = [X,Y], \quad X,\, Y \in \Gamma\big(\mathbf{T}(\Sigma)\big),$$

where $[X,Y] \equiv XY - YX$ is the commutator of X and Y, and

$$\text{(6.2.2)}\qquad X\big(Y|Z\big) = \big(\nabla_X Y|Z\big) + \big(Y|\nabla_X Z\big) \quad \text{for } X,\, Y,\, Z \in \Gamma\big(\mathbf{T}(\Sigma)\big).$$

In addition, we will use $\operatorname{grad}\phi \in \Gamma\big(\mathbf{T}(\Sigma)\big)$ and $\operatorname{div} X \in C^\infty(\Sigma;\mathbb{R})$ to denote the gradient of $\phi \in C^\infty(\Sigma;\mathbb{R})$ and the divergence of $X \in \Gamma\big(\mathbf{T}(\Sigma)\big)$. Thus, for $X \in \Gamma\big(\mathbf{T}(\Sigma)\big)$:

$$\text{(6.2.3)}\qquad X\phi = \big(X|\operatorname{grad}\phi\big), \quad \phi \in C^\infty(\Sigma;\mathbb{R}),$$

and

$$\int_\Sigma X\phi\, d\lambda = -\int_\Sigma \phi \operatorname{div} X\, d\lambda \quad \text{for } \phi \in C_c^\infty(\Sigma;\mathbb{R}), \tag{6.2.4}$$

where $C_c^\infty(\Sigma;\mathbb{R})$ denotes the class of $\phi \in C^\infty(\Sigma;\mathbb{R})$ which have compact support. In particular, with the use of normal coordinates, one can easily check that

$$\operatorname{div} X(\sigma) = \sum_{k=1}^N \big(E_k|\nabla_{E_k}X\big)(\sigma), \quad X \in \Gamma\big(\mathbf{T}(\Sigma)\big), \tag{6.2.5}$$

if $\{E_k\}_1^N \subset \Gamma\big(\mathbf{T}(\Sigma)\big)$ is orthonormal at σ. Finally, we will use Δ to denote the **Laplace–Beltrami operator** given by

$$\Delta\phi = \operatorname{div}\big(\operatorname{grad}\phi\big), \quad \phi \in C^\infty(\Sigma;\mathbb{R}).$$

The reason for our introducing the preceding terminology is that we are going to be dealing with diffusions on Σ corresponding to an operator L of the form

$$[L\phi] = \frac{e^U}{2}\operatorname{div}\big(e^{-U}\operatorname{grad}\phi\big) = \frac{1}{2}\Big([\Delta\phi] - (\operatorname{grad} U|\operatorname{grad}\phi)\Big) \tag{6.2.6}$$

for $\phi \in C^\infty(\Sigma;\mathbb{R})$, where U is a fixed element of $C^\infty(\Sigma;\mathbb{R})$ which satisfies

$$\int_\Sigma e^{-U}\, d\lambda = 1. \tag{6.2.7}$$

(Note that Example 6.1.11 corresponds to $\Sigma = \mathbb{R}$ with the standard EUCLIDean structure and $U(x) = \big(x^2 - \log 2\pi\big)/2$.) Our first step will be to make sure that such a diffusion exists and that the measure $m \in \mathbf{M}_1(\Sigma)$ given by

$$m(d\sigma) = e^{-U(\sigma)}\,\lambda(d\sigma) \tag{6.2.8}$$

is reversing for the corresponding transition probability function. To be precise, we will prove the following.

6.2.9 Theorem. *Set $\Omega = C\big([0,\infty);\Sigma\big)$ and define the evaluation map $\Sigma_t : \Omega \longrightarrow \Omega$ and the σ-algebra $\mathcal{B}_t$ for $t \in [0,\infty)$ accordingly. Then, for each $\sigma \in \Sigma$, there is precisely one $P_\sigma \in \mathbf{M}_1(\Omega)$ with the property that*

$$\left(\phi\big(\Sigma_t(\omega)\big) - \phi(\sigma) - \int_0^t [L\phi]\big(\Sigma_s(\omega)\big)\, ds,\ \mathcal{B}_t,\ P_\sigma\right) \tag{6.2.10}$$

is a mean-zero martingale for every $\phi \in C_c^\infty(\Sigma;\mathbb{R})$. *Moreover, the map* $\sigma \in \Sigma \longmapsto P_\sigma \in \mathbf{M}_1(\Omega)$ *is continuous and the family* $\{P_\sigma : \sigma \in \Sigma\}$ *is (time-homogeneous)* MARKOV. *Finally, let* $P(t,\sigma,\cdot)$ *denote the associated transition probability function (i.e.,* $P(t,\sigma,\Gamma) = P_\sigma(\{\omega : \Sigma_t(\omega) \in \Gamma\})$*). Then the measure* m *in (6.2.8) is* $P(t,\sigma,\cdot)$*-reversing. In fact, the corresponding* DIRICHLET *form* $\mathcal{E}$ *is given by*

$$\mathcal{E}(\phi,\phi) = \frac{1}{2}\int_\Sigma |\mathrm{grad}\,\phi|^2\,dm \tag{6.2.11}$$

for $\phi \in L^2(m) \cap C^\infty(\Sigma;\mathbb{R})$ *with* $|\mathrm{grad}\,\phi| \in L^2(m)$*; and* $\mathcal{E}$ *is the closure of its own restriction to* $C_c^\infty(\Sigma;\mathbb{R})$ *in the sense that* $\phi \in L^2(m)$ *is an element of* **Dom**$(\mathcal{E})$ *(i.e., satisfies* $\mathcal{E}(\phi,\phi) < \infty$*) if and only if* ϕ *is the limit in* $L^2(m)$ *of a sequence* $\{\phi_n\}_1^\infty \subseteq C_c^\infty(\Sigma;\mathbb{R})$ *with the property that*

$$\lim_{m\to\infty} \sup_{n\ge m} \mathcal{E}\big(\phi_n - \phi_m, \phi_n - \phi_m\big) = 0,$$

in which case $\mathcal{E}(\phi,\phi) = \lim_{n\to\infty} \mathcal{E}\big(\phi_n,\phi_n\big)$. *In particular, if* $\{\overline{P}_t : t > 0\}$ *is the semigroup on* $L^2(m)$ *determined by* $P(t,\sigma,\cdot)$*, then for every* $\phi \in L^2(m)$*,* $[\overline{P}_t\phi] \longrightarrow \int_\Sigma \phi\,dm$ *in* $L^2(m)$ *as* $t \longrightarrow \infty$.

Aside from rather mundane probabilistic considerations, the proof of Theorem 6.2.9 comes down to showing that the diffusion "generated" by L does not explode (cf. Chapter 10 of [**104**]); and the key to checking this is contained in the following variant of a lemma due to M. GAFFNEY [**52**], which shows how to utilize the completness assumption that we have made about the RIEMANNian structure on Σ. (For the required standard facts about RIEMANNian geometry, the reader might want to consult MILNOR's marvelous [**76**].)

6.2.12 Lemma. (GAFFNEY) *There exists a* $\psi \in C^\infty\big(\Sigma;[0,\infty)\big)$ *with the properties that the level set* $\{\sigma : \psi(\sigma) \le R\}$ *is compact for each* $R \in (0,\infty)$ *and that* $|\mathrm{grad}\,\psi|$ *is bounded. In particular, there exists a non-decreasing sequence* $\{\eta_n\}_1^\infty \subseteq C_c^\infty\big(\Sigma;[0,1]\big)$ *with the properties that*

$$\{\sigma : \eta_n(\sigma) = 1\} \nearrow \Sigma \quad \text{and} \quad \big\|\, |\mathrm{grad}\,\eta_n|\,\big\|_B \longrightarrow 0 \quad \text{as } n \longrightarrow 0.$$

PROOF: Choose and fix a reference point $\sigma_0 \in \Sigma$, and set

$$\phi(\sigma) = \mathrm{dist}(\sigma,\sigma_0), \quad \sigma \in \Sigma,$$

where "distance" is being measured with respect to the RIEMANNian distance function on Σ. Because Σ is connected, $\Sigma = \{\sigma : \phi(\sigma) < \infty\}$; and by the triangle inequality, it is obvious that ϕ is LIPSCHITZ continuous with LIPSCHITZ constant 1. Moreover, because the RIEMANNian structure on Σ is complete, the level sets $K(R) \equiv \{\sigma : \phi(\sigma) \le R\}$ are compact, and clearly they exhaust Σ. Thus, we can find an open cover $\{U_m\}_1^\infty$ and an atlas $\{(W_m,\Phi_m)\}_1^\infty$ with the following properties:

(i) Every pair of points in W_m are joined by a unique geodesic which lies entirely inside of W_m.

(ii) $\overline{U}_m \subset\subset W_m$.

(iii) For every $R \in (0,\infty)$ there are only finitely many $m \in \mathbb{Z}^+$ with $W_m \cap K(R) \neq \emptyset$; and if $W_m \cap K(R) \neq \emptyset$, then $\overline{W}_m \subseteq K(R+1)$.

Finally, choose $\alpha_m \subseteq C_c^\infty\big(\Sigma;[0,1]\big)$ to be a partition of unity which is subordinate to $\{U_m\}_1^\infty$.

For $0 < \epsilon < r_m \equiv \big|\Phi_m(U_m) - \Phi_m(W_m)^c\big|_{\mathbb{R}^N}$, define

$$\phi_{m,\epsilon}(\sigma) = \int_{\Phi_m(W_m)} \phi \circ \Phi_m^{-1}(y)\rho_\epsilon\big(\Phi_m(\sigma) - y\big)\,dy, \quad \sigma \in U_m,$$

where $\rho_\epsilon(y) = \epsilon^{-N}\rho(y/\epsilon)$ and $\rho \in C^\infty\big(\mathbb{R}^N;[0,\infty)\big)$ is compactly supported in the unit ball and has total (LEBESGUE) integral 1. Clearly, $\phi_{m,\epsilon} \in C^\infty\big(U_m;[0,\infty)\big)$. In addition, for every $\sigma \in U_m$,

$$\begin{aligned}\big|\phi_{m,\epsilon}(\sigma) - \phi(\sigma)\big| &\leq \int_{\{y\in\mathbb{R}^N:|y|<\epsilon\}} \Big|\phi \circ \Phi_m^{-1}\big(\Phi_m(\sigma) - y\big) - \phi(\sigma)\Big|\rho_\epsilon(y)\,dy \\ &\leq \sup_{|y|<\epsilon} \operatorname{dist}\big(\Psi_{m,y}(\sigma),\sigma\big),\end{aligned}$$

where, for a fixed $|y| < \epsilon$, $\Psi_{m,y}$ is the diffeomorphism on U_m into W_m given by $\sigma \in U_m \longmapsto \Phi_m^{-1}\big(\Phi_m(\sigma) - y\big)$. In particular,

$$A_m(\epsilon) \equiv \sup_{\sigma\in U_m} \big|\phi_{m,\epsilon}(\sigma) - \phi(\sigma)\big| \longrightarrow 0 \quad \text{as } \epsilon \longrightarrow 0.$$

Similarly, for all σ, $\tau \in U_m$,

$$\big|\phi_{m,\epsilon}(\tau) - \phi_{m,\epsilon}(\sigma)\big| \leq \sup_{|y|<\epsilon} \operatorname{dist}\big(\Psi_{m,y}(\tau), \Psi_{m,y}(\sigma)\big).$$

Using $D\Psi_{m,y}$ to denote the JACOBian matrix of $\Psi_{m,y}$ and noting that there is a $C_m \in (0,\infty)$ such that

$$(1 - C_m\epsilon)I_\sigma \leq \big(D\Psi_{m,y}(\sigma)\big)^* D\Psi_{m,y}(\sigma) \leq (1 + C_m\epsilon)I_\sigma$$

for $|y| < \epsilon$ and $\sigma \in U_m$ (where I_σ is used here to denote the identity map on $\mathbf{T}_\sigma(\Sigma)$ and the asterisk indicates the adjoint relative to the RIEMANNian metric) one easily sees that

$$\sup_{|y|<\epsilon} \big|\Psi_{m,y}(\tau) - \Psi_{m,y}(\sigma)\big| \leq \big(1 + \delta_m(\epsilon)\big)\operatorname{dist}(\tau,\sigma), \quad \sigma, \tau \in U_m,$$

where $\delta_m(\epsilon) \longrightarrow 0$ as $\epsilon \longrightarrow 0$. We can therefore choose $0 < \epsilon_m < r_m$ so that

$$\sup_{\sigma \in U_m} \left|\phi_m(\sigma) - \phi(\sigma)\right| \left[1 + \left|\operatorname{grad} \alpha_m(\sigma)\right|\right] \leq \frac{1}{2^{m+1}}$$

and

$$\left|\operatorname{grad} \phi_m(\sigma)\right| \leq \frac{3}{2}, \quad \sigma \in U_m,$$

when

$$\phi_m \equiv \phi_{m,\epsilon_m} \text{ on } U_m. \tag{6.2.13}$$

We now set

$$\psi = \sum_{m=1}^{\infty} \alpha_m \phi_m,$$

where the ϕ_m 's are the ones defined in (6.2.13). Obviously,

$$\psi \in C^{\infty}\big(\Sigma; [0, \infty)\big).$$

In addition, $|\psi - \phi| \leq 1$ on Σ; and therefore the level sets of ψ are compact. Finally,

$$\operatorname{grad} \psi = \sum_{m=1}^{\infty} \alpha_m \operatorname{grad} \phi_m + \sum_{m=1}^{\infty} \phi_m \operatorname{grad} \alpha_m.$$

Clearly, the first sum contributes at most $\frac{3}{2}$ to $|\operatorname{grad} \psi|$. To estimate the contribution of the second sum, note that

$$\sum_{m=1}^{\infty} \operatorname{grad} \alpha_m = 0;$$

and therefore that the length of the second sum is dominated by

$$\sum_{m=1}^{\infty} |\phi_m - \phi| \left|\operatorname{grad} \alpha_m\right| \leq \frac{1}{2}.$$

Hence, $|\operatorname{grad} \psi| \leq 2$.

To complete the proof, choose $\eta \in C^{\infty}\big(\mathbb{R}, [0,1]\big)$ so that $\eta \equiv 1$ on $(-\infty, 1]$, $\eta \equiv 0$ on $[2, \infty)$, and $|\eta'| \leq 1$ everywhere. It is then clear that the functions η_n given by

$$\eta_n(\sigma) = \eta\left(\frac{\psi(\sigma)}{n}\right)$$

have the desired properties. ∎

PROOF OF THEOREM 6.2.9: Let $\hat{\Sigma} = \Sigma \cup \{p\}$ be the one-point compactification of Σ, and set $\hat{\Omega} = C\big([0,\infty);\hat{\Sigma}\big)$. Then, by standard diffusion theoretic techniques (see Chapter 10 of [**104**] or [**51**]) one can show that, for each $\sigma \in \Sigma$, there is a unique $P_\sigma \in \mathbf{M}_1(\hat{\Sigma})$ with the properties that the expression in (6.2.10) is a mean-zero martingale for all $\phi \in C_c^\infty(\Sigma;\mathbb{R})$ and

$$P_\sigma\Big(\big\{\omega : \hat{\Sigma}_s(\omega) = p \text{ and } \hat{\Sigma}_t(\omega) \neq p \text{ for some } t \in (s,\infty)\big\}\Big) = 0$$

for $s \in [0,\infty)$. (We extend compactly supported functions on Σ to $\hat{\Sigma}$ by taking them equal to 0 at p.) Moreover, $\sigma \in \Sigma \longmapsto P_\sigma \in \mathbf{M}_1(\hat{\Omega})$ is continuous; and if P_p is the measure concentrated on the constant path at p, then $\{P_{\hat{\sigma}} : \hat{\sigma} \in \hat{\Sigma}\}$ forms a time-homogeneous MARKOV family. Hence, if

$$\zeta(\omega) \equiv \inf\big\{t \in [0,\infty) : \hat{\Sigma}_t(\omega) = p\big\}$$

and

$$[Q_t\phi](\sigma) = \int_{\{\omega:\,\zeta(\omega)>t\}} \phi\big(\hat{\Sigma}_t(\omega)\big)\, P_\sigma(d\omega), \quad (t,\sigma) \in (0,\infty)\times\Sigma,$$

for every $\phi \in B(\Sigma;\mathbb{R})$, then $\{Q_t : t > 0\}$ forms a FELLER-continuous, sub-MARKOVian semigroup which is weakly continuous at 0.

What is not so obvious, but is nonetheless true (cf. [**51**]), is the fact that the symmetry of $L|_{C_c^\infty(\Sigma;\mathbb{R})}$ in $L^2(m)$:

$$\text{(6.2.14)} \quad -\big(\phi, L\psi\big)_{L^2(m)} = \frac{1}{2}\int_\Sigma (\operatorname{grad}\phi|\operatorname{grad}\psi)\, dm, \quad \phi,\ \psi \in C_c^\infty(\Sigma;\mathbb{R}),$$

implies that the Q_t 's are also symmetric on $L^2(m)$. Thus each Q_t admits a unique extension as a self-adjoint contraction operator $\overline{Q}_t$ on $L^2(m)$ and the semigroup $\{\overline{Q}_t : t > 0\}$ is strongly continuous. In fact, $\{\overline{Q}_t : t > 0\}$ is the semigroup which is generated by the FRIEDRICHS extension $\overline{L}$ of $L|_{C_c^\infty(\Sigma;\mathbb{R})}$. Using $\{E_\lambda : \lambda \in [0,\infty)\}$ to denote the spectral resolution of $-\overline{L}$, we have the representation

$$\overline{Q}_t = \int_{[0,\infty)} e^{-\lambda t}\, dE_\lambda.$$

In particular, if

$$\mathcal{E}(\phi,\phi) \equiv \int_{[0,\infty)} \lambda\, d\big(E_\lambda\phi,\phi\big)_{L^2(m)}, \quad \phi \in L^2(m),$$

then (6.2.14) leads to

$$\mathcal{E}(\phi,\phi) = \frac{1}{2}\int_\Sigma |\mathrm{grad}\,\phi|^2\,dm, \quad \phi \in C_c^\infty(\Sigma;\mathbb{R}), \tag{6.2.15}$$

and, for $\phi \in L^2(m)$,

$$-\frac{d}{dt}\big(\phi, [\overline{Q}_t\phi]\big)_{L^2(m)} = \int_{[0,\infty)} \lambda e^{-\lambda t}\, d\big(E_\lambda\phi,\phi\big)_{L^2(m)} \le \mathcal{E}(\phi,\phi) \tag{6.2.16}$$

for $t \in (0,\infty)$. A basic fact about the FRIEDRICHS extension of a non-negative operator is that its DIRICHLET form is the closure of its quadratic form. Thus, in the present situation, $\mathcal{E}$ is the closure of its restriction to $C_c^\infty(\Sigma;\mathbb{R})$.

We next want to prove (6.2.11). To this end, let $\phi \in C^\infty(\Sigma;\mathbb{R}) \cap L^2(m)$ with $|\mathrm{grad}\,\phi| \in L^2(m)$ be given, and observe that, by (6.2.15) and the fact that $\mathcal{E}$ is closed, all that we have to do is produce a sequence $\{\phi_n\}_1^\infty \subseteq C_c^\infty(\Sigma;\mathbb{R})$ such that $\phi_n \longrightarrow \phi$ in $L^2(m)$ and

$$\int_\Sigma |\mathrm{grad}\,\phi_n - \mathrm{grad}\,\phi|^2\,dm \longrightarrow 0 \quad \text{as } n \longrightarrow 0.$$

To this end, choose the functions η_n as in the last part of Lemma 6.2.12 and simply take $\phi_n = \eta_n\phi$.

As an immediate consequence of (6.2.11) and (6.2.16) with $\phi = 1$, we see that

$$\int_\Sigma [Q_t 1]\,dm \ge 1 \quad \text{for all } t \in (0,\infty);$$

and because $[Q_t 1]$ is continuous and dominated by 1, this proves that $[Q_t 1] \equiv 1$. Equivalently, we now know that $P_\sigma\big(\{\omega : \zeta(\omega) \le t\}\big) = 0$ for every $(t,\sigma) \in [0,\infty) \times \Sigma$; and therefore the measures P_σ are actually concentrated on Ω. In particular, $\{P_\sigma : \sigma \in \Sigma\}$ is itself a FELLER-continuous time-homogeneous MARKOV family of probability measures on Ω; and all of the statements which we have made about the Q_t's immediately become statements about the semigroup $\{P_t : t > 0\}$ determined by $\{P_\sigma : \sigma \in \Sigma\}$.

We still have to prove the final assertion of the theorem. Using the spectral representation of $\overline{P}_t = \overline{Q}_t$, one sees that it is sufficient to show that the range of the projection E_0 is the constant functions. Equivalently, this comes down to checking that ϕ is constant if $\phi \in L^2(m)$ with $\mathcal{E}(\phi,\phi) = 0$. To this end, assume that $\mathcal{E}(\phi,\phi) = 0$. One then has that $\mathcal{E}(\phi,\psi) = 0$ for every $\psi \in \mathbf{Dom}(\mathcal{E})$ and therefore that

$$\int_\Sigma \phi[L\psi]\,dm = \lim_{t\searrow 0}\frac{1}{t}\big(\phi, [P_t\psi] - \psi\big)_{L^2(m)} = -\mathcal{E}(\phi,\psi) = 0$$

for every $\psi \in C_c^\infty(\Sigma;\mathbb{R})$. But this means that $[L\phi] = 0$ in the sense of distributions and therefore, by standard elliptic regularity theory, that $\phi \in C^\infty(\Sigma;\mathbb{R})$. In particular, this now leads to the conclusion that

$$-\frac{1}{2}\int_\Sigma (\operatorname{grad}\phi|\operatorname{grad}\psi)\,dm = \int_\Sigma \phi[L\psi]\,dm = 0, \quad \psi \in C_c^\infty(\Sigma;\mathbb{R}),$$

and, therefore, that $\operatorname{grad}\phi = 0$ everywhere. Clearly the constancy of ϕ follows from this and the connectedness of Σ. ∎

From now on m will be the probability measure in (6.2.8) and we will use $\langle\phi\rangle_m$ to denote the m-integral of a $\phi \in L^1(m)$. Also, $P(t,\sigma,\cdot)$ will be the transition probability function for the MARKOV family $\{P_\sigma : \sigma \in \Sigma\}$ produced in Theorem 6.2.9, and $\{P_t : t > 0\}$ will be the corresponding FELLER-continuous semigroup.

Before proceeding, we will need the following technical addendum to Theorem 6.2.9.

6.2.17 Lemma. *Set*

(6.2.18) $$\mathcal{F} = \Big\{f \in C_b(\Sigma;\mathbb{R}) \cap C^\infty(\Sigma;\mathbb{R}) : [Lf] \in L^2(m)\Big\}.$$

Then, for each $f \in \mathcal{F}$, $(t,\sigma) \in (0,\infty)\times\Sigma \longmapsto [P_tf](\sigma)$ is smooth,

(6.2.19) $$\frac{d}{dt}[P_tf] = [LP_tf] = [P_tLf], \quad t \in [0,\infty),$$

and $|\operatorname{grad} f| \in L^2(m)$. In fact,

(6.2.20) $$-\Big(g,[Lf]\Big)_{L^2(m)} = \frac{1}{2}\Big\langle(\operatorname{grad} f|\operatorname{grad} g)\Big\rangle_m \quad \text{for } f,\, g \in \mathcal{F}.$$

Finally, $\mathcal{F}$ is $\{P_t : t > 0\}$-invariant.

PROOF: Let $f \in \mathcal{F}$ and $\psi \in C_c^\infty(\Sigma;\mathbb{R})$ be given. Then,

$$\begin{aligned}\frac{d}{dt}\Big(\psi,[P_tf]\Big)_{L^2(m)} &= \frac{d}{dt}\Big([P_t\psi],f\Big)_{L^2(m)} \\ &= \Big([P_tL\psi],f\Big)_{L^2(m)} = \Big([L\psi],[P_tf]\Big)_{L^2(m)}.\end{aligned}$$

Thus, $(t,\sigma) \in (0,\infty)\times\Sigma \longmapsto [P_tf](\sigma)$ satisfies the first equality in (6.2.19) is the sense of distributions; and therefore, by elliptic regularity theory, it is a smooth function which satisfies this equality in the classical sense.

Before attempting to check the second inequality in (6.2.19), we will prove $|\mathrm{grad}\, f| \in L^2(m)$, $f \in \mathcal{F}$, and (6.2.20). To this end, choose $\{\eta_n\}_1^\infty$ as in the last part of Lemma 6.2.12. Then

$$-\Big(\eta_n^2 f, [Lf]\Big)_{L^2(m)} = \Big\langle \eta_n f \big(\mathrm{grad}\, \eta_n | \mathrm{grad}\, f\big)\Big\rangle_m + \frac{1}{2}\Big\langle \eta_n^2\, |\mathrm{grad}\, f|^2\Big\rangle_m$$
$$\geq -\big\|\eta_n |\mathrm{grad}\, f|\big\|_{L^2(m)} \big\|f\, |\mathrm{grad}\, \eta_n|\big\|_{L^2(m)} + \frac{1}{2}\big\|\eta_n\, |\mathrm{grad}\, f|\big\|_{L^2(m)}^2,$$

from which it is a simple matter to estimate $\big\|\, |\mathrm{grad}\, f|\, \big\|_{L^2(m)}^2$ in terms of $\|f\|_{L^2(m)}\|Lf\|_{L^2(m)}$. Thus, we now know that $|\mathrm{grad}\, f| \in L^2(m)$ for all $f \in \mathcal{F}$; and once one knows this, the proof of (6.2.20) is easy:

$$\begin{aligned}
-2\Big(g, [Lf]\Big)_{L^2(m)} &= -2 \lim_{n\to\infty} \Big(\eta_n g, [Lf]\Big)_{L^2(m)} \\
&= \lim_{n\to\infty} \Big[\Big\langle \eta_n \big(\mathrm{grad}\, g | \mathrm{grad}\, f\big)\Big\rangle_m + \Big\langle g \big(\mathrm{grad}\, \eta_n | \mathrm{grad}\, f\big)\Big\rangle_m\Big] \\
&= \Big\langle \big(\mathrm{grad}\, g | \mathrm{grad}\, f\big)\Big\rangle_m .
\end{aligned}$$

Returning to the proof of the the second equality in (6.2.19), note that we already know that (6.2.19) holds for elements of $C_c^\infty(\Sigma; \mathbb{R})$; and therefore, if $\psi \in C_c^\infty(\Sigma; \mathbb{R})$, then

$$\begin{aligned}
\big(\psi, [LP_t f]\big)_{L^2(m)} &= \frac{d}{dt}\big(\psi, [P_t f]\big)_{L^2(m)} \\
&= \frac{d}{dt}\big([P_t \psi], f\big)_{L^2(m)} = \big([LP_t\psi], f\big)_{L^2(m)} \\
&= \big([P_t\psi], [Lf]\big)_{L^2(m)} = \big(\psi, [P_t Lf]\big)_{L^2(m)},
\end{aligned}$$

where, in the passage from the first to the second lines, we have used the facts that $[P_t\psi] \in \mathcal{F}$, $t \in [0, \infty)$, and therefore that (6.2.20) applies. Clearly the second equality in (6.2.19) follows from the above. Moreover, we now see that $\mathcal{F}$ is $\{P_t : t > 0\}$-invariant, since the only thing that we had left to check is that $[LP_t f] \in L^2(m)$, and this is obvious from the second equality in (6.2.19). ∎

Our goal now is to find conditions which will tell us when the results in Sections 5.3 and 5.4 apply to the processes described in Theorem 6.2.9. We begin with the following.

6.2.21 Theorem. *Set* $V = \frac{1}{4}|\mathrm{grad}\, U|^2 - \frac{1}{2}\Delta U$ *and assume that the level sets* $\big\{\sigma \in \Sigma : V(\sigma) \leq R\big\}$, $R \in [0, \infty)$ *are compact. Then* $J_{\mathcal{E}}$ *is a good rate* *function and*

$$
(6.2.22)\quad
\begin{aligned}
-\inf_{\Gamma^\circ} J_{\mathcal{E}} &\le \inf_{\nu\in\mathbf{M}_1(\Sigma)} \underline{\lim_{t\to\infty}} \frac{1}{t}\log\Big(P_\nu\big(\{\omega : \mathbf{L}_t(\omega)\in\Gamma\}\big)\Big)\\
&\le \sup_{\nu\in\mathbf{M}_1(\Sigma)} \overline{\lim_{t\to\infty}} \frac{1}{t}\log\Big(P_\nu\big(\{\omega : \mathbf{L}_t(\omega)\in\Gamma\}\big)\Big) \le -\inf_{\overline{\Gamma}} J_{\mathcal{E}}
\end{aligned}
$$

for every measurable $\Gamma\subseteq\mathbf{M}_1(\Sigma)$.

PROOF: Recall that, for any $R\in[0,\infty)$, the set

$$
\left\{\phi\in C_c^\infty(B;\mathbb{R}) : \int_B \sum_{j=1}^N \left|\frac{\partial\phi}{\partial x_j}\right|^2 dx \le R\right\}
$$

is relatively compact in $L^2(\mathbb{R}^N)$, where B is the open unit ball in $\mathbb{R}^N$. Hence, with the use of a partition of unity, one can easily check that, for any relatively compact open set $G\subseteq\Sigma$,

$$
\left\{\phi\in C_c^\infty(G;\mathbb{R}) : \int_G |\operatorname{grad}\phi|^2\,dm \le R\right\}
$$

is a relatively compact subset of $L^2(m)$ for every $R\in[0,\infty)$. Knowing this and using, once again, the functions η_n from Lemma 6.2.12, one concludes that, for each $R\in[0,\infty)$,

$$
\Phi(R)\equiv\left\{\phi\in C_c^\infty(\Sigma;\mathbb{R}) : \mathcal{E}(\phi,\phi)\le R\right\}
$$

is relatively compact in $L^2_{\text{loc}}(m)$. That is, every sequence $\{\phi_n\}\subseteq\Phi(R)$ contains a subsequence $\{\phi_{n'}\}$ which is $L^2(m)$-convergent on each compact subset of Σ. Thus, we would know that $\Phi(R)$ is relatively compact in $L^2(m)$ if we could produce a sequence $\{K_\ell\}_1^\infty$ of compact subsets in Σ such that

$$
(6.2.23)\qquad \lim_{\ell\to\infty}\ \sup_{\phi\in\Phi(R)} \int_{K_\ell^c} \phi^2\,dm = 0.
$$

To prove (6.2.23) under the stated hypothesis on V, note that if $\phi\in C_c^\infty(\Sigma;\mathbb{R})$ and $\psi=e^{-U/2}\phi$, then

$$
\begin{aligned}
0 &\le \int_\Sigma |\operatorname{grad}\psi|^2\,d\lambda\\
&= 2\mathcal{E}(\phi,\phi) - \frac{1}{2}\int_\Sigma \big(\operatorname{grad}U\big|\operatorname{grad}\phi^2\big)\,dm + \frac{1}{4}\int_\Sigma |\operatorname{grad}U|^2\,\phi^2\,dm\\
&= 2\mathcal{E}(\phi,\phi) + \int_\Sigma [LU]\phi^2\,dm + \frac{1}{4}\int_\Sigma |\operatorname{grad}U|^2\phi^2\,dm\\
&= 2\mathcal{E}(\phi,\phi) - \int_\Sigma V\phi^2\,dm;
\end{aligned}
$$

and therefore

$$\int_{\Sigma} V\phi^2\,dm \leq 2\mathcal{E}(\phi,\phi).$$

Since the level sets of V are compact, it is clear from this how to choose the sets K_ℓ.

To complete the proof that $J_{\mathcal{E}}$ is good, remember that $\mathcal{E}$ is the closure of its restriction to $C_c^\infty(\Sigma:\mathbb{R})$ and conclude that

$$\begin{aligned}&\Big\{\nu \in \mathbf{M}_1(\Sigma) : J_{\mathcal{E}}(\nu) \leq R\Big\}\\ &\qquad\subseteq \Big\{\nu \in \mathbf{M}_1(\Sigma) : d\nu = \phi^2\,dm \text{ for some } \phi \in \overline{\Phi(R)}\Big\},\end{aligned}$$

where $\overline{\Phi(R)}$ is the closure of $\Phi(R)$ in $L^2(m)$. Thus, if $\{\nu_n\}_1^\infty \subseteq \mathbf{M}_1(\Sigma)$ with $J_{\mathcal{E}}(\nu_n) \leq R$, $n \in \mathbb{Z}^+$, then $d\nu_n = \phi_n^2\,dm$, where $\{\phi_n\}_1^\infty \subseteq \overline{\Phi(R)}$. Now choose a subsequence $\{\phi_{n'}\}$ which converges in $L^2(m)$ to an element ϕ of $\overline{\Phi(R)}$. It is then clear that $\nu_{n'} \Longrightarrow \nu$, where $d\nu = \phi^2\,dm$. Moreover, since (cf. (4.2.54))

$$J_{\mathcal{E}}(\nu) = \mathcal{E}(|\phi|,|\phi|) \leq \mathcal{E}(\phi,\phi) \leq R,$$

it is also clear that $J_{\mathcal{E}}(\nu) \leq R$.

The rest of the proof is nothing but an application of elliptic regularity theory and Exercise 5.3.14. Indeed, elliptic regularity theory assures us that $P(t,\sigma,d\tau) = p(t,\sigma,\tau)\,m(d\tau)$, where $p \in C^\infty\big((0,\infty)\times\Sigma\times\Sigma;(0,\infty)\big)$. ∎

Having found a condition which enables us to apply the results in Section 5.3, we next want to see what we can do to bring the results in Section 5.4 to bear. As we pointed out in Remark 6.1.10, the strong form of ($\tilde{\mathbf{U}}$) in (**SU**) is more than enough to guarantee that the semigroup $\{P_t : t > 0\}$ is hypercontractive. Of course, at least from the standpoint of large deviation theory, this is not a very useful observation since (**SU**) itself implies far stronger large deviation results than does hypermixing. On the other hand, Example 6.1.11 clearly demonstrates that there are interesting situations in which (**SU**) fails to hold but $\{P_t : t > 0\}$ is nonetheless hypercontractive; and what we want to do now is develop machinery for recognizing such situations.

Thus, we are about to embark on a program which will eventually give us a criterion with which to determine when $\{P_t : t > 0\}$ is hypercontractive even though (**SU**) may fail. The program which we have in mind is based on the work of Bakry and Emery [**3**] and entails the analysis of the function

$$H(t) \equiv \langle f_t \log f_t \rangle_m, \quad t \in [0,\infty), \tag{6.2.24}$$

where f is a uniformly positive element of $\mathcal{F}$ (cf. (6.2.18)) and we use f_t to denote $[P_t f]$. Using Lemma 6.1.17, one can easily justify the steps:

$$H'(t) = \langle [Lf_t] \rangle_m + \langle [Lf_t] \log f_t \rangle_m = -\frac{1}{2} \left\langle \frac{|\text{grad } f_t|^2}{f_t} \right\rangle_m .$$

Thus, since, by the last part of Theorem 6.2.9,

$$f_t \log f_t \longrightarrow \langle f \rangle_m \log \langle f \rangle_m \quad \text{in } L^1(m),$$

$$\left\langle f \log \frac{f}{\langle f \rangle_m} \right\rangle_m = \frac{1}{2} \int_0^\infty \left\langle \frac{|\text{grad } f_t|^2}{f_t} \right\rangle_m dt. \tag{6.2.25}$$

Clearly (6.2.25) is potentially related to a logarithmic SOBOLEV inequality. In particular, it indicates that we would be well-advised to study quantities related to the integrand on the right hand side. With this in mind, we introduce, for $\delta \in (0, \infty)$, the function

$$u_\delta(t, \sigma) \equiv \left(\left| \text{grad } f_t(\sigma) \right|^2 + \delta \right)^{1/2}, \quad (t, \sigma) \in [0, \infty) \times \Sigma. \tag{6.2.26}$$

By straight-forward computation (ITÔ's transformation rule for second order operators), one can show that

$$[Lu_\delta](t, \sigma) - \frac{\partial u_\delta}{\partial t}(t, \sigma) = \frac{v(t, \sigma) - w_\delta(t, \sigma)}{2u_\delta(t, \sigma)}, \tag{6.2.27}$$

where

$$v(t, \sigma) \equiv \left[L\left(|\text{grad } f_t|^2 \right) \right](\sigma) - 2\left(\text{grad } f_t | \text{grad } [Lf_t] \right)(\sigma) \tag{6.2.28}$$

and

$$w_\delta(t, \sigma) \equiv \frac{\left| \text{grad } \left(|\text{grad } f_t|^2 \right) \right|^2 (\sigma)}{4u_\delta(t, \sigma)^2}. \tag{6.2.29}$$

Our next goal is to interpret the quantity $v(t, \sigma)$ in (6.2.28). In doing so, it will be necessary to recall some more notions from RIEMANNian geometry. In the first place, if $g \in C^\infty(\Sigma; \mathbb{R})$, then the **Hessian**, $\text{Hess}\, g$, is the element of $\Gamma\left(\mathbf{T}^*(\Sigma) \otimes \mathbf{T}^*(\Sigma)\right)$ given by $\text{Hess}\, g(X, Y) = XYg - \nabla_X Y g$ for X, $Y \in \Gamma\left(\mathbf{T}(\Sigma)\right)$. Note that, because the LEVI–CIVITA connection is torsion free, $\text{Hess}\, g$ is symmetric. Also, an elementary calculation leads to

$$\text{Hess}\, g(X, Y) = \left(\nabla_X \text{grad}\, g | Y \right), \quad X, Y \in \Gamma\left(\mathbf{T}(\Sigma)\right). \tag{6.2.30}$$

A second notion which we will need is that of the RICCI curvature tensor. For this purpose, recall that the **Riemann curvature** is the tensor $R \in \Gamma\big(\mathbf{T}^*(\Sigma)^{\otimes 4}\big)$ defined by

$$R(X,V,Y,W) = -\big(\nabla_X \circ \nabla_V Y - \nabla_V \circ \nabla_X Y - \nabla_{[X,V]} Y \big| W\big)$$

for X, Y, V, $W \in \Gamma\big(\mathbf{T}(\Sigma)\big)$, and that the **Ricci curvature** is the tensor $\mathrm{Ric} \in \Gamma\big(\mathbf{T}^*(\Sigma)^{\otimes 2}\big)$ such that

$$\text{(6.2.31)} \qquad \mathrm{Ric}(X,Y)(\sigma) = \sum_{k=1}^{N} R(X,E_k,Y,E_k)(\sigma), \quad X,\, Y \in \Gamma\big(\mathbf{T}(\Sigma)\big),$$

as long as $\{E_k\}_1^N \subseteq \Gamma\big(\mathbf{T}(\Sigma)\big)$ is orthonormal at σ.

We will now show that

$$\text{(6.2.32)} \qquad v(t,\cdot) = \big(\mathrm{Ric} + \mathrm{Hess}\, U\big)(\mathrm{grad}\, f_t, \mathrm{grad}\, f_t) + \big\|\mathrm{Hess}\, f_t\big\|_{\mathrm{H.S.}}^2,$$

where, for any $\{E_k\}_1^N \subseteq \Gamma\big(\mathbf{T}(\Sigma)\big)$ which is orthonormal at σ,

$$\big\|\mathrm{Hess}\, f_t(\sigma)\big\|_{\mathrm{H.S.}} = \left(\sum_{k,\ell=1}^{N} \Big[\mathrm{Hess}\, f_t(E_k,E_\ell)(\sigma)\Big]^2\right)^{1/2}$$

is the HILBERT-SCHMIDT norm of $\mathrm{Hess}\, f_t(\sigma)$. In the derivation of (6.2.32), a central role will be played by the identity

$$\text{(6.2.33)} \qquad \mathrm{grad}\,(\mathrm{grad}\, u|\mathrm{grad}\, v) = \nabla_{\mathrm{grad}\, u}\mathrm{grad}\, v + \nabla_{\mathrm{grad}\, v}\mathrm{grad}\, u$$

for u, $v \in C^\infty(\Sigma;\mathbb{R})$. To prove (6.2.33), set $Y = \mathrm{grad}\, u$ and $Z = \mathrm{grad}\, v$. Then, for $X \in \Gamma\big(\mathbf{T}(\Sigma)\big)$:

$$\begin{aligned}
\big(X|\nabla_Y Z + \nabla_Z Y\big) &= Y(X|Z) - \big(\nabla_Y X|Z\big) + Z(X|Y) - \big(\nabla_Z X|Y\big) \\
&= YXv - \big(\nabla_X Y|Z\big) + ZXu - \big(Y|\nabla_X Z\big) - \big([Y,X]|Z\big) - \big([Z,X]|Y\big) \\
&= XYv + XZu - X(Y|Z) = X(Y|Z) = \big(X|\mathrm{grad}\,(Y|Z)\big),
\end{aligned}$$

where we made use of the torsion free nature of ∇.

Turning to the proof of (6.2.32), note that

$$\begin{aligned}
v(t,\cdot) &= \frac{1}{2}\Delta(\mathrm{grad}\, f_t|\mathrm{grad}\, f_t) - \frac{1}{2}\Big(\mathrm{grad}\, U\Big|\mathrm{grad}\,(\mathrm{grad}\, f_t|\mathrm{grad}\, f_t)\Big) \\
&\qquad - \big(\mathrm{grad}\,\Delta f_t\big|\mathrm{grad}\, f_t\big) + \big(\mathrm{grad}\,(\mathrm{grad}\, U|\mathrm{grad}\, f_t)|\mathrm{grad}\, f_t\big) \\
&\equiv v^0(t,\cdot) - \frac{1}{2}(\mathrm{grad}\, U|\mathrm{grad}\,(\mathrm{grad}\, f_t|\mathrm{grad}\, f_t)) \\
&\qquad + \big(\mathrm{grad}\,(\mathrm{grad}\, U|\mathrm{grad}\, f_t)|\mathrm{grad}\, f_t\big).
\end{aligned}$$

At the same time, by (6.2.33) (with $u = U$ and $v = f_t$), and (6.2.30):

$$\begin{aligned}
-\frac{1}{2}\big(\operatorname{grad} U|\operatorname{grad}(\operatorname{grad} f_t|\operatorname{grad} f_t)\big) &+ \big(\operatorname{grad}(\operatorname{grad} U|\operatorname{grad} f_t)|\operatorname{grad} f_t\big)\\
&= -\frac{1}{2}\operatorname{grad} U(\operatorname{grad} f_t|\operatorname{grad} f_t) + \big(\nabla_{\operatorname{grad} U}\operatorname{grad} f_t|\operatorname{grad} f_t\big)\\
&\qquad\qquad + \operatorname{Hess} U(\operatorname{grad} f_t, \operatorname{grad} f_t)\\
&= \operatorname{Hess} U(\operatorname{grad} f_t|\operatorname{grad} f_t).
\end{aligned}$$

Thus, all that remains is to show that

$$v^0(t,\cdot) = \operatorname{Ric}(\operatorname{grad} f_t, \operatorname{grad} f_t) + \big\|\operatorname{Hess} f_t\big\|^2_{\mathrm{H.S.}}. \tag{6.2.34}$$

In order to check (6.2.34), it will be convenient to fix a $\sigma \in \Sigma$ and to choose $\{E_k\} \subseteq \Gamma\big(\mathbf{T}(\Sigma)\big)$ so that $\{E_k(\sigma)\}_1^N$ is orthonormal and $\nabla_X E_k(\sigma) = 0$ for $1 \le k \le N$ and $X \in \Gamma\big(\mathbf{T}(\Sigma)\big)$. For example, one can choose a normal coordinate system $(x^1, \dots, x^N)$ in a neighborhood $\mathcal{O}$ of σ and arrange that $E_k = \frac{\partial}{\partial x^k}$ in $\mathcal{O}$. By (6.2.33), one then has

$$\begin{aligned}
\frac{1}{2}\Delta\big(|\operatorname{grad} f_t|^2\big)(\sigma) &= \operatorname{div}\big(\nabla_{\operatorname{grad} f_t}\operatorname{grad} f_t\big)(\sigma)\\
&= \sum_{k=1}^N \big(\nabla_{E_k}\nabla_{\operatorname{grad} f_t}\operatorname{grad} f_t\big|E_k\big)(\sigma)
\end{aligned}$$

and, by (6.2.2),

$$\begin{aligned}
\big(\operatorname{grad}\Delta f_t\big|\operatorname{grad} f_t\big)(\sigma) &= \sum_{k=1}^N \Big(\operatorname{grad}\big(\nabla_{E_k}\operatorname{grad} f_t\big|E_k\big)\Big|\operatorname{grad} f_t\Big)(\sigma)\\
&= \sum_{k=1}^N \big(\nabla_{\operatorname{grad} f_t}\nabla_{E_k}\operatorname{grad} f_t\big|E_k\big)(\sigma) + \sum_{k=1}^N \big(\nabla_{E_k}\operatorname{grad} f_t\big|\nabla_{\operatorname{grad} f_t}E_k\big)(\sigma)\\
&= \sum_{k=1}^N \big(\nabla_{\operatorname{grad} f_t}\nabla_{E_k}\operatorname{grad} f_t\big|E_k\big)(\sigma).
\end{aligned}$$

Thus, after subtracting the second of these from the first, we arrive at

$$v^0(t,\sigma) = \operatorname{Ric}\big(\operatorname{grad} f_t|\operatorname{grad} f_t\big)(\sigma) + \sum_{k=1}^N \Big(\nabla_{[E_k,\operatorname{grad} f_t]}\operatorname{grad} f_t\big|E_k\Big)(\sigma).$$

Finally, note that, because the HESSian is symmetric and ∇ is torsion free,

$$\begin{aligned}
\Big(\nabla_{[E_k,\operatorname{grad} f_t]}&\operatorname{grad} f_t\big|E_k\Big)(\sigma)\\
&= \operatorname{Hess} f_t\big([E_k, \operatorname{grad} f_t], E_k\big)(\sigma) = \big(\nabla_{E_k}\operatorname{grad} f_t\big|[E_k, \operatorname{grad} f_t]\big)(\sigma)\\
&= \big(\nabla_{E_k}\operatorname{grad} f_t|\nabla_{E_k}\operatorname{grad} f_t\big)(\sigma) - \big(\nabla_{E_k}\operatorname{grad} f_t|\nabla_{\operatorname{grad} f_t}E_k\big)(\sigma)\\
&= \big(\nabla_{E_k}\operatorname{grad} f_t|\nabla_{E_k}\operatorname{grad} f_t\big)(\sigma).
\end{aligned}$$

Thus, (6.2.34) follows after summing the preceding over $1 \le k \le N$.

Having dealt with $v(t,\sigma)$, we next want to estimate $w_\delta(t,\sigma)$ in (6.2.29). Remembering that the square of the HILBERT-SCHMIDT norm dominates the square of the largest eigenvalue of a symmetric matrix, use (6.2.33) to check that

$$\begin{aligned}\frac{1}{4}\Big|\operatorname{grad}\big(|\operatorname{grad} f_t|^2\big)\Big|^2 &= \big|\nabla_{\operatorname{grad} f_t}\operatorname{grad} f_t\big|^2\\ &= \big(\operatorname{Hess} f_t\big)^2\big(\operatorname{grad} f_t, \operatorname{grad} f_t\big)\\ &\le \big\|\operatorname{Hess} f_t\big\|_{\text{H.S.}}^2\,\big|\operatorname{grad} f_t\big|^2;\end{aligned}$$

and therefore

$$w_\delta(t,\cdot) \le \|\operatorname{Hess} f_t\|_{\text{H.S.}}^2. \tag{6.2.35}$$

By combining (6.2.27), (6.2.32), and (6.2.35), we arrive at the important relation

$$\big[Lu_\delta\big](t,\sigma) - \frac{\partial u_\delta}{\partial t}(t,\sigma) \ge \frac{\big(\operatorname{Ric} + \operatorname{Hess} U\big)\big(\operatorname{grad} f_t, \operatorname{grad} f_t\big)(\sigma)}{2u_\delta(t,\sigma)}. \tag{6.2.36}$$

In particular, if we now make the assumption that

$$\operatorname{Ric} + \operatorname{Hess} U \ge 2\epsilon I, \tag{\textbf{B\&E}}$$

for some $\epsilon > 0$, then

$$\frac{\partial u_\delta}{\partial t}(t,\cdot) \le \big[Lu_\delta\big](t,\cdot) - \epsilon\,\frac{|\operatorname{grad} f_t|^2}{u_\delta(t,\cdot)}, \quad t \in [0,\infty). \tag{6.2.37}$$

6.2.38 Lemma. *Let $T \in (0,\infty)$ and $v \in C^\infty\big([0,T]\times\Sigma;[0,\infty)\big)$ be given, and assume that $t \in [0,T] \longmapsto \|v(t,\cdot)\|_{L^2(m)}$ is bounded. If*

$$\frac{\partial v}{\partial t}(t,\cdot) \le \big[Lv(t,\cdot)\big], \quad t \in [0,T],$$

then

$$\int_0^T \big\|\,|\operatorname{grad} v(t,\cdot)|\,\big\|_{L^2(m)}^2\,dt \le \|v(0,\cdot)\|_{L^2(m)}^2 - \|v(T,\cdot)\|_{L^2(m)}^2.$$

PROOF: Choose $\{\eta_n\}_1^\infty$ as in the last part of Lemma 6.2.12 and set

$$\beta_n = \sup_{t\in[0,T]} \|v(t,\cdot)\|_{L^2(m)}\,\big\|\,|\operatorname{grad}\eta_n|\,\big\|_B.$$

Then

$$\Big\langle \eta_n^2\big(v(T,\cdot)^2 - v(0,\cdot)^2\big)\Big\rangle_m \le 2\int_0^T \Big(\eta_n^2 v(t,\cdot),[Lv(t,\cdot)]\Big)_{L^2(m)}\,dt$$

$$= -\int_0^T \Big\langle\Big(\operatorname{grad}\big(\eta_n^2 v(t,\cdot)\big)\Big|\operatorname{grad} v(t,\cdot)\Big)\Big\rangle_m\,dt$$

$$\le -\int_0^T \Big\|\eta_n|\operatorname{grad} v(t,\cdot)|\Big\|_{L^2(m)}^2\,dt$$

$$+ 2\int_0^T \Big\|\eta_n|\operatorname{grad} v(t,\cdot)|\Big\|_{L^2(m)}\Big\|v(t,\cdot)|\operatorname{grad}\eta_n|\Big\|_{L^2(m)}\,dt$$

$$\le -\int_0^T \Big\|\eta_n|\operatorname{grad} v(t,\cdot)|\Big\|_{L^2(m)}^2\,dt$$

$$+ 2T^{1/2}\beta_n\left(\int_0^T \Big\|\eta_n|\operatorname{grad} v(t,\cdot)|\Big\|_{L^2(m)}^2\,dt\right)^{1/2}$$

$$= -\left[\left(\int_0^T \Big\|\eta_n|\operatorname{grad} v(t,\cdot)|\Big\|_{L^2(m)}^2\,dt\right)^{1/2} - T^{1/2}\beta_n\right]^2 + T\beta_n^2,$$

from which the desired inequality follows after one takes the limit as $n \longrightarrow \infty$. ∎

With the preceding preparations, we are at last ready to prove the estimate toward which our efforts have been directed.

6.2.39 Lemma. *Assume that* **(B&E)** *holds for some* $\epsilon > 0$. *Then, for every uniformly positive element* f *of* $\mathcal{F}$,

$$\Big|\operatorname{grad}[P_T f]\Big| \le e^{-\epsilon T}\Big[P_T\big|\operatorname{grad} f\big|\Big], \quad T\in[0,\infty). \tag{6.2.40}$$

PROOF: Define u_δ as in (6.2.26). Then, by Lemma 6.2.17, (6.2.37), and Lemma 6.2.38, we know that

$$\int_0^T \Big\|\,\big|\operatorname{grad} u_\delta(t,\cdot)\big|\,\Big\|_{L^2(m)}^2\,dt < \infty, \quad T\in[0,\infty). \tag{6.2.41}$$

Now let $\phi,\ \psi \in C_c^\infty\big(\Sigma;[0,\infty)\big)$ be given and set

$$F(t) = \Big(\phi, \big[P_t\big(\psi u_\delta(T-t,\cdot)\big)\big]\Big)_{L^2(m)} \quad \text{for } t\in(0,T).$$

Then, by (6.2.37),

$$F'(t) - \epsilon F(t) + \epsilon\delta^{1/2}\big([P_t\phi], \psi\big)_{L^2(m)}$$

$$\geq \Big(\big[LP_t\phi\big], \psi u_\delta(T-t,\cdot)\Big)_{L^2(m)} - \Big(\big[P_t\phi\big], \psi\big[Lu_\delta(T-t,\cdot)\big]\Big)_{L^2(m)}$$

$$= \Big(\big[P_t\phi\big], \big[L\big(\psi u_\delta(T-t,\cdot)\big)\big]\Big)_{L^2(m)} - \Big(\big[P_t\phi\big], \psi\big[Lu_\delta(T-t,\cdot)\big]\Big)_{L^2(m)}$$

$$= \Big([P_t\phi], u_\delta(T-t,\cdot)[L\psi]\Big)_{L^2(m)} + \Big\langle [P_t\phi]\big(\operatorname{grad} u_\delta(T-t,\cdot)\big|\operatorname{grad}\psi\big)\Big\rangle_m$$

$$= -\frac{1}{2}\Big\langle u_\delta(T-t,\cdot)\big(\operatorname{grad}[P_t\phi]\big|\operatorname{grad}\psi\big)\Big\rangle_m + \frac{1}{2}\Big\langle [P_t\phi]\big(\operatorname{grad} u_\delta(T-t,\cdot)\big|\operatorname{grad}\psi\big)\Big\rangle_m;$$

and therefore

$$e^{-\epsilon T}\Big(\phi, \big[P_T\big(\psi u_\delta(0,\cdot)\big)\big]\Big)_{L^2(m)} - \big(\phi, \psi u_\delta(T,\cdot)\big)_{L^2(m)}$$

$$\geq \frac{1}{2}\int_0^T e^{-\epsilon t}\Big[\Big\langle [P_t\phi]\big(\operatorname{grad} u_\delta(T-t,\cdot)\big|\operatorname{grad}\psi\big)\Big\rangle_m - \Big\langle u_\delta(T-t,\cdot)\big(\operatorname{grad}[P_t\phi]\big|\operatorname{grad}\psi\big)\Big\rangle_m\Big]\,dt$$

$$- \epsilon\delta^{1/2}\int_0^T e^{-\epsilon t}\big([P_t\phi], \psi\big)_{L^2(m)}\,dt.$$

Now let $\{\eta_n\}_1^\infty$ be the sequence produced in Lemma 6.2.12, replace ψ in the preceding by η_n, let $n \longrightarrow \infty$ and $\delta \searrow 0$, and use the above together with (6.2.41) to conclude that

$$\big(\phi, \big|\operatorname{grad}[P_T f]\big|\big)_{L^2(m)} \leq e^{-\epsilon T}\Big(\phi, \big[P_T|\operatorname{grad} f|\big]\Big)_{L^2(m)}.$$

Finally, because this is true for an arbitrary $\phi \in C_c^\infty\big(\Sigma; [0,\infty)\big)$, it obviously implies (6.2.40). ∎

6.2.42 Theorem. *Assume that* **(B&E)** *holds for some* $\epsilon > 0$. *Then, for all* $1 < p \leq q < \infty$,

$$\|P_t\|_{L^p(m)\to L^q(m)} = 1 \quad \text{for } t \in (0,\infty) \text{ with } e^{2\epsilon t} \geq \frac{q-1}{p-1}. \tag{6.2.43}$$

In particular, $\{P_t : t > 0\}$ *is hypercontractive at time* $(\log 3)/2\epsilon$ *and*

$$\inf\Big\{\mathcal{E}(\phi,\phi) : \big\|\phi - \langle\phi\rangle_m\big\|_{L^2(m)} = 1\Big\} \geq \epsilon. \tag{6.2.44}$$

PROOF: Let f be a uniformly positive element of $\mathcal{F}$. Then, from (6.2.40), we have that

$$\begin{aligned}|\mathrm{grad}\,[P_T f]|^2 &\le e^{-2\epsilon T}\left[P_T\left(f^{1/2}\frac{|\mathrm{grad}\,f|}{f^{1/2}}\right)\right]^2 \\ &\le e^{-2\epsilon T}[P_T f]\left[P_T\left(\frac{|\mathrm{grad}\,f|^2}{f}\right)\right];\end{aligned}$$

and so, by (6.2.25),

$$\left\langle f\log\frac{f}{\langle f\rangle_m}\right\rangle_m \le \frac{1}{4\epsilon}\left\langle\frac{|\mathrm{grad}\,f|^2}{f}\right\rangle_m.$$

Next, let $q\in(1,\infty)$ and a uniformly positive $\phi\in\mathcal{F}$ be given. Choosing $\{\eta_n\}_1^\infty$ as in Lemma 6.2.12, set $f_n = \left(\eta_n\,\phi^{q/2}+1/n\right)^2$. Plugging this f_n into the above, noting that

$$\frac{1}{4}\left\langle\frac{|\mathrm{grad}\,f_n|^2}{f_n}\right\rangle_m = 2\mathcal{E}\left(f_n^{1/2},f_n^{1/2}\right),$$

and then letting $n\longrightarrow\infty$, we arrive at

$$\int_\Sigma \phi^q\log\frac{\phi^q}{\|\phi^{q/2}\|^2_{L^2(m)}}\,dm \le \frac{2}{\epsilon}\mathcal{E}\left(\phi^{q/2},\phi^{q/2}\right).$$

Since $\phi_t\equiv[P_t\phi]$ is a uniformly positive element of $\mathcal{F}$ whenever ϕ itself is, we can use this in (6.1.16) with $q(t)=1+(p-1)e^{2\epsilon t}$ to conclude that

$$t\in[0,\infty)\longmapsto\|P_t\phi\|_{L^{q(t)}(m)}$$

is non-increasing; and from this point it is an easy step to (6.2.43). Finally, (6.2.44) follows from (6.2.43) together with Theorem 6.1.14 and Corollary 6.1.17. ∎

6.2.45 Corollary. *Assume that there is a bounded $V\in C^\infty(\Sigma;\mathbb{R})$ with the property that*

$$\mathrm{Ric}+\mathrm{Hess}\,(U+V)\ge\epsilon I \tag{6.2.46}$$

for some $\epsilon>0$. Then $\{P_t : t>0\}$ is hypercontractive.

PROOF: Without loss of generality, we will assume that

$$\int_\Sigma e^V\,dm=\int_\Sigma e^{V+U}\,d\lambda=1.$$

Define $m' \in \mathbf{M}_1(\Sigma)$ and the DIRICHLET form $\mathcal{E}'$ relative to $U + V$. By Theorem 6.1.14 and Theorem 6.2.42,

$$\int_\Sigma \phi^2 \log \frac{|\phi|^2}{\|\phi\|^2_{L^2(m')}}\, dm' \le \frac{2}{\epsilon}\mathcal{E}'(\phi,\phi), \quad \phi \in L^2(m').$$

Using the technique in part (**i**) of Exercise 6.1.27, one sees that

$$\int_\Sigma \phi^2 \log \frac{|\phi|^2}{\|\phi\|^2_{L^2(m)}}\, dm \le \frac{2\|e^V\|_B}{\epsilon}\mathcal{E}'(\phi,\phi), \quad \phi \in L^2(m).$$

At the same time, by (6.2.11),

$$\mathcal{E}'(\phi,\phi) \le \|e^{-V}\|_B \mathcal{E}(\phi,\phi), \quad \phi \in \mathcal{F};$$

and therefore

$$\int_\Sigma \phi^2 \log \frac{|\phi|^2}{\|\phi\|^2_{L^2(m)}}\, dm \le \frac{2\|e^V\|_B \|e^{-V}\|_B}{\epsilon}\mathcal{E}(\phi,\phi), \quad \phi \in \mathcal{F}.$$

Thus, we find ourselves at the same place as we were when we started the second paragraph in the proof of Theorem 6.2.42; and therefore the same argument applies here. ∎

6.2.47 Exercise.

Let $\Sigma = \mathbb{R}^N$ and give $\mathbb{R}^N$ the standard EUCLIDean structure. Then the RIEMANNian measure is LEBESGUE's measure and Δ is the standard EUCLIDean LAPLACE operator. Let $U \in C^\infty(\mathbb{R}^N;\mathbb{R})$ be a function which is bounded below and satisfies (6.2.7), and define $m \in \mathbf{M}_1(\mathbb{R}^N)$ and L on $C_c^\infty(\mathbb{R}^N;\mathbb{R})$ accordingly. Finally, let $\mathcal{E}$ be the corresponding DIRICHLET form described in Theorem 6.2.9, and define V as in Theorem 6.2.21.

(**i**) It is interesting to see that, at least for the setting just described, Theorem 6.2.21 is quite sharp. To see this, suppose that there is an $r \in (0,\infty)$ and a sequence $\sigma_n \longrightarrow \infty$ with the property that

$$\sup_{n\in\mathbb{Z}^+} \sup_{\tau\in B(\sigma_n,r)} V(\tau) < \infty,$$

where $B(\sigma, r)$ denotes the open EUCLIDean ball with center σ and radius r. Choose

$$\psi \in C_c^\infty\big(B(0,r);[0,\infty)\big) \quad \text{with} \int_{\mathbb{R}^N} \psi\, dx = 1,$$

and set $\phi_n = \exp\big(U/2\big)\psi_n$ where $\psi_n(\tau) \equiv \psi(\tau + \sigma_n)$, $\tau \in \mathbb{R}^N$. Show that

$$\|\phi_n\|_{L^2(m)} = 1 \text{ for all } n \in \mathbb{Z}^+ \quad \text{and} \sup_{n\in\mathbb{Z}^+} \mathcal{E}\big(\phi_n,\phi_n\big) < \infty;$$

and conclude from this that the associated $J_{\mathcal{E}}$ cannot be good.

(**ii**) Assume that

$$U(\sigma) = \left(1 + |\sigma|^2\right)^{\alpha/2} - \log c_\alpha, \quad \sigma \in \mathbb{R}^N,$$

where $\alpha \in (0, \infty)$ and c_α is chosen so that the normalization condition is satisfied. Show that $J_{\mathcal{E}}$ is good if and only if $\alpha \in (1, \infty)$ and that the associated semigroup $\{P_t : t > 0\}$ is hypercontractive if $\alpha \in [2, \infty)$. Finally, if $\alpha \in (1, 2)$, show that **(LS)** fails and therefore that $\{P_t : t > 0\}$ is not hypercontractive. (**Hint:** Try test functions of the form $e^{\beta U}$ with $\beta \in (0, 1/2)$.)

(**iii**) The preceding result showed that the ORNSTEIN–UHLENBECK semigroup in Exercise 6.1.11 (i.e., the case when $\alpha = 2$) is at the borderline of hypercontractivity. By a remarkable coincidence, it turns out that Theorem 6.2.42 predicts the optimal hypercontractive result for this semigroup. To see this, check that in this case **(B&E)** holds with $\epsilon = \frac{1}{2}$ and therefore that

$$\|P_t\|_{L^p(m) \to L^q(m)} \quad \text{for } e^t \geq \frac{q-1}{p-1}.$$

Using the fact (cf. the last part of Example 6.1.11) that

$$\inf\left\{\mathcal{E}(\phi, \phi) : \|\phi - <\phi>_m\|_{L^2(m)} = 1\right\} = \frac{1}{2},$$

show that

$$\|P_t\|_{L^2(m) \to L^4(m)} > 1 \quad \text{if } e^t < 3;$$

and therefore that the predicted result is optimal. Actually, one can do even better. Namely, by considering the functions $\phi_\gamma(x) = \exp\left(\gamma|x|^2\right)$, one can show that for any $1 < p < q < \infty$ and $t \in (0, \infty)$ with $e^t < (q-1)/(p-1)$,

$$\|P_t\|_{L^p(m) \to L^q(m)} = \infty.$$

The facts contained in this exercise were first obtained by E. NELSON [**79**] and constitute the origins of all hypercontractivity considerations.

6.2.48 Exercise.

It is interesting to look at the BAKRY–EMERY argument when Σ is compact; even though, in that case, we already know that **(SU)** holds and therefore that $\{P_t : t > 0\}$ is more than hypercontractive. In this exercise we outline the argument for the compact case and point out that the argument is not only simpler but also leads to a slightly sharper statement. Observe that the key to the simplification is hidden entirely in the fact that the space $C^\infty(\Sigma; \mathbb{R})$ is invariant under both L and $\{P_t : t > 0\}$.

(**i**) Let $f \in C^\infty(\Sigma;\mathbb{R})$ be uniformly positive and set $H(t) = \langle f_t \log f_t \rangle_m$, where, once again, $f_t \equiv [P_t f]$. First show that

$$-H'(t) = \frac{1}{2}\Big\langle f_t \big|\mathrm{grad}\,\psi_t\big|^2\Big\rangle_m,$$

where $\psi_t \equiv \log f_t$, and second that

$$H''(t) = \frac{1}{2}\Big\langle f_t\Big[\|\mathrm{Hess}\,\psi_t\|^2_{\mathrm{H.S.}} + \big(\mathrm{Ric} + \mathrm{Hess}\,U\big)\big(\mathrm{grad}\,\psi_t, \mathrm{grad}\,\psi_t\big)\Big]\Big\rangle_m.$$

Now conclude that the condition

$$(\mathbf{B\&E'}) \qquad \begin{aligned} &\Big\langle e^\psi\Big[\|\mathrm{Hess}\,\psi\|^2_{\mathrm{H.S.}} + \big(\mathrm{Ric} + \mathrm{Hess}\,U\big)\big(\mathrm{grad}\,\psi, \mathrm{grad}\,\psi\big)\Big]\Big\rangle_m \\ &\geq 2\epsilon\Big\langle e^\psi|\mathrm{grad}\,\psi|^2\Big\rangle_m \qquad \text{for } \psi \in C^\infty(\Sigma;\mathbb{R}) \end{aligned}$$

implies (6.2.43).

(**ii**) The major advantage that (**B&E**$'$) has over (**B&E**) is that it leaves open the possibility of applying it even when no point-wise estimate holds. For example, consider the case when Σ is the flat N-torus ($= (\mathbb{R}/\mathbb{Z})^N$) and $U \equiv 0$. Then, since the RICCI curvature vanishes, the left hand side of (**B&E**$'$) becomes

$$\Big\langle e^\psi \|\mathrm{Hess}\,\psi\|^2_{\mathrm{H.S.}}\Big\rangle_\lambda,$$

which is easily seen to dominate

$$\sum_{j=1}^N \left\langle e^\psi \left(\frac{\partial^2\psi}{\partial\theta_j^2}\right)^2\right\rangle_\lambda,$$

where $(\theta_1, \ldots, \theta_N)$ is the standard coordinate system on Σ. Thus, in this case, (**B&E**$'$) holds for all $N \in \mathbb{Z}^+$ with a given ϵ if it holds when $N = 1$ for that ϵ. Therefore, assume that $N = 1$, and observe that when $h = e^{\psi/2}$ then the preceding dominates $4\|h''\|^2_{L^2(\lambda)}$, whereas the factor to be estimated on the right hand side of (**B&E**$'$) becomes $4\|h'\|^2_{L^2(\lambda)}$. Use these observations to show that (**B&E**$'$) holds with $\epsilon = \frac{1}{2}$.

6.3 Hypoelliptic Diffusions on a Compact Manifold

In this section we will describe a particularly good situation to which the results in Section 4.2 apply and will attempt to give a more pleasing expression for the associated rate function, even when the process involved is not symmetric.

The general setting in which we will be working is as follows. The space Σ will be a connected, compact, N-dimensional differentiable manifold; and λ will denote a fixed probability measure on Σ which is "smooth" in the sense that, for any coordinate chart (W, Φ), there is an $\alpha \in C^\infty\big(W; (0,\infty)\big)$ for which

$$\lambda(\Gamma) = \int_{\Phi(\Gamma)} \alpha(x)\, dx, \quad \Gamma \in \mathcal{B}_W.$$

In particular, for any $X \in \Gamma\big(\mathbf{T}(\Sigma)\big)$, there is a (unique) $g_X \in C^\infty(\Sigma;\mathbb{R})$ with the property that

$$\big(\psi, X\phi\big)_{L^2(\lambda)} = \big(X^*\psi, \phi\big)_{L^2(\lambda)}, \quad \phi, \psi \in C^\infty(\Sigma;\mathbb{R}),$$

where

(6.3.1) $$X^*\psi \equiv -X\psi + g_X\psi, \quad \psi \in C^\infty(\Sigma;\mathbb{R}).$$

Now suppose that $X_1, \ldots, X_d$, and Y are given elements of $\Gamma\big(\mathbf{T}(\Sigma)\big)$ and define the operator

$$L^Y = -\sum_{k=1}^{d} X_k^* \circ X_k + Y \quad \text{on } C^\infty(\Sigma;\mathbb{R}).$$

The following theorem contains a few important facts about the diffusion determined by L^Y.

6.3.2 Theorem. *Let* $\Omega = C\big([0,\infty);\Sigma\big)$, $\omega \in \Omega \longmapsto \Sigma_t(\omega) \in \Sigma$, $t \in [0,\infty)$, *and* $\{\mathcal{B}_t : t > 0\}$ *be as in* Theorem *6.2.9. Then, for each* $\sigma \in \Sigma$, *there is a unique* $P_\sigma \in \mathbf{M}_1(\Omega)$ *for which*

$$\left(\phi(\Sigma_t(\omega)) - \phi(\sigma) - \int_0^t \big[L^Y\phi\big]\big(\Sigma_s(\omega)\big)\, ds,\ \mathcal{B}_t,\ P_\sigma^Y\right)$$

is a mean-zero martingale. In addition, $\big\{P_\sigma^Y : \sigma \in \Sigma\big\}$ *is a* FELLER-*continuous* MARKOV *family. Finally, let* $\big\{P_t^Y : t > 0\big\}$ *denote the associated* MARKOV *semigroup. Then, for each* $\phi \in C^\infty(\Sigma;\mathbb{R})$, *the function*

$(t,\sigma) \in [0,\infty) \times \Sigma \longmapsto [P_t^Y \phi](\sigma) \in \mathbb{R}$ *is an element of* $C^\infty([0,\infty) \times \Sigma; \mathbb{R})$ *which satisfies*

$$(6.3.3) \quad \frac{\partial u}{\partial t}(t,\sigma) = [L^Y u](t,\sigma), \quad (t,\sigma) \in [0,\infty) \times \Sigma, \quad \text{with } u(0,\cdot) = \phi;$$

λ *is* $\{P_t^Y : t > 0\}$*-invariant if and only if* $g_Y = 0$*; and* λ *is* $\{P_t^Y : t > 0\}$*-reversing if and only if* $Y = 0$.

PROOF: There are many ways in which one can prove each of these facts. For the sake of completeness, we will outline a proof which should be pleasing to the probabilists, if no one else.

Without loss of generality, we assume that Σ is an embedded submanifold of $\mathbb{R}^n$ for a suitably large $n \in \mathbb{Z}^+$ and that the vector fields $X_1, \dots, X_d$, and Y are the restrictions to Σ of vector fields $\hat{X}_1, \dots, \hat{X}_d$, and $\hat{Y}$ on $\mathbb{R}^n$ with coefficients in $C_b^\infty(\mathbb{R}^n; \mathbb{R})$ (i.e., bounded continuous derivatives of all orders). At the same time, we think of each of the functions g_{X_k} as the restriction to Σ of some $\hat{g}_{X_k} \in C^\infty(\mathbb{R}^n; \mathbb{R})$, and then set $\widehat{X_k^*} = -\hat{X}_k + \hat{g}_{X_k}$. Hence, if $\hat{\Omega} \equiv C^\infty([0,\infty); \mathbb{R}^n)$, then one can use ITÔ's theory of stochastic integral equations to construct a FELLER-continuous, MARKOV family $\{\hat{P}_x : x \in \mathbb{R}^n\} \subseteq \mathbf{M}_1(\hat{\Omega})$ with the property that, for every $x \in \mathbb{R}^n$,

$$\left(\hat{\phi}(\hat{\Sigma}_t(\hat{\omega})) - \hat{\phi}(x) - \int_0^t [\hat{L}\hat{\phi}](\hat{\Sigma}_s(\hat{\omega}))\, ds,\, \hat{\mathcal{B}}_t,\, \hat{P}_x \right)$$

is a mean-zero martingale for every $\hat{\phi} \in C_b^\infty(\mathbb{R}^n; \mathbb{R})$, where $\hat{\omega} \in \hat{\Omega} \longmapsto \hat{\Sigma}_t(\hat{\omega}) \in \mathbb{R}^n$ and $\hat{\mathcal{B}}_t$ are defined by analogy to their "unhatted" counterparts, and

$$\hat{L} = -\sum_{k=1}^d \widehat{X_k^*} \circ \hat{X}_k + \hat{Y}.$$

In fact, one knows that it is possible to differentiate the solution to ITÔ's equations as a function of the starting point x. As a consequence, one finds first that the associated semigroup $\{\hat{P}_t : t > 0\}$ maps $C_b^\infty(\mathbb{R}^n; \mathbb{R})$ into itself and then that $(t,x) \in [0,\infty) \times \Sigma \longmapsto [\hat{P}_t\hat{\phi}](x) \in \mathbb{R}$ is a smooth solution to

$$\frac{\partial \hat{u}}{\partial t}(t,x) = [\hat{L}\hat{u}](t,x), \quad t \in [0,\infty) \times \Sigma \quad \text{with } \hat{u}(0,\cdot) = \hat{\phi}$$

for each $\hat{\phi} \in C_b^\infty(\mathbb{R}^n; \mathbb{R})$. Finally, if $x = \sigma \in \Sigma$, then one can easily show that $\hat{P}_\sigma(\Omega) = 1$; and so we get all the required existence results by simply taking $P_\sigma = \hat{P}_\sigma\big|_{\mathcal{B}_\Omega}$, $\sigma \in \Sigma$. Furthermore, the asserted uniqueness statement follows easily (cf. Theorem 6.3.2 in [**104**]) from the fact that we now also know how to find a smooth solution to (6.3.3) for every smooth ϕ; namely, one simply chooses $\hat{\phi} \in C^\infty(\mathbb{R}^n; \mathbb{R})$ so that $\hat{\phi}|_\Sigma = \phi$ and then takes $u(t,\sigma) = [\hat{P}_t\hat{\phi}](\sigma)$.

To complete the proof, let $\phi,\ \psi \in C^\infty(\Sigma;\mathbb{R})$ be given and note that, for any $T \in (0,\infty)$,

$$\frac{d}{dt}\int_\Sigma [P^Y_{T-t}\psi](\sigma)[P^Y_t\phi](\sigma)\,\lambda(d\sigma)$$
$$= -2\int_\Sigma [YP^Y_{T-t}\psi](\sigma)[P^Y_t\phi](\sigma)\,\lambda(d\sigma)$$
$$+\int_\Sigma g_Y(\sigma)[P^Y_{T-t}\psi](\sigma)[P^Y_t\phi](\sigma)\,\lambda(d\sigma)$$

for $t \in [0,T]$. Hence, with $\psi = 1$, we see that λ is $\{P^Y_t : t > 0\}$-invariant if and only if $g_Y = 0$. At the same time, if $Y = 0$, then

$$\Big(\psi, [P^0_T\phi]\Big)_{L^2(\lambda)} = \Big([P^0_T\psi],\phi\Big)_{L^2(\lambda)};$$

whereas, if λ is $\{P^Y_t : t > 0\}$-reversing, then $(Y\psi,\phi)_{L^2(\lambda)} = 0$. ∎

6.3.4 Remark.

Note that if $U \in C^\infty(\Sigma;\mathbb{R})$ and $Y^U \in \Gamma(\mathbf{T}(\Sigma))$ and $m_U \in \mathbf{M}_1(\Sigma)$ are defined by

$$Y^U = -\sum_{k=1}^d (X_kU)X_k \quad\text{and}\quad m_U(d\sigma) = \frac{e^{-U(\sigma)}}{Z_U}\lambda(d\sigma), \tag{6.3.5}$$

where $Z_U \equiv \int_\Sigma e^{-U}\,d\lambda$, then m_U is $\{P^Y_t : t > 0\}$-reversing if and only if $Y = Y^U$. Indeed, for any $X \in \Gamma(\mathbf{T}(\Sigma))$, one can easily check that

$$\int_\Sigma \psi\, X\phi\, dm_U = \int_\Sigma \big(X^*\psi + (XU)\psi\big)\,\phi\, dm_U, \qquad \phi,\ \psi \in C^\infty(\Sigma;\mathbb{R});$$

from which it is clear that the reasoning used to prove the last part of the preceding theorem applies with m_U replacing λ and $X^*_k + (X_kU)$ replacing X^*_k.

As yet we have not made any assumptions which would guarantee the sort of conditions required to make the results in Section 4.2 applicable. For this reason, we will now add the following hypothesis:

$$\text{Lie}(X_1,\dots,X_d) = \mathbf{T}(\Sigma), \tag{H}$$

where $\text{Lie}(X_1,\dots,X_d)$ denotes the LIE sub-algebra of $\Gamma(\mathbf{T}(\Sigma))$ generated by $\{X_1,\dots,X_d\}$ and the equality means that, at each $\sigma \in \Sigma$,

$$\Big\{X(\sigma) : X \in \text{Lie}(X_1,\dots,X_d)\Big\} = \mathbf{T}_\sigma(\Sigma).$$

According to HÖRMANDER's famous theorem (see [**63**]), the hypothesis (**H**) is more than enough to guarantee that, for any $Y \in \Gamma\big(\mathbf{T}(\Sigma)\big)$, the operator

$$\frac{\partial}{\partial t} + L^Y$$

is "hypoelliptic." In particular, this means that

$$P^Y(t,\sigma,d\tau) = p^Y(t,\sigma,\tau)\,\lambda(d\tau),$$

where the function p^Y is a non-negative element of $C^\infty\big((0,\infty)\times\Sigma\times\Sigma;\mathbb{R}\big)$. In addition, (**H**) is sufficient to guarantee that p^Y must be everywhere strictly positive. To see this, one can either invoke BONY's strong maximum principle (see [**13**]) or one can use the "support theorem" in [**103**]. Thus, with (**H**), we have more than enough information to see that not only does the condition ($\tilde{\mathbf{U}}$) hold but even that, for every $t \in (0,\infty)$,

$$\text{(6.3.6)} \qquad \frac{1}{M_t}\lambda \le P^Y(t,\sigma,\cdot) \le M_t\lambda, \quad \sigma \in \Sigma,$$

for some $M_t \in [1,\infty)$.

In view of the preceding, we now know that (**H**) allows us to apply the results of Section 4.2, and the following lemma summarizes what we can say immediately on the basis of those results.

6.3.7 Lemma. *Assume that* (**H**) *holds, and define* $\omega \in \Omega \longmapsto \mathbf{L}_t(\omega) \in \mathbf{M}_1(\Sigma)$, $t \in (0,\infty)$ *as in* Remark 4.2.2. *Then, for every* $\Gamma \in \mathcal{B}_{\mathbf{M}_1(\Sigma)}$,

$$\text{(6.3.8)} \qquad \begin{aligned} -\inf_{\Gamma^\circ} J^Y &\le \varliminf_{t\to\infty} \frac{1}{t}\log\Big[\inf_\sigma P_\sigma\Big(\big\{\omega : \mathbf{L}_t(\omega)\in\Gamma\big\}\Big)\Big] \\ &\le \varlimsup_{t\to\infty} \frac{1}{t}\log\Big[\sup_\sigma P_\sigma\Big(\big\{\omega : \mathbf{L}_t(\omega)\in\Gamma\big\}\Big)\Big] \le -\inf_{\overline{\Gamma}} J^Y, \end{aligned}$$

where

$$\text{(6.3.9)} \quad J^Y(\nu) \equiv \sup\left\{-\int_\Sigma \frac{L^Y u}{u}\,d\nu : u \in C^\infty\big(\Sigma;[1,\infty)\big)\right\}, \quad \nu \in \mathbf{M}_1(\Sigma).$$

Moreover, if $\mathcal{E}$ *denotes the* DIRICHLET *form corresponding to* $(t,\sigma) \in (0,\infty)\times\Sigma \longmapsto P^0(t,\sigma,\cdot) \in \mathbf{M}_1(\Sigma)$ *and* λ, *then*

$$\text{(6.3.10)} \qquad J^0(\nu) = J_{\mathcal{E}}(\nu) = \begin{cases} \mathcal{E}\big(f^{1/2}, f^{1/2}\big) & \text{if } d\nu = f\,d\lambda \\ \infty & \text{otherwise ,} \end{cases}$$

where $J^0 \equiv J^Y$ *with* $Y = 0$.

PROOF: Let L be the operator defined in the discussion preceding Lemma 4.2.31 and define $\mathbf{D}_c$ as in Lemma 4.2.35. In view of Theorem 4.2.43, the first assertion will be proved once we note that $\mathbf{D}_c \subseteq C^\infty(\Sigma;\mathbb{R})$, $Lu = L^Y u$ for $u \in C^\infty(\Sigma;\mathbb{R})$, and that, for every $u \in \mathbf{D}_c$ there is a sequence $\{u_n\}_1^\infty \subseteq C^\infty(\Sigma;\mathbb{R})$ such that $(u_n, L^Y u_n) \longrightarrow (u, Lu)$ uniformly as $n \longrightarrow \infty$. Clearly the only one of these needing comment is the last. But, for every $u \in \mathbf{D}_c$, $u_n \equiv [P^Y_{1/n}u] \in C^\infty(\Sigma;\mathbb{R})$ and $L^Y u_n = [P^Y_{1/n}Lu]$.

Finally, since $(\tilde{\mathbf{U}})$ holds, the second assertion is an immediate consequence of Theorem 4.2.58. ∎

6.3.11 Remark.

In connection with Remark 6.3.4, one should notice that the last part of Lemma 6.3.8 can be immediately modified to say that $J^Y = J_{\mathcal{E}^U}$ when Y is the Y^U in (6.3.5) and $\mathcal{E}^U$ is the DIRICHLET form associated with the corresponding symmetric MARKOV semigroup on $L^2(m_U)$.

Our main goal in the rest of this section will be to obtain a better expression for the rate function J^Y, even in cases when Remark 6.3.11 does not apply. In particular, what we are seeking is an expression in which one can clearly see the distinct contributions made to J^Y by the "symmetrizable" and "non-symmetrizable" parts of L^Y.

In order to carry out our program, it will be useful to introduce the following notions. In the first place, for $\phi \in C^\infty(\Sigma;\mathbb{R})$ define $\mathbf{X}\phi \in C^\infty(\Sigma;\mathbb{R}^d)$ by

$$\mathbf{X}\phi = \begin{bmatrix} X_1\phi \\ \vdots \\ X_d\phi \end{bmatrix}.$$

Next, for $p \in [1,\infty)$, define $W_p^{(1)}(\mathbf{X},\lambda)$ to be the space of $\phi \in L^p(\lambda)$ for which there exists a sequence $\{\phi_n\}_1^\infty \subseteq C^\infty(\Sigma;\mathbb{R})$ with the properties that

$$\text{(6.3.12)} \qquad \phi_m \longrightarrow \phi \text{ in } L^p(\lambda) \quad \text{and} \quad \sup_{n\geq m} \left\|\mathbf{X}\phi_n - \mathbf{X}\phi_m\right\|_{L^p(\lambda;\mathbb{R}^d)} \longrightarrow 0$$

as $m \longrightarrow \infty$.

6.3.13 Lemma. *For any* $p \in [1,\infty)$, *there is a unique continuous linear mapping*

$$\overline{\mathbf{X}}^{(p)} : W_p^{(1)}(\mathbf{X},\lambda) \longrightarrow L^p(\lambda;\mathbb{R}^d)$$

for which $\overline{\mathbf{X}}^{(p)}\phi = \mathbf{X}\phi$ *whenever* $\phi \in C^\infty(\Sigma;\mathbb{R})$. *In fact,* $\overline{\mathbf{X}}^{(p)}\phi$ *is the unique element of* $L^p(\lambda;\mathbb{R}^d)$ *with the property that*

$$\sum_{k=1}^d \int_\Sigma \Psi_k \left(\overline{\mathbf{X}}^{(p)}\phi\right)_k d\lambda = \sum_{k=1}^d \int_\Sigma \phi\left(X_k^*\Psi_k\right) d\lambda \quad \text{for every } \Psi \in C^\infty\left(\Sigma;\mathbb{R}^d\right);$$

and therefore, $\overline{\mathbf{X}}^{(p)}\phi = \overline{\mathbf{X}}^{(q)}\phi$ λ*-almost everywhere when* $\phi \in W_p^{(1)}(\mathbf{X},\lambda) \cap W_q^{(1)}(\mathbf{X},\lambda)$. *Moreover, if* $\eta \in C^1(\mathbb{R};\mathbb{R})$ *and* ϕ *is an element of* $W_p^{(1)}(\mathbf{X},\lambda)$ *which satisfies* $\eta\circ\phi \in L^q(\lambda)$ *and* $(\eta'\circ\phi)\overline{\mathbf{X}}^{(p)}\phi \in L^q(\lambda;\mathbb{R}^d)$ *for some* $q \in [1,\infty)$, *then* $\eta\circ\phi \in W_q^{(1)}(\mathbf{X},\lambda)$ *and*

$$\overline{\mathbf{X}}^{(q)}(\eta\circ\phi) = (\eta'\circ\phi)\,\overline{\mathbf{X}}^{(p)}\phi.$$

PROOF: We first note that, for any $\phi \in L^p(\lambda)$, there is at most one $\Phi \in L^p(\lambda;\mathbb{R}^d)$ with the property that

$$\sum_{k=1}^d \int_\Sigma \Psi_k\,\Phi_k\,d\lambda = \sum_{k=1}^d \int_\Sigma \phi\left(X_k^*\Psi_k\right) d\lambda \tag{6.3.14}$$

for every $\Psi \in C^\infty\left(\Sigma;\mathbb{R}^d\right)$. Second, we observe that if $\{\phi_n\}_1^\infty \subseteq C^\infty(\Sigma;\mathbb{R})$ satisfies (6.3.12), then $\mathbf{X}\phi_n$ converges in $L^p(\lambda;\mathbb{R}^d)$ to a $\Phi \in L^p(\lambda;\mathbb{R}^d)$ for which (6.3.14) holds. Thus, both the existence and uniqueness statements follow immediately, and all the other statements are easy applications of these. ∎

Because the program which we have in mind rests on L^Y being a compact perturbation of L^0, we will have to assume that

$$Y = \sum_{k=1}^d a_k X_k \quad \text{for some } \{a_k\}_1^d \subseteq C^\infty(\Sigma;\mathbb{R}). \tag{6.3.15}$$

The importance of (6.3.15) is already apparent in the next result.

6.3.16 Lemma. *Assume that* (**H**) *holds. Then* $W_2^{(1)}(\mathbf{X},\lambda) = \mathbf{Dom}(\mathcal{E})$ *and*

$$\mathcal{E}(\phi,\phi) = \left\|\overline{\mathbf{X}}^{(2)}\phi\right\|^2_{L^2(\lambda;\mathbb{R}^d)} \quad \text{for } \phi \in \mathbf{Dom}(\mathcal{E}).$$

If, in addition, Y *is given by* (6.3.15), *then* $J^Y(\nu) < \infty$ *if and only if* $d\nu = f\,d\lambda$, *where* f *is non-negative and* $f^{1/2} \in W_2^{(1)}(\mathbf{X},\lambda)$.

PROOF: To prove the first part, note that

$$\mathcal{E}(\phi,\phi) = -\int_\Sigma \phi\,L^0\phi\,d\lambda = \left\|\mathbf{X}\phi\right\|^2_{L^2(\lambda;\mathbb{R}^d)}$$

for $\phi \in C^\infty(\Sigma;\mathbb{R})$. Thus, since $\phi \in \mathbf{Dom}(\mathcal{E})$ and

$$\mathcal{E}(\phi,\phi) = \lim_{n\to\infty} \mathcal{E}\big(\phi_n,\phi_n\big) \quad \text{if } \phi = \lim_{n\to\infty} \phi_n \text{ in } L^2(\lambda)$$

when $\{\phi_n\}_1^\infty \subseteq \mathbf{Dom}(\mathcal{E})$ satisfies

$$\lim_{m\to\infty} \sup_{n\geq m} \mathcal{E}\big(\phi_n - \phi_m, \phi_n - \phi_m\big) = 0,$$

we see that $W_2^{(1)}(\mathbf{X},\lambda) \subseteq \mathbf{Dom}(\mathcal{E})$ and that $\mathcal{E}(\phi,\phi) = \big\|\overline{\mathbf{X}}^{(2)}\phi\big\|^2_{L^2(\lambda;\mathbb{R}^d)}$ for $\phi \in W_2^{(1)}(\mathbf{X},\lambda)$. To prove the opposite inclusion, let $\phi \in \mathbf{Dom}(\mathcal{E})$ be given and set $\phi_n = \big[P^0_{1/n}\phi\big]$, $n \in \mathbb{Z}^+$. Then, because of $(\mathbf{H})$, $\{\phi_n\}_1^\infty \subseteq C^\infty(\Sigma;\mathbb{R})$, and clearly $\phi_n \longrightarrow \phi$. At the same time, by the Spectral Theorem,

$$\sup_{n\geq m} \big\|\mathbf{X}\phi_n - \mathbf{X}\phi_m\big\|_{L^2(\lambda;\mathbb{R}^d)} = \sup_{n\geq m} \mathcal{E}\big(\phi_n - \phi_m, \phi_n - \phi_m\big) \longrightarrow 0$$

as $m \longrightarrow \infty$.

Turning to the second part, note that (cf. Theorem 4.2.58) there is nothing to do when $Y = 0$. On the other hand, if Y is given by (6.3.15), then, after writing $u \in C^\infty\big(\Sigma;[1,\infty)\big)$ as $e^{-\phi}$, we see that

$$J^Y(\nu) = \sup\left\{\int_\Sigma L^Y\phi\,d\nu - \big\|\mathbf{X}\phi\big\|^2_{L^2(\lambda;\mathbb{R}^d)} : \phi \in C^\infty(\Sigma;\mathbb{R})\right\}. \tag{6.3.17}$$

Hence, if we take

$$\mathbf{A} = \begin{bmatrix} a_1 \\ \vdots \\ a_d \end{bmatrix}, \tag{6.3.18}$$

then we find that

$$\begin{aligned} J^Y(\nu) \leq \sup\Bigg\{\int_\Sigma L^0\phi\,d\nu &- \frac{3}{4}\big\|\mathbf{X}\phi\big\|^2_{L^2(\lambda;\mathbb{R}^d)} + \big\|\mathbf{A}\big\|^2_{L^2(\lambda;\mathbb{R}^d)} \\ &- \Big\|\mathbf{A} - \frac{1}{2}\mathbf{X}\phi\Big\|^2_{L^2(\lambda;\mathbb{R}^d)} : \phi \in C^\infty(\Sigma;\mathbb{R})\Bigg\} \\ \leq \frac{4}{3}J^0(\nu) + \big\||\mathbf{A}|\big\|^2_B. \end{aligned}$$

By reversing the preceding argument, we also find that

$$J^0(\nu) \leq \frac{4}{3}J^Y(\nu) + \big\||\mathbf{A}|\big\|^2_B;$$

and so we now see that $J^Y(\nu) < \infty$ if and only if $J^0(\nu) < \infty$. ∎

In order to complete our program, let Y be given by (6.3.15), define $\mathbf{A}$ as in (6.3.18); and, for $\nu \in \mathbf{M}_1(\Sigma)$, define $\mathbf{A}_\nu$ to be the orthogonal projection in $L^2(\nu;\mathbb{R}^d)$ of $\mathbf{A}$ onto

$$\overline{\{\mathbf{X}\phi : \phi \in C^\infty(\Sigma;\mathbb{R})\}}^{L^2(\nu;\mathbb{R}^d)},$$

and set

$$P(\mathbf{A};\nu) = \left\|\mathbf{A}_\nu\right\|^2_{L^2(\nu;\mathbb{R}^d)}.$$

Since

$$P(\mathbf{A};\nu) = \sup\Big\{2\big(\mathbf{A},\mathbf{X}\phi\big)_{L^2(\nu;\mathbb{R}^d)} - \left\|\mathbf{X}\phi\right\|^2_{L^2(\nu;\mathbb{R}^d)} : \phi \in C^\infty(\Sigma;\mathbb{R})\Big\},$$

it is clear that $\nu \in \mathbf{M}_1(\Sigma) \longmapsto P(\mathbf{A},\nu)$ is lower semi-continuous and convex.

6.3.19 Theorem. *Assume that* (**H**) *holds and that* Y *is given by* (6.3.15). *Then*

$$J^Y(\nu) = J_{\mathcal{E}}(\nu) + \frac{1}{4}P(\mathbf{A};\nu) + \frac{1}{2}\int_\Sigma R_Y\, d\nu, \tag{6.3.20}$$

where $\mathbf{A}$ *is defined as in* (6.3.18) *and*

$$R_Y = -\sum_{k=1}^d X_k^* a_k.$$

PROOF: In view of Lemma 6.3.16, we need only consider $\nu \in \mathbf{M}_1(\Sigma)$ for which $d\nu = f\,d\lambda$ for some non-negative f with $f^{1/2} \in W_2^{(1)}(\mathbf{X},\lambda)$. In addition, since both sides of (6.3.20) are lower semi-continuous and convex, we may and will assume that $f \geq \epsilon$ for some $\epsilon > 0$. (Otherwise, set $\nu_\epsilon = \epsilon\lambda + (1-\epsilon)\nu$ and let $\epsilon \searrow 0$.)

We begin by proving that

$$J^Y(\nu) = J_{\mathcal{E}}(\nu) + \frac{1}{4}P(\mathbf{A};\nu) - \left(\mathbf{A}_\nu, \frac{\overline{\mathbf{X}}^{(2)} f^{1/2}}{f^{1/2}}\right)_{L^2(\nu;\mathbb{R}^d)}. \tag{6.3.21}$$

To this end, choose $\{\phi_n\}_1^\infty \subseteq C^\infty(\Sigma;\mathbb{R})$ so that

$$\left\|\mathbf{X}\phi_n - \frac{1}{2}\mathbf{A}_\nu\right\|_{L^2(\nu;\mathbb{R}^d)} \leq \frac{1}{n} \quad \text{for } n \in \mathbb{Z}^+,$$

and set $\Phi_n = \mathbf{X}\phi_n - \frac{1}{2}\mathbf{A}_\nu$. Then (cf. (6.3.17)) $J^Y(\nu)$ equals

$$\sup\left\{\int_\Sigma L^Y(\phi_n + \psi)\, d\nu - \left\|\mathbf{X}(\phi_n+\psi)\right\|^2_{L^2(\nu;\mathbb{R}^d)} : \psi \in C^\infty(\Sigma;\mathbb{R})\right\};$$

and, for any given $\psi \in C^\infty(\Sigma;\mathbb{R})$,

$$\begin{aligned}
&\int_\Sigma L^Y(\phi_n+\psi)\,d\nu - \left\|\mathbf{X}(\phi_n+\psi)\right\|^2_{L^2(\nu;\mathbb{R}^d)} \\
&= \int_\Sigma L^0(\phi_n+\psi)\,d\nu + \Big(\mathbf{A}_\nu, \mathbf{X}(\phi_n+\psi)\Big)_{L^2(\nu;\mathbb{R}^d)} \\
&\qquad - \left\|\frac{1}{2}\mathbf{A}_\nu + \mathbf{X}\psi\right\|^2_{L^2(\nu;\mathbb{R}^d)} - (\Phi_n, \mathbf{A}_\nu + 2\mathbf{X}\psi)_{L^2(\nu;\mathbb{R}^d)} - \left\|\Phi_n\right\|^2_{L^2(\nu;\mathbb{R}^d)} \\
&= \int_\Sigma L^0\psi\,d\nu - \left\|\mathbf{X}\psi\right\|^2_{L^2(\nu;\mathbb{R}^d)} + \frac{1}{4}P(\mathbf{A};\nu) \\
&\qquad - 2\big(\Phi_n, \mathbf{X}\psi\big)_{L^2(\nu;\mathbb{R}^d)} - \left\|\Phi_n\right\|^2_{L^2(\nu;\mathbb{R}^d)} + \int_\Sigma L^0\phi_n\,d\nu.
\end{aligned}$$

At the same time,

$$\begin{aligned}
\int_\Sigma L^0\phi_n\,d\nu &= -\sum_{k=1}^d \int_\Sigma \left(\overline{\mathbf{X}}^{(1)} f\right)_k X_k\phi_n\,d\lambda = -2\left(\frac{\overline{\mathbf{X}}^{(2)} f^{1/2}}{f^{1/2}}, \mathbf{X}\phi_n\right)_{L^2(\nu;\mathbb{R}^d)} \\
&= -\left(\frac{\overline{\mathbf{X}}^{(2)} f^{1/2}}{f^{1/2}}, \mathbf{A}_\nu\right)_{L^2(\nu;\mathbb{R}^d)} - 2\left(\frac{\overline{\mathbf{X}}^{(2)} f^{1/2}}{f^{1/2}}, \Phi_n\right)_{L^2(\nu;\mathbb{R}^d)}.
\end{aligned}$$

Hence, since $\frac{\overline{\mathbf{X}}^{(2)} f^{1/2}}{f^{1/2}} \in L^2\big(\nu;\mathbb{R}^d\big)$,

$$\begin{aligned}
&\int_\Sigma L^Y(\phi_n+\psi)\,d\nu - \left\|\mathbf{X}(\phi_n+\psi)\right\|^2_{L^2(\nu;\mathbb{R}^d)} \\
&\quad = \int_\Sigma L^0\psi\,d\nu - \left\|\mathbf{X}\psi\right\|^2_{L^2(\nu;\mathbb{R}^d)} + \frac{1}{4}P(\mathbf{A};\nu) \\
&\qquad\qquad - \left(\frac{\overline{\mathbf{X}}^{(2)} f^{1/2}}{f^{1/2}}, \mathbf{A}_\nu\right)_{L^2(\nu;\mathbb{R}^d)} + E_n(\psi),
\end{aligned}$$

where

$$\left|E_n(\psi)\right| \le \frac{C}{n}\left\|\mathbf{X}\psi\right\|_{L^2(\nu;\mathbb{R}^d)}, \quad n \in \mathbb{Z}^+,$$

for some $C \in (0,\infty)$ depending only on $\mathbf{A}$ and f. Clearly, by using (6.3.17) with $Y = 0$ to compute $J_\mathcal{E}(\nu)$, one can easily use the preceding to get (6.3.21).

To prove (6.3.20) from (6.3.21), all that we have to do is check that

$$\left(\frac{\overline{\mathbf{X}}^{(2)} f^{1/2}}{f^{1/2}}, \mathbf{A}_\nu\right)_{L^2(\nu;\mathbb{R}^d)} = -\frac{1}{2}\int_\Sigma R_Y\,d\nu;$$

and this comes down to showing that there exists a sequence $\{g_n\}_1^\infty \subseteq C^\infty(\Sigma;\mathbb{R})$ such that

$$\mathbf{X}g_n \longrightarrow \frac{\overline{\mathbf{X}}^{(2)} f^{1/2}}{f^{1/2}} \text{ in } L^2(\nu;\mathbb{R}^d).$$

For this purpose, choose $\{u_n\}_1^\infty \subseteq C^\infty\big(\Sigma;[1,\infty)\big)$ so that

$$-\int_\Sigma \frac{L^0 u_n}{u_n}\,d\nu \longrightarrow J^0(\nu) = \left\|\overline{\mathbf{X}}^{(2)} f^{1/2}\right\|^2_{L^2(\nu;\mathbb{R}^d)} \quad \text{as } n \longrightarrow \infty;$$

and set $g_n = \log u_n$. One then has that

$$\begin{aligned}
&\left\|\frac{\overline{\mathbf{X}}^{(2)} f^{1/2}}{f^{1/2}} - \mathbf{X}g_n\right\|^2_{L^2(\nu;\mathbb{R}^d)} \\
&\quad = J^0(\nu) - 2\Big(\overline{\mathbf{X}}^{(2)} f^{1/2}, f^{1/2}\mathbf{X}g_n\Big)_{L^2(\lambda;\mathbb{R}^d)} + \|\mathbf{X}g_n\|^2_{L^2(\nu;\mathbb{R}^d)} \\
&\quad = J^0(\nu) - \Big(\overline{\mathbf{X}}^{(1)} f, \mathbf{X}g_n\Big)_{L^2(\lambda;\mathbb{R}^d)} + \|\mathbf{X}g_n\|^2_{L^2(\nu;\mathbb{R}^d)} \\
&\quad = J^0(\nu) + \int_\Sigma \frac{L^0 u_n}{u_n}\,d\nu \longrightarrow 0.
\end{aligned}$$

We can now complete the proof by simply noting that

$$\begin{aligned}
&\left(\frac{\overline{\mathbf{X}}^{(2)} f^{1/2}}{f^{1/2}}, \mathbf{A}_\nu\right)_{L^2(\nu;\mathbb{R}^d)} = \lim_{n\to\infty} \big(\mathbf{A}_\nu, \mathbf{X}g_n\big)_{L^2(\nu;\mathbb{R}^d)} \\
&\quad = \lim_{n\to\infty} \big(\mathbf{A}, \mathbf{X}g_n\big)_{L^2(\nu;\mathbb{R}^d)} = \left(\frac{\overline{\mathbf{X}}^{(2)} f^{1/2}}{f^{1/2}}, \mathbf{A}\right)_{L^2(\nu;\mathbb{R}^d)} \\
&\quad = -\frac{1}{2}\Big(\mathbf{A}, \overline{\mathbf{X}}^{(1)} f\Big)_{L^2(\lambda;\mathbb{R}^d)} = -\frac{1}{2}\int_\Sigma R_Y\,d\nu \;\blacksquare.
\end{aligned}$$

6.3.22 Exercise.

Let $X_1, \ldots, X_d$, and Y be smooth vector fields on the connected, compact manifold Σ; and assume that the X_k 's satisfy (**H**). Next, set $\hat{\Sigma} = \Sigma \times \mathbb{R}$ and define the vector fields $\hat{X}_1, \ldots, \hat{X}_d$, and $\hat{Y}$ on $\hat{\Sigma}$ by

$$\big[\hat{X}_k\hat{\phi}\big](\hat{\sigma}) = \big[X_k\hat{\phi}(\cdot,\xi)\big](\sigma) + b_k(\sigma)\frac{\partial\hat{\phi}}{\partial\xi}(\sigma,\xi)$$

and

$$[\hat{Y}\hat{\phi}](\hat{\sigma}) = [Y\hat{\phi}(\cdot,\xi)](\sigma) + c(\sigma)\frac{\partial\hat{\phi}}{\partial\xi}(\sigma,\xi)$$

for $\hat{\phi} \in C^\infty(\hat{\Sigma};\mathbb{R})$, where $b_1,\dots,b_d$, and c are given elements of $C^\infty(\Sigma;\mathbb{R})$. Finally, define

$$\hat{L} = \sum_{k=1}^{d} X_k^2 + \hat{Y} \quad \text{on } C^\infty(\hat{\Sigma};\mathbb{R}).$$

One can then show that $\hat{L}$ determines a (unique) FELLER-continuous, MARKOV family

$$\{\hat{P}_{\hat{\sigma}} : \sigma \in \hat{\Sigma}\}$$

of probability measures on $\Omega = C([0,\infty);\hat{\Sigma})$ with the property that

$$\left(\hat{\phi}(\hat{\Sigma}_t(\omega)) - \phi(\hat{\sigma}) - \int_0^t [\hat{L}\hat{\phi}](\hat{\Sigma}_s(\omega))\,ds,\ \mathfrak{M}_t,\ \hat{P}_{\hat{\sigma}}\right)$$

is a mean-zero martingale for every $\hat{\sigma} \in \hat{\Sigma}$ and all $\hat{\phi} \in C^\infty(\hat{\Sigma};\mathbb{R})$. In fact, as aficionados of stochastic differential equations will easily verify, if $\hat{\sigma} = (\sigma,\xi) \in \Sigma \times \mathbb{R}$, then $\hat{P}_{\hat{\sigma}}$ is the joint distribution under d-dimensional WIENER's measure $\mathcal{W}$ of the solution to

$$d\,\Xi_t(\theta) = 2^{1/2}\sum_{k=1}^{d} X_k(\Xi_t(\theta)) \circ d\theta_k(t) + Y(\Xi_t(\omega))\,dt \quad \text{with } \Xi_0(\omega) = \sigma$$

together with

$$\xi_t(\theta) = \xi + 2^{1/2}\sum_{k=1}^{d}\int_0^t b_k(\Xi_s(\theta)) \circ d\theta_k(s) + \int_0^t c(\Xi_s(\theta))\,ds.$$

(In both of these expressions, the stochastic integrals are taken in the sense of STRATONOVICH.)

(**i**) Write

$$\hat{\Sigma}_t(\omega) = (\Sigma_t(\omega), \mathbf{S}_t(\omega)) \in \Sigma \times \mathbb{R},$$

show that the hypotheses at the beginning of Section 4.2 are met by the measures $\hat{P}_\sigma \equiv \hat{P}_{(\sigma,0)}$, $\sigma \in \Sigma$, and check that the condition $(\tilde{\mathbf{U}})$ is satisfied. Conclude that, for every $\beta \in \mathbb{R}$, the limit

$$\Lambda(\beta) \equiv \lim_{t\to\infty}\frac{1}{t}\log\left[\int_\Omega \exp\left[\beta\mathbf{S}_t(\omega)\right]\hat{P}_\sigma(d\omega)\right]$$

(this would have been denoted by $\Lambda_{\hat{P}}(\beta)$ in Section 4.2) exists uniformly in $\sigma \in \Sigma$ and that the the large deviations of $\overline{\mathbf{S}}_t(\omega) \equiv \frac{\mathbf{S}_t(\omega)}{t}$ under $\{P_\sigma : \sigma \in \Sigma\}$ are uniformly governed by the good rate function I given by

$$I(\alpha) = \sup\Big\{\alpha\beta - \Lambda(\beta) : \beta \in \mathbb{R}\Big\}, \quad \alpha \in \mathbb{R}.$$

(ii) In order to get a handle on $\Lambda(\beta)$, define

$$w_\beta(t,\sigma) = \int_\Omega \exp\Big[\beta \mathbf{S}_t(\omega)\Big]\, \hat{P}_\sigma(d\omega), \quad (t,\sigma) \in [0,\infty) \times \Sigma,$$

and show that w_β is the smooth solution to

$$\frac{\partial w_\beta}{\partial t} = \big[L^\beta w_\beta\big] + Q^\beta w_\beta \quad \text{on } [0,\infty) \times \Sigma \quad \text{with } w_\beta(0,\cdot) = 1,$$

where

$$L^\beta = \sum_{k=1}^d X_k^2 + Y^\beta \quad \text{with} \quad Y^\beta = Y + 2\beta \sum_{k=1}^d b_k X_k$$

and

$$Q^\beta = \beta c + \sum_{k=1}^d \Big(\beta\big[X_k b_k\big] + (\beta b_k)^2\Big).$$

Hint: Consider the function

$$\hat{w}_\beta(t,\hat{\sigma}) = e^{\beta\xi} w_\beta(t,\sigma), \quad t \in [0,\infty) \text{ and } \hat{\sigma} = (\sigma,\xi) \in \Sigma \times \mathbb{R},$$

and apply the FEYNMAN–KAC formula to see that

$$\frac{\partial \hat{w}_\beta}{\partial t}(t,\hat{\sigma}) = \big[\hat{L}\hat{w}_\beta\big](t,\hat{\sigma}) + \beta\xi\hat{w}_\beta(t,\hat{\sigma}).$$

Now apply Theorem 4.2.43 to the MARKOV process determined by L^β and conclude that

$$\Lambda(\beta) = \sup\left\{\int_\Sigma Q^\beta\, d\nu - J^\beta(\nu) : \nu \in \mathbf{M}_1(\Sigma)\right\},$$

where

$$J^\beta(\nu) = \sup\left\{-\int_\Sigma \frac{L^\beta u}{u}\, d\nu : u \in C^\infty\big(\Sigma;[1,\infty)\big)\right\}.$$

(iii) Finally, we add the assumption that

$$Y = \sum_{k=1}^d a_k X_k \quad \text{for some } \{a_k\}_1^d \subseteq C^\infty(\Sigma;\mathbb{R}),$$

introduce a smooth probability measure λ on Σ, and, using X^* to denote the λ-adjoint of X, define g_X by (6.3.1). Applying Theorem 4.3.19, check that

$$J^\beta(\nu) = J_{\mathcal{E}}(\nu) + \frac{1}{4}P\big(\mathbf{A}^\beta;\nu\big) + \frac{1}{2}\int_\Sigma R^\beta\,d\nu,$$

where $\mathcal{E}$ is the DIRICHLET form obtained by closing

$$\phi \in C^\infty(\Sigma;\mathbb{R}) \longmapsto \sum_{k=1}^d \int_\Sigma \big(X_k\phi\big)^2\,d\lambda$$

in $L^2(\lambda)$,

$$\mathbf{A}^\beta = \begin{bmatrix} a_1 + g_{X_1} + 2\beta b_1 \\ \vdots \\ a_d + g_{X_d} + 2\beta b_d \end{bmatrix},$$

and

$$R^\beta = -\sum_{k=1}^d X_k^*\big(a_k + g_{X_k} + 2\beta b_k\big).$$

After combining the preceding with part **(ii)**, show that

$$\lim_{|\beta|\to\infty} \frac{\Lambda(\beta)}{\beta^2} = \inf\Big\{\big\|\,|\mathbf{B} - \mathbf{X}\phi|\,\big\|_B^2 : \phi \in C^\infty(\Sigma;\mathbb{R})\Big\},$$

where

$$\mathbf{B} = \begin{bmatrix} b_1 \\ \vdots \\ b_d \end{bmatrix} \quad \text{and } \mathbf{X}\phi = \begin{bmatrix} X_1\phi \\ \vdots \\ X_d\phi \end{bmatrix}.$$

A more complete discussion of these, and related, matters can be found in [**102**] and [**9**].

Historical Notes and References

Disclaimer

The authors make no claims for the completeness and few claims for the accuracy or value of what follows. In other words, we are doing no more than perpetuating the mathematical tradition of covering up the tracks which we and others may or may not have made. For example, we are aware that there is a huge body of excellent work in the statistics literature on the subject of large deviations and that we have done little more than acknowledge the statisticians' just claim of paternity. In addition, we have given rather short shift to some very beautiful mathematics in which large deviation theory is employed to find the asymptotics of heat kernels. In particular, we have essentially ignored the important contributions of MOLCHANOV [**77**], AZENCOTT [**1**], BISMUT [**10**], and their students to this topic. Thus, the interested reader should consult VENTCEL and FREIDLIN's book [**111**] for applications to dynamical systems. In another direction, ELLIS's book [**39**] describes connections with statistical mechanics.

Chapter I

§**1.1** & §**1.2** Historically, the theory of large deviations emerged as an attempt to carry the well known CENTRAL LIMIT THEOREM one step further. To be more precise, suppose that $\mu \in \mathbf{M}_1(\mathbb{R})$ has mean 0 and variance 1. Then the classical CENTRAL LIMIT THEOREM says that, for each $x \in \mathbb{R}$,

$$\mu^n\left(\frac{1}{\sqrt{n}}\sum_{i=1}^{n} x_i > x\right)$$

converges as $n \longrightarrow \infty$ to

$$\frac{1}{\sqrt{2\pi}} \int_{[x,\infty)} e^{t^2/2}\, dt.$$

The problem in which people became interested was that of determining just how fast the tails of the approximents were approaching those of the GAUSSian. As early as 1928 KHINCHIN [**69**], followed by SMIRNOFF [**98**], studied this question by replacing the fixed x by a function $x(n)$ which tends to ∞ as $n \longrightarrow \infty$. By restricting their attention to BERNOULLI random variables, they were able to give a very precise answer in the case when $\lim_{n\to\infty} \frac{x(n)}{n^{1/2}} = 0$. Although KHINCHIN said that he was studying "große Abweichungen," from the standpoint the theory presented in this book the deviations with which he was dealing would have to be considered "moderate deviations." For related work and other references, see LINNIK [**75**], RICHTER [**90**] and PETROV [**87**].

The archetype for large deviation results of the sort dealt with in this book is the one which we have called CRAMÈR's Theorem. In [**20**] CRAMÈR proved this theorem for distributions μ on $\mathbb{R}$ which are not singular to LEBESGUE's measure. From the viewpoint of later developments, the most significant idea introduced by CRAMÈR was that of transforming the given measure (apparently the transformation itself antedates CRAMÈR's use of it and goes back to ESSCHER [**43**]). As distinguished from the use to which we put it in the proof of (1.2.7), CRAMÈR uses the transformation as an initial step in a program which eventually enabled him to bring to bear refined estimates about the CENTRAL LIMIT THEREM. The first proof of the general statement in Theorem 1.2.6 is due to CHERNOFF [**17**]. CHERNOFF uses CHEBYSHEV's inequality (in exactly the same way as we) to obtain the upper bound, but he gets the lower bound via approximation by discrete distributions and a clever application of STIRLING's formula.

CHERNOFF's motivation came from statistics. In particular, he was interested in questions about the asymptotic efficiency of statistical tests and initiated a program which has been carried further by several statisticians: BAHADUR [**4**] and [**5**], BAHADUR and RANGA RAO [**6**], BARNDORFF-NIELSEN [**8**], and DACUNHA-CASTELLE [**22**].

§1.3 & **§1.4** Theorem 1.3.27, which appeared in SCHILDER's thesis [**95**], is the first example of a large deviation result for measures on a function space. At the time, SCHILDER was a student of M. DONSKER, and it seems clear from DONSKER's earlier work that the idea for such a result should be credited to him. Be that as it may, what we have called SCHILDER's

Theorem contains only the first step of a program in which it was envisioned that function-space integral techniques could be used to provide an entire asymptotic expansion for the quantities under consideration. Although SCHILDER's thesis contains the first examples of this line of reasoning, the real breakthrough came in the article of VARADHAN [**106**] where the foundations were laid for large deviation theory as we have presented it here.

It seems that BOROVKOV [**14**] should also be cited as one of the first to study large deviation theory in a function space context, although his work does not appear to have had a great deal of influence even in Russia. Also, slightly later and apparently independently, VENTCEL and FREIDLIN in [**109**], [**110**], and [**111**] started to use essentially the same function-space integral ideas to analyze randomly perturbed dynamical systems. Our brief presentation of their estimates as an application of SCHILDER's Theorem is based on the ideas of AZENCOTT [**1**] who is also responsible for much of the recent progress toward the completion of SCHILDER's program.

CHAPTER II

§**2.1** The formulation given for the principle of large deviations as well as the LAPLACE asymptotic result contained in Theorem 2.1.10 appear for the first time in VARADHAN's pioneering work [**106**]. Lemma 2.1.4 is simply an abstraction of ideas which had already been used by several authors, in particular, by AZENCOTT in his treatment of the VENTCEL-FREIDLIN estimates.

Part **ii**) of Exercise 2.1.13 stems from a problem posed by G. STEIN (and answered independently, with entirely different methods, by E.M. STEIN); projective limits of large deviation principles (cf. Exercise 2.1.21) play a prominent role in DAWSON and GÄRTNER [**24**]; and Exercise 2.1.24 is an adaptation of a technique used by extensively by ELLIS in [**39**].

§**2.2** The basic relation between large deviations and the LEGENDRE transform of the logarithmic moment generating function is already present in CRAMÈR's and CHERNOFF's papers; and the role that convex analysis has to play in the theory became increasingly evident in the work of several authors (cf. especially J. GÄRTNER [**53**] and also the comments below on §3.1).

The systematic use of the LEGENDRE transformation as a tool for identifying rate functions is an underlying principle in the second half of [**101**].

Chapter III

The general outline of this chapter is taken from Azencott's excellent treatment in [**1**].

§**3.1** The use of sub-additivity to show that limits like those in 3.1.4 exist was introduced by Ruelle in [**92**] and [**93**] and systematically exploited by Landford in [**73**]. At the time, they were dealing with the problem of thermodynamical limits and the characterization of specific entropy in terms of Gibbs' variational principle. The first authors to apply this technique specifically to large deviation theory were Bahadur and Zabell [**7**] who, after significantly generalizing the idea, used it to derive both Sanov's Theorem and the Banach space case of Cramèr's Theorem (cf. Theorems 3.2.17 and Theorem 3.3.11).

§**3.2** Sanov's elegant result was at first so surprising that several authors expressed doubts about its veracity! The theorem, which Sanov proved only for empirical distributions of $\mathbb{R}$-valued random variables, was extended by many authors: Hoadley [**60**], Hoeffding [**61**], and at last achieved the form stated here in Donsker and Varadhan [**33**]. The first statement of Sanov's Theorem relative to the strong topology (cf. Theorem 3.2.21) appeared in Groeneboom, Oosterhoff and Ruymgaart [**55**].

The relative entropy function had been introduced into statistics by Kullback and Leibner [**71**], and its properties were investigated by Csiszàr [**21**]. Lemma 3.2.13 (which is implicit in the cited work by Ruelle and Lanford on Gibbs' variational principle) owes its present form to Donsker and Varadhan [**30**]. Finally, the estimate (3.2.25) is taken from Kemperman [**68**].

§**3.3** & §**3.4** The large deviation results as they are stated in these two sections were first obtained by Donsker and Varadhan [**33**]. However, Theorem 3.4.5 was proved earlier by Freidlin [**49**] and Ventcel [**108**] for both Hilbert space as well as $C([0,1])$. The important idea of obtaining the required exponential tightness in the Banach space setting from the corresponding result for the empirical distributions (cf. Lemma 3.3.10) is due to Donsker and Varadhan but was, to some extent, anticipated by Ranga Rao [**89**] in his elegant proof of the Strong Law (cf. Theorem 3.3.4).

The applications to Gaussian measures (in particular, Corollary 3.4.6) given in Section 3.4 are again basically due to Donsker and Varadhan, although the outline of our treatment follows that of Azencott [**1**] and

Corollary 3.4.6 itself should be viewed as the culmination of the program initiated by LANDAU and SHEPP [**72**].

Exercise 3.3.12 is taken from FÖLLMER [**47**], and more recent results about BANACH spaces can be found in BOLTHAUSEN [**11**].

CHAPTER IV

Credit for the theory of large deviations of the occupation time functional for a MARKOV process should unquestionably go to DONSKER and VARADHAN; even though similar ideas and results were formulated and proved a little later by GÄRTNER [**53**]. The moving force behind DONSKER and VARADHAN's investigation was DONSKER's stubborn conviction that something deep must underlie KAC's formula [**66**] for the smallest eigenvalue of a SCHRÖDINGER operator; and it is this force which led first to [**29**] and eventually to [**30**], [**31**], [**33**], [**34**], [**36**], and [**107**].

§**4.1** & §**4.2** The results contained herein are, more or less, the same as those in [**30**]. However, in addition to some technical improvements in the hypotheses under which they worked, our presentation is entirely different from DONSKER and VARADHAN's. Indeed, their derivation is much more a direct application of the principles underlying the proof in Section 1.2 of CRAMÈR's theorem, whereas ours (which is a slight embellishment of the one adopted in [**101**]) is an immediate descendent of BAHADUR and ZABELL's approach to CRAMÈR's Theorem. Moreover, our procedure for identifying the rate function is quite different from DONSKER and VARADHAN's and was influenced by the heuristic exposition given in [**67**] by KAC. Finally, the possibility of working in the strong topology as well as that of dispensing with FELLER continuity were first considered, in this setting, by BOLTHAUSEN [**12**].

The extent to which these results extend to general irreducible Markov chains has been investigated by various authors: DE ACOSTA [**25**] and [**26**], N.C. JAIN [**64**] and [**65**], NEY [**80**], and NEY and NUMMELIN [**81**].

§**4.3** Apart for minor changes in the ingredients, the recipe which we have used to cook the WIENER sausage is the same as the original one in [**31**] and [**32**]. In [**35**], DONSKER and VARADHAN applied the same ideas to random walks; and recently SZNITMAN [**99**] has carried out the analogous computation for hyperbolic spaces.

§**4.4** DONSKER and VARADHAN in [**36**] were the first to formulate and prove the large deviation principle at the level of processes. Aside from the intrinsic aesthetic appeal of this formulation (in particular, the reappearance of entropy as the rate function), it was only after introducing this formulation that they were able to solve the so-called Polaron problem in [**37**].

It should be remarked that some of the arguments in this section are intimately related to similar results in information theory. In particular, the lower bound at the process level can be viewed as an application of SHANNON-MC MILLAN technology; and this is the way it was developed first in MOY [**78**] and later in FÖLLMER [**46**] and [**47**], and OREY [**83**] and [**84**]. However, direct application of SHANNON-MC MILLAN ideas do not prove the lower bound except at ergodic points; and there is work to be done before one can handle the general case.

In an attempt to get away from processes and to handle random fields, several authors, have carried out a version of the DONSKER and VARADHAN program in connection with GIBBS measures for lattice systems (cf. COMETS [**19**], FÖLLMER and OREY [**48**] and OLLA [**82**]).

ELLIS was the first to suggest that the process level result ought to be obtainable from a multi-dimensional position level result, and he carried out such a program first for independent variables (cf. [**39**]) and, more recently, for MARKOV chains (cf. [**40**] and [**41**]). Our own treatment in Section 4.4 is based on the same idea.

CHAPTER V

§**5.1** This section basically reproduces the first part of Chapter 8 of [**101**]. The criterion in EXERCISE 5.1.17 **iii**) already appeared in [**33**].

§**5.2** So far as we know, the first person to see the Maximal Ergodic Inequality as an easy corollary of the Sunrise Lemma was P. HARTMAN [**59**]. The history of the Ergodic Decompostion Theorem starts with the paper [**70**] by KRYLOV and BOGOLIOUBOFF. Their results were re-worked by OXTOBY in [**86**], and it is on OXTOBY's ideas that our own proof is based.

§**5.3** This section is taken from the second part of Chapter 8 of [**101**]. The motivation here is to provide conditions which have a chance of holding even in an infinite dimensional context. See [**62**] for an example of this sort.

§**5.4** & §**5.5** These two sections are based on CHIYONOBU and KUSUOKA [**18**]. Our proof of the upper bound of Theorem 5.4.27 takes into account the results of 5.1 but otherwise differs very little from theirs. On the other hand, our proof of the lower bound is based on the methods of Section 3.1 and, as such, is quite different from theirs. The hypermixing property was formulated and used (in the context of constructive quantum field theory) by GUERRA, ROSEN and SIMON [**57**]. It is clear that the notion is intimately related to NELSON's ideas about hypercontractive semigroups (cf. [**79**] and the discussion for Chapter VI below). Lemma 5.5.9 is due to SIMON [**96**], and Corollary 5.5.16 was first derived in [**57**].

Other references to large deviations for non-MARKOV processes are: DONSKER and VARADHAN [**38**] for GAUSSian processes and OREY [**84**], OREY and PELIKAN [**85**], and TAKAHASHI [**105**] for dynamical systems.

CHAPTER VI

§**6.1** This section is an expanded version of the contents of Chapter 9 of [**101**]. Hypercontractivity was introduced by NELSON [**79**] in connection with his construction of a two-dimensional quantum field where he proved it for the ORNSTEIN-UHLENBECK semigroup and used it in the form that it appears in (6.1.20). A precursor of the logarithmic Sobolev inequality can be found in the article [**44**] by FEDERBUSH, but a its systematic exploitation appears for the first time in GROSS's [**56**], and [**101**] may be the first place where it is stated in full generality. The second part of LEMMA 6.1.5 is due to GLIMM [**54**].

§**6.2** The key to our handling of diffusions on a non-compact RIEMANNian manifold is contained in Lemma 6.2.12 which, in turn, is taken from GAFF -NEY [**52**]. In particular, it is GAFFNEY's result which tells us how to exploit completeness.

BAKRY and EMERY [**3**] were the first ones to provide the local condition for hypercontractivity given in Theorem 6.2.42. Although our treatment is derived from their ideas, it is not clear from their presentation when one can work in a non-compact setting. On the other hand, their formulation is couched in more abstract terms and therefore may be applied in situations which are not covered by us. The formula (6.2.32) is familiar to differential geometers, who think of it as an application of the BOCHNER-LICHNEROWICZ-WEITZENBÖCK formula. Closely related topics are treated in YAU [**113**], DAVIES and SIMON [**23**], and BAKRY [**2**]. Finally, CARLEN and STROOCK [**16**] apply BAKRY and EMERY's criterion to certain infinite dimensional diffusion processes.

The idea outlined in Exercise 6.2.48 comes from EMERY and YUKICH [**42**].

§**6.3** The contents of this section are taken from [**9**], where they are used to address the sort of question raised in Exercise 6.3.20. Related computations and ideas appear in the article [**88**] by R. PINSKY.

REFERENCES

[1] R. Azencott, *Grandes déviations et applications*, in "Ecoles d'Eté de Probabilités de Saint-Flour VIII-1978," edited by P.L. Hennequin. Lecture Notes in Mathematics **774**, Springer, Berlin, 1980, pp. 1–176.

[2] D. Bakry, *Un critère de non-explosion pour certaines diffusions sur une variété riemannienne complète*, C.R. Acad. Sc. Paris Série I **303** (1986), 23–25.

[3] D. Bakry and M. Emery, *Diffusions hypercontractives*, in "Séminaire de probabilités XIX," Lecture Notes in Mathematics **1123**, Springer, Berlin, 1985, pp. 179–206.

[4] R.R. Bahadur, *Rates of convergence of estimates and test statistics*, Ann. Math. Statist. **38** (1967), 303–324.

[5] R.R. Bahadur, "Some Limit Theorems in Statistics," Society for Industrial and Applied Mathematics, Philadelphia, 1971.

[6] R.R. Bahadur and R. Ranga Rao, *On deviations of the sample mean*, Ann. Math. Statist. **31** (1960), 1015–1027.

[7] R.R. Bahadur and S.L. Zabell, *Large deviations of the sample mean in general vector spaces*, Ann. Probab. **7** (1979), 587–621.

[8] O. Barndorff-Nielsen, "Information and Exponential Families in Statistical Theory," Wiley, Chichester, 1978.

[9] P.H. Baxendale and D.W. Stroock, *Large deviations and stochastic flows of diffeomorphisms*, Probab. Th. and rel. Fields **80** (1988), 169-216.

[10] J-M. Bismut, "Large Deviations and the Malliavin Calculus," Birkhäuser, Basel, 1984.

[11] E. Bolthausen, *On the probability of large deviations in Banach spaces*, Ann. Probab. **12** (1984), 427–435.

[12] E. Bolthausen, *Markov process large deviations in the τ-topology*, Stoch. Proc. and Appl. **25** (1987), 95–108.

[13] J.-M. Bony, *Principe du maximum, inégalité de Harnack et unicité du problème de Cauchy pour les operateurs elliptiques dégénérés*, Ann. Inst. Fourier **XIX no. 1** (1969), 277–304.

[14] A.A. Borovkov, *Boundary-value problems for random walks and large deviations in function spaces*, Th. Prob. Appl. **12** (1967), 575–595.

[**15**] R.H. Cameron and W.T. Martin, *Transformations of Wiener integrals under translations*, Ann. Math. **45** (1955), 386–396.

[**16**] E.A. Carlen and D.W. Stroock, *An application of the Bakry- Emery criterion to infinite dimensional diffusions*, in "Séminaire de probabilités XX," Lecture Notes in Mathematics **1204**, Springer, Berlin, 1986, pp. 341–347.

[**17**] H. Chernoff, *A measure of asymptotic efficiency for tests of a hypothesis based on the sum of observations*, Ann. Math. Statist. **23** (1952), 493–507.

[**18**] T. Chiyonobu and S. Kusuoka, *The large deviation principle for hypermixing processes*, Probab. Th. and Rel. Fields **78** (1988), 627–649.

[**19**] F. Comets, *Grandes déviations pour des champs de Gibbs sur* $\mathbb{Z}^d$, C.R. Acad. Sc. Paris, Série I **303** (1986), p. 511.

[**20**] H. Cramèr, *Sur un nouveau théorème-limite de la théorie des probabilités*, Actualités Scientifiques et Industrielles **736** (1938), 5–23. Colloque consacré à la théorie des probabilités, Vol. 3, Hermann, Paris.

[**21**] Csiszàr, *I-divergence geometry of probability distributions and minimization problems*, Ann. Probab. **3** (1975), 146–158.

[**22**] D. Dacunha-Castelle, *Formule de Chernoff pour une suite de variables réelles*, in "Grandes Deviations et Applications Statistiques," Astérisque **68**, Société Mathématique de France, Paris, 1979, pp. 19–24.

[**23**] E.B. Davies and B. Simon, *Ultracontractivity and the heat kernel for Schrödinger operators and Dirichlet Laplacians*, J. Func. Anal. **59** (1984), 335- 395.

[**24**] D.W. Dawson and J. Gärtner, *Long time fluctuations of weakly interacting diffusions*, Stochastics **20** (1987), 247–308.

[**25**] A. de Acosta, *Upper bounds for large deviations of dependent random vectors*, Z. Wahrsch. verw. Geb. **69** (1985), 551–565.

[**26**] A. de Acosta, *Large deviations for vector valued functionals of a Markov chain: lower bounds*, Ann. Probab. **16** (1988), 925–960.

[**27**] J.D. Deuschel and D.W. Stroock, *A function space large deviation principle for certain stochastic integrals*, Probab. Th. Rel. Fields (to

appear).

[28] W. Doeblin, *Élément d'une théorie générale des chaines simples constantes de Markoff*, Ann. Sc. École Norm. Sup. **57** (1940), 61–111.

[29] M.D. Donsker and S.R.S. Varadhan, in "Functional Integration and Its Applications," Proceedings of the International Conference Held at Cumberland Lodge, Winstor Great Park, London, April 1974, Edited by A.M. Arthurs, Clarenton, Oxford, pp. 15–33.

[30] M.D. Donsker and S.R.S. Varadhan, *Asymptotic evaluation of certain Markov process expectations for large time,I*, Comm. Pure Appl. Math. **28** (1975), 1–47.

[31] M.D. Donsker and S.R.S. Varadhan, *Asymptotic evaluation of certain Markov process expectations for large time,II*, Comm. Pure Appl. Math. **28** (1975), 279–301.

[32] M.D. Donsker and S.R.S. Varadhan, *Asymptotics for the Wiener sausage*, Comm. Pure Appl. Math. **28** (1975), 525–565.

[33] M.D. Donsker and S.R.S. Varadhan, *Asymptotic evaluation of certain Markov process expectations for large time,III*, Comm. Pure Appl. Math. **29** (1976), 389–461.

[34] M.D. Donsker and S.R.S. Varadhan, *On the principal eigenvalue of second-order elliptic differential operators*, Comm. Pure Appl. Math. **29** (1976), 595–621.

[35] M.D. Donsker and S.R.S. Varadhan, *On the number of distinct sites visited by a random walk*, Comm. Pure Appl. Math. **32** (1979), 721–747.

[36] M.D. Donsker and S.R.S. Varadhan, *Asymptotic evaluation of certain Markov process expectations for large time,IV*, Comm. Pure Appl. Math. **36** (1983), 183–212.

[37] M.D. Donsker and S.R.S. Varadhan, *Asymptotics for the polaron*, Comm. Pure Appl. Math. **36** (1983), 505–528.

[38] M.D. Donsker and S.R.S. Varadhan, *Large deviations for stationary Gaussian processes*, Comm. Math. Phys. **97** (1985), 187–210.

[39] R.S. Ellis, "Entropy, Large Deviations and Statistical Mechanics," Springer, Berlin, 1985.

[40] R.S. Ellis, *Large deviation for the empirical measure of a Markov*

chain with an application to the multivariate empirical measure, Ann. Probab. **16** (1988), 1496–1508.

[41] R.S. Ellis and A. Wyner, *Uniform large deviation property of the empirical measure of a Markov chain*, Ann. Probab. (to appear).

[42] M. Emery and J.E. Yukich, *A simple proof of the logarithmic inequality on the circle*, in "Séminaire de Probabilités XXI," Lecture Notes in Mathematics **1247**, Springer, Berlin, 1987, pp. 173- 176.

[43] F. Esscher, *On the probability function in the collective theory of risk*, Skandinavisk Aktuarietidskrift **15** (1932), 175–195.

[44] P. Federbush, *Partially alternative derivation of a result of Nelson*, J. Math. Phys. **10 no. 1** (1969), 50-52.

[45] X. Fernique, *Régularité des trajectoires des fonctions aléatoires gaussiennes*, in "Ecoles d'Eté de Probabilités de Saint-Flour IV-1974," edited by P.L. Hennequin. Lecture Notes in Mathematics **480**, Springer, Berlin, 1975, pp. 1–97.

[46] H. Föllmer, *On entropy and information gain in random fields*, Z. Wahrsch. verw. Geb. **26** (1973), 207–217.

[47] H. Föllmer, *Random fields and diffusion processes*, in "Ecoles d'Eté de Probabilités de Saint-Flour XVI- 1986" (to appear).

[48] H. Föllmer and S. Orey, *Large deviations for the empirical field of a Gibbs measure*, Ann. Probab. **16** (1988), 961–977.

[49] M.I. Freidlin, *Action functional for a class of stochastic processes*, Th. Prob. Appl. **17** (1972), 511–515.

[50] M. Fukushima, "Dirichlet Forms and Markov Processes," NorthHolland, Amsterdam, 1980.

[51] M. Fukushima and D.W. Stroock, *Reversibility of solutions to martingale problems*, in "Probability, Statistical Mechanics, and Number Theory Advances in Mathematics Supplemental Studies, Vol.9," Academic Press, New York, 1986, pp. 107–123.

[52] M.P. Gaffney, *The conservation property of the heat equation on riemannian manifolds*, Comm. Pure Appl. Math. **12** (1959), 1–11.

[53] J. Gärtner, *On large deviations from the invariant measure*, Th. Prob. Appl. **22** (1977), 24–39.

[54] J. Glimm, *Boson fields with nonlinear self-interaction in two dimen-*

sions, Comm. Math. Phys. **8** (1968), 12–25.

[**55**] P. Groeneboom, J. Oosterhoff and F.H. Ruymgaart, *Large deviation theorems for empirical probability measures*, Ann. Probab. **7** (1979), 553–586.

[**56**] L. Gross, *Logarithmic Sobolev inequalities*, Amer. J. Math. **97** (1976), 1061–1083.

[**57**] F. Guerra, L. Rosen and B. Simon, *The $P(\Phi)_2$ Euclidean quantum field theory as classical statistical mechanics*, Ann. of Math. **101** (1975), 111–259.

[**58**] G.H. Hardy and J.E. Littlewood, *A maximal theorem with function-theoretic applications*, Acta Math. **5** (1930), 81–116.

[**59**] P. Hartman, *On the ergodic theorem*, Am. J. Math. **69** (1947), 193–199.

[**60**] A.B. Hoadley, *On the probability of large deviations of functions of several empirical cdf's*, Ann. Math. Stat. **38** (1967), 360–382.

[**61**] W. Hoeffding, *On probabilities of large deviations*, in "Proceedings of the Fifth Berkeley Symposium on Mathematical Statistics and Probability," Univ. of California Press, Berkeley, 1965, pp. 203–219.

[**62**] R. Holley and D.W. Stroock, *Logarithmic Sobolev inequalities and stochastic Ising models*, J. Stat. Physics **46** (1987), 1159–1194.

[**63**] L. Hörmander, *Hypoelliptic second order differential equations*, Acta Math. **119** (1967), 147–171.

[**64**] N.C. Jain, *A Donsker-Varadhan type invariance principle*, Z. Wahrsch. verw. Geb. **59** (1982), 117–138.

[**65**] N.C. Jain, *Large deviation lower bounds for additive functionals of Markov processes*, Ann. Prob. (to appear).

[**66**] M. Kac, *On some connections between probability theory and differential and integral equations*, in "Proceedings of the Second Berkeley Symposium on Mathematical Statistics and Probability," Univ. of California Press, Berkeley, 1950, pp. 189–215.

[**67**] M. Kac, "Integration in Function Spaces," Fermi Lectures, Academia Nazionale dei Lincei Scuola Normale Superiore, Pisa, 1980.

[**68**] J.H.B. Kemperman, *On the optimum rate of transmitting information*, in "Probability and Information Theory," Lecture Notes in Math.**89**,

Springer, Berlin, 1967, pp. 120–169.

[**69**] A.I. Khinchin, *Über einen neuen Grenzwertsatz der Wahrscheinlichkeitsrechnung*, Math. Annalen **101** (1929), 745-752.

[**70**] N.Krylov and N. Bogolioubov, *La théorie générale de la mesure dans son application à l'étude des systèmes de la mécanique non linéaires*, Ann. of Math. **38** (1937), 65–113.

[**71**] S. Kullback and R.A. Leibler, *On information and sufficiency*, Ann. Math. Statist. **22** (1951), 79–86.

[**72**] H.J Landau and L.A. Shepp, *On the supremum of a Gaussian process*, Sankhyā **Ser. A no. 32** (1970), 369-378.

[**73**] O.E. Landford, *Entropy and equilibrium states in classical statistical mechanics*, in "Statistical Mechanics and Mathematical Problems," Edited by A. Lenard. Lecture Notes in Physics **20**, Springer, Berlin, 1973, pp. 1–113.

[**74**] E. Lieb, *Existence and uniqueness of the minimizing solution of Choquard's non-linear equation*, Studies in Appl. Math. **57** (1977), 93–105.

[**75**] Y.V. Linnik, *On the probability of large deviations for the sums of independent variables*, in "Proceedings of the Fourth Berkeley Symposium on Mathematical Statistics and Probability," Univ. of California Press, Berkeley, 1961, pp. 289–306.

[**76**] J. Milnor, "Morse Theory," Princeton Univ. Press, Princeton, 1969.

[**77**] S. Molchanov, *Diffusion processes and Riemannian geometry*, Russian Math. Surveys **30** (1975), 1- 53.

[**78**] S.T. Moy, *Generalisation of Shannon-Mc Millan Theorem*, Pacific J. Math. **11** (1960), 1371–1438.

[**79**] E. Nelson, *The free Markov field*, J. Func. Anal. **12** (1973), 211–227.

[**80**] P. Ney, *Dominating points and the asymptotics of large deviations for random walk on* $\mathbb{R}^d$, Ann. Probab. **11** (1983), 158–167.

[**81**] P. Ney and E. Nummelin, *Markov additive processes II: large deviations*, Ann. Probab. **15** (1987), 593–609.

[**82**] S. Olla, *Large deviations for Gibbs random fields*, Prob. Th. Rel. Fields **77** (1988), 343–357.

[83] S. Orey, *On the Shannon-Perez- Moy theorem*, in "Proceedings on Particle systems, random media and large deviations (New Brunswick, Maine)," Contemp. Math. **41**, A.M.S., Providence R.I., 1985, pp. 319–327.

[84] S. Orey, *Large deviations in ergodic theory*, in "Seminar on Stochastic Processes," Edited by E.Çinlar, K.L. Chung and R.K. Getoor, Birkhäuser, Basel, 1985, pp. 195–249.

[85] S. Orey and S. Pelikan, *Large deviation principles for stationary processes*, Ann. Probab. **16** (1988), 1481–1496.

[86] J.C. Oxtoby, *Ergodic sets*, Bull. Amer. Math. Soc. **58** (1952), 116–136.

[87] V.V. Petrov, "Sums of Independent Random Variables," translated by A.A. Brown, Springer, Berlin, 1975.

[88] R. Pinsky, *On evaluating the Donsker-Varadhan I-functional*, Ann. Probab. **13** (1985), 342-362.

[89] R. Ranga Rao, *Relations between weak and uniform convergence of measures with applications*, Ann. Math. Statis. **33** (1962), 659–680.

[90] V. Richter, *Local limit theorems for large deviations*, Th. Prob. Appl. **2** (1957), 206–220.

[91] F. Riesz, *Sur un théorème de maximum de MM. Hardy et Littlewood*, J. London Math. Soc. **7** (1931), 10–13.

[92] D. Ruelle, *Correlation functionals*, J. Math. Physics **6** (1965), 201 –220.

[93] D. Ruelle, *A variational formulation of equilibrium statistical mechanics and the Gibbs phase rule*, Comm. Math. Phys. **5** (1967), 324–329.

[94] I.N. Sanov, *On the probability of large deviations of random variables*, (in Russian), Mat. Sb. **42** (1957), 11–44. (English translation in *Selected Translations in Mathematical Statistics and Probability I* (1961) pp. 213- 244.)

[95] M. Schilder, *Some asymptotics formulae for Wiener integrals*, Trans. Amer. Math. Soc. **125** (1966), 63–85.

[96] B. Simon, "The $P(\Phi)_2$ Euclidian (Quantum) Field Theory," Princeton Univ. Press, Princeton, 1974.

[97] A.V. Skorokhod, *Limit theorems for stochastic processes*, Th. Prob.

and Appl. **I** (1956), 261–290.

[**98**] N. Smirnoff, *Über Wahrscheinlichkeiten grosser Abweichungen*, Rec. Soc. Math. Moscou **40** (1933), 441–455.

[**99**] A.S. Sznitman, *Lipschitz tail and Wiener sausage on hyperbolic space*, Comm. Pure Appl. Math. (to appear).

[**100**] V. Strassen, *An invariance principle for the law of the iterated logarithm*, Z. Wahrsch. verw. Geb. **3** (1964), 211–226.

[**101**] D.W. Stroock, "An Introduction to the Theory of Large Deviations," Springer, Berlin, 1984.

[**102**] D.W. Stroock, *On the rate at which a homogeneous diffusion approaches a limit, an application of the large deviation theory of certain stochastic integrals*, Ann. Probab. **14** (1986), 840–859.

[**103**] D.W. Stroock and S.R.S. Varadhan, *On the support of diffusion processes, with applications to the strong maximum principle*, in "Proceedings of the Sixth Berkeley Symposium on Mathematical Statistics and Probability," Univ. California Press, Berkeley, 1970, pp. 333–360.

[**104**] D.W. Stroock and S.R.S. Varadhan, "Multidimensional Diffusion Processes," Springer, Berlin, 1979.

[**105**] Y. Takahashi, *Entropy function (free energy) for dynamical systems and their random perturbations*, in "Proceedings Taniguchi Symposium on Stochastic Analysis at Katata and Kyoto," edited by K. Itô, Kinokuniya and North Holland, Tokyo, 1982.

[**106**] S.R.S. Varadhan, *Asymptotic probabilities and differential equations*, Comm. Pure Appl. Math. **19** (1966), 261–286.

[**107**] S.R.S. Varadhan, "Large Deviations and Applications," Society for Industrial and Applied Mathematics, Philadelphia, 1984.

[**108**] A.D. Ventcel, *Action functional for gaussian random function*, Th. Prob. Appl. **17** (1972), 515–517.

[**109**] A.D. Ventcel and M.I. Freidlin , *On small perturbations of dynamical systems*, Russian Math. Surveys **25** (1970), 1– 55.

[**110**] A.D. Ventcel and M.I. Freidlin , *Some problems concerning stability under small random perturbations*, Th. Prob. Appl. **17** (1972), 269–283.

[**111**] A.D. Ventcel and M.I. Freidlin, "Random Perturbations of Dynam-

ical Systems," translated by J. Szücs, Springer, Berlin, 1984.

[**112**] N. Wiener, *Differential spaces*, J. Math. Phys. **2** (1923), 131–174.

[**113**] S.T. Yau, *On the heat kernel of a complete Riemannian manifold*, J. Math. Pures et Appl. **57** (1978), 191–201.

Frequently Used Notation

Symbols	Description	Page
$\mathcal{B}_E$	BOREL *field over* E	8
$B(\Sigma;\mathbb{R})$	*bounded, measurable functions on* Σ	101
$\\| \\|_B$	*uniform norm on* $B(\Sigma;\mathbb{R})$	101
$B(x,r)$	*open ball of radius* r *around* x	12
(**B&E**)	BAKRY–EMERY *hypercontractivity criterion*	264
C_c^∞	*smooth functions with compact support*	12, 147
χ_Γ	*indicator function of* Γ	65
$\subset\subset$	*compact subset of*	11
$\mathbf{D}$ & $\mathbf{D}^{\mathbf{V}}$	*domains of* L & L^V	123
$\mathbf{D}_c$	FELLER *domain of* L	125
$\mathbf{D}(\mathcal{E})$	*domain of* $\mathcal{E}$	129
$\delta_{\omega_T} \otimes_T Q$	*splice of measures*	171
$D([0,T];\Sigma)$	SKOROKHOD'*s path space*	169
Δ	LAPLACE–BELTRAMI *operator*	251
$divX$	*divergence of* X	250
$\mathcal{E}$	DIRICHLET *form*	129
E_I	*conditional expectation operator*	231
(**EM**)	ϵ-MARKOV *property*	232
$F^{(\delta)}$	*open* δ*-hull of* F	31
$\|\Gamma\|$	LEBESGUE *measure of* Γ	146
$\overline{\Gamma}^\tau$ & $\Gamma^{\circ\,\tau}$	*closure & interior in the* τ*-topology*	73
γ_ϵ	GAUSS*ian measure with variance* ϵ	2
$grad\,\phi$	RIEMANN*ian gradient of* ϕ	250
$\Gamma(\mathbf{T}(\Sigma))$	*smooth sections of* $\mathbf{T}(\Sigma)$	250
$\mathbf{H}(\nu\|\mu)$	*relative entropy functional*	68
$\mathbf{H}_I(Q\|P)$	*relative entropy restricted to the interval* I	215
$\mathbf{H}(Q)$	*specific entropy function*	215
$H^1([0,\infty);\mathbb{R}^d)$	*underlying* HILBERT *space for* $\mathcal{W}$	11
$\\| \\|_{H^1}$	*associated norm*	11
$H^1(\mathbb{R}^N)$	SOBOLEV *space*	154
$\\| \\|_{H^1(\mathbb{R}^N)}$	*associated norm*	154
(**H-1**) & (**H-2**)	*hypermixing conditions*	214
$Hess f$	HESS*ian of* f	261

$\langle\phi\rangle_m$	*mean of* ϕ *under* m	*257*
$\mu \otimes_d \Pi$	*splice of* μ *with* Π	*165*
$\mu_{\hat{\sigma},n}$	*distribution of* $\overline{\mathbf{S}}_n$ *under* $\hat{P}_{\hat{\sigma}}$	*92*
$\mu_{\sigma,n}$	*distribution of* $\mathbf{L}_n$ *under and* P_σ	*92*
$\tilde{\mu}_n$	*distribution of* $\mathbf{L}_n$ *under* μ^n	*68*
$\mu_{\sigma,t}$	*distribution of* $\overline{\mathbf{S}}_t$ *under* P_σ	*111*
$\nabla_X Y$	*covariant derivative of* Y *relative to* X	*250*
$\nu \otimes_T P_*$	*splice of* ν *with* P_σ	*177*
$\omega_T \otimes \omega'$	*splice of paths*	*170*
$\hat{\Pi}(\hat{\sigma},\cdot)$ & $\Pi(\sigma,\cdot)$	*transition probability functions (discrete-time)*	*91, 92*
Π_V	FEYNMAN–KAC *kernel (discrete-time)*	*101*
$\hat{P}_{\hat{\sigma}}$ & P_σ	*law of the* MARKOV *chain starting from* $\hat{\sigma}$ & σ	*91, 92*
$\hat{P}(t,\sigma,\cdot)$	*continuous-time transition probability function*	*110*
$P(t,\sigma,\cdot)$	*variant of the preceding*	*110*
$\{P_t^V : t > 0\}$	FEYNMAN–KAC *semigroup*	*121*
$P^V(t,\sigma,\cdot)$	FEYNMAN–KAC *kernel (continuous time)*	*122*
$\{\overline{P}_t : t > 0\}$	$L^2(m)$*-extension of* $\{P_t : t > 0\}$	*129*
P_σ	*law of the* MARKOV *process starting from* σ	*111*
P_ω^I	*r.c.p.d. of* P *given* $\mathcal{B}_I$	*231*
$P \otimes_0 \Pi$	*splice of* P *with* Π	*168*
$\mathbf{R}_n$ & $\mathbf{R}_t$	*empirical process measures*	*161, 171*
$\mathbf{R}_T$	*variant of the preceding*	*214*
r.c.p.d.	*regular conditional probability distribution*	*163*
Ric	RICCI *curvature tensor*	*262*
$\mathbf{S}_n$ & $\mathbf{S}_n^m$	*partial sums*	*59, 91, 93*
$\overline{\mathbf{S}}_n$ & $\overline{\mathbf{S}}_n^m$	*normalized partial sums*	*59, 91, 93*
$\mathbf{S}_t$ & $\overline{\mathbf{S}}_t$	*additive & normalized additive functionals*	*110*
$\mathbf{S}^1$	*unit circle*	*202*
$\mathfrak{S}_t^{(\epsilon)}(\theta)$	ϵ*-sausage around* $\theta\|_{[o,t]}$	*146*
$\mathcal{T}$	*tail* σ*-field*	*205*
Θ & Θ^*	$\mathbb{R}^d$*-valued* WIENER *paths and dual*	*8*
$\\| \\|_\Theta$	*associated norm*	*8*
${}_{\Theta^*}\langle\cdot,\cdot\rangle_\Theta$	*duality relation between* Θ^* *and* Θ	*8*
$\mathbf{T}(\Sigma)$	*tangent bundle over* Σ	*250*
$(\hat{\mathbf{U}})$	*a uniform ergodic condition for* $\hat{\Pi}(\hat{\sigma},\cdot)$	*95*
$(\mathbf{U})$	*a uniform ergodic condition for* $\Pi(\sigma,\cdot)$	*100*
$(\tilde{\mathbf{U}})$	*a uniform ergodic condition for* $\hat{P}(t,\sigma,\cdot)$	*113*
$(\mathbf{SU})$	*a variant of the preceding*	*240*
$\\| \\|_{\text{var}}$	*(total) variation norm*	*64*
$\mathcal{W}$	WIENER*'s measure*	*8*

Subject Index

PURE AND APPLIED MATHEMATICS

Vol. 1 Arnold Sommerfeld, *Partial Differential Equations in Physics*

Vol. 2 Reinhold Baer, *Linear Algebra and Projective Geometry*

Vol. 3 Herbert Busemann and Paul Kelly, *Projective Geometry and Projective Metrics*

Vol. 4 Stefan Bergman and M. Schiffer, *Kernel Functions and Elliptic Differential Equations in Mathematical Physics*

Vol. 5 Ralph Philip Boas, Jr., *Entire Functions*

Vol. 6 Herbert Busemann, *The Geometry of Geodesics*

Vol. 7 Claude Chevalley, *Fundamental Concepts of Algebra*

Vol. 8 Sze-Tsen Hu, *Homotopy Theory*

Vol. 9 A. M. Ostrowski, *Solution of Equations in Euclidean and Banach Spaces, Third Edition of Solution of Equations and Systems of Equations*

Vol. 10 J. Dieudonné, *Treatise on Analysis: Volume I, Foundations of Modern Analysis; Volume II; Volume III; Volume IV; Volume V; Volume VI; Volume VII*

Vol. 11* S. I. Goldberg, *Curvature and Homology*

Vol. 12* Sigurdur Helgason, *Differential Geometry and Symmetric Spaces*

Vol. 13 T. H. Hildebrandt, *Introduction to the Theory of Integration*

Vol. 14 Shreeram Abhyankar, *Local Analytic Geometry*

Vol. 15* Richard L. Bishop and Richard J. Crittenden, *Geometry of Manifolds*

Vol. 16* Steven A. Gaal, *Point Set Topology*

Vol. 17 Barry Mitchell, *Theory of Categories*

Vol. 18 Anthony P. Morse, *A Theory of Sets*

Vol. 19 Gustave Choquet, *Topology*

Vol. 20 Z. I. Borevich and I. R. Shafarevich, *Number Theory*

Vol. 21 José Luis Massera and Juan Jorge Schaffer, *Linear Differential Equations and Function Spaces*

Vol. 22 Richard D. Schafer, *An Introduction to Nonassociative Algebras*

Vol. 23* Martin Eichler, *Introduction to the Theory of Algebraic Numbers and Functions*

Vol. 24 Shreeram Abhyanker, *Resolution of Singularities of Embedded Algebraic Surfaces*

* Presently out of print

Vol. 25 François Treves, *Topological Vector Spaces, Distributions, and Kernels*

Vol. 26 Peter D. Lax and Ralph S. Phillips, *Scattering Theory*

Vol. 27 Oystein Ore, *The Four Color Problem*

Vol. 28* Maurice Heins, *Complex Function Theory*

Vol. 29 R. M. Blumenthal and R. K. Getoor, *Markov Processes and Potential Theory*

Vol. 30 L. J. Mordell, *Diophantine Equations*

Vol. 31 J. Barkley Rosser, *Simplified Independence Proofs: Boolean Valued Models of Set Theory*

Vol. 32 William F. Donoghue, Jr., *Distributions and Fourier Transforms*

Vol. 33 Marston Morse and Stewart S. Cairns, *Critical Point Theory in Global Analysis and Differential Topology*

Vol. 34* Edwin Weiss, *Cohomology of Groups*

Vol. 35 Hans Freudenthal and H. De Vries, *Linear Lie Groups*

Vol. 36 Laszlo Fuchs, *Infinite Abelian Groups*

Vol. 37 Keio Nagami, *Dimension Theory*

Vol. 38 Peter L. Duren, *Theory of H^p Spaces*

Vol. 39 Bodo Pareigis, *Categories and Functors*

Vol. 40* Paul L. Butzer and Rolf J. Nessel, *Fourier Analysis and Approximation: Volume I, One-Dimensional Theory*

Vol. 41* Eduard Prugovečki, *Quantum Mechanics in Hilbert Space*

Vol. 42 D. V. Widder, *An Introduction to Transform Theory*

Vol. 43 Max D. Larsen and Paul J. McCarthy, *Multiplicative Theory of Ideals*

Vol. 44 Ernst-August Behrens, *Ring Theory*

Vol. 45 Morris Newman, *Integral Matrices*

Vol. 46 Glen E. Bredon, *Introduction to Compact Transformation Groups*

Vol. 47 Werner Greub, Stephen Halperin, and Ray Vanstone, *Connections, Curvature, and Cohomology:*
Volume I, De Rham Cohomology of Manifolds and Vector Bundles
Volume II, Lie Groups, Principal Bundles, and Characteristic Classes
Volume III, Cohomology of Principal Bundles and Homogeneous Spaces

Vol. 48 Xia Dao-xing, *Measure and Integration Theory of Infinite-Dimensional Spaces: Abstract Harmonic Analysis*

Vol. 49 Ronald G. Douglas, *Banach Algebra Techniques in Operator Theory*
Vol. 50 Willard Miller, Jr., *Symmetry Groups and Theory Applications*
Vol. 51 Arthur A. Sagle and Ralph E. Walde, *Introduction to Lie Groups and Lie Algebras*
Vol. 52 T. Benny Rushing, *Topological Embeddings*
Vol. 53* James W. Vick, *Homology Theory: An Introduction to Algebraic Topology*
Vol. 54 E. R. Kolchin, *Differential Algebra and Algebraic Groups*
Vol. 55 Gerald J. Janusz, *Algebraic Number Fields*
Vol. 56 A. S. B. Holland, *Introduction to the Theory of Entire Functions*
Vol. 57 Wayne Roberts and Dale Varberg, *Convex Functions*
Vol. 58 H. M. Edwards, *Riemann's Zeta Function*
Vol. 59 Samuel Eilenberg, *Automata, Languages, and Machines: Volume A, Volume B*
Vol. 60 Morris Hirsch and Stephen Smale, *Differential Equations, Dynamical Systems, and Linear Algebra*
Vol. 61 Wilhelm Magnus, *Noneuclidean Tesselations and Their Group*
Vol. 62 François Treves, *Basic Linear Partial Differential Equations*
Vol. 63* William M. Boothby, *An Introduction to Differentiable Manifolds and Riemannian Geometry*
Vol. 64 Brayton Gray, *Homotopy Theory: An Introduction to Algebraic Topology*
Vol. 65 Robert A. Adams, *Sobolev Spaces*
Vol. 66 John J. Benedetto, *Spectral Synthesis*
Vol. 67 D. V. Wilder, *The Heat Equation*
Vol. 68 Irving Ezra Segal, *Mathematical Cosmology and Extragalactic Astronomy*
Vol. 69 I. Martin Isaacs, *Character Theory of Finite Groups*
Vol. 70 James R. Brown, *Ergodic Theory and Topological Dynamics*
Vol. 71 C. Truesdell, *A First Course in Rational Continuum Mechanics: Volume 1, General Concepts*
Vol. 72 K. D. Stroyan and W. A. J. Luxemburg, *Introduction to the Theory of Infinitesimals*
Vol. 73 B. M. Puttaswamaiah and John D. Dixon, *Modular Representations of Finite Groups*
Vol. 74 Melvyn Berger, *Nonlinearity and Functional Analysis: Lectures on Nonlinearity Problems in Mathematical Analysis*
Vol. 75 George Gratzer, *Lattice Theory*

Vol. 76 Charalambos D. Aliprantis and Owen Burkinshaw, *Locally Solid Riesz Spaces*
Vol. 77 Jan Mikusinski, *The Bochner Integral*
Vol. 78 Michiel Hazelwinkel, *Formal Groups and Applications*
Vol. 79 Thomas Jech, *Set Theory*
Vol. 80 Sigurdur Helgason, *Differential Geometry, Lie Groups, and Symmetric Spaces*
Vol. 81 Carl L. DeVito, *Functional Analysis*
Vol. 82 Robert B. Burckel, *An Introduction to Classical Complex Analysis*
Vol. 83 C. Truesdell and R. G. Muncaster, *Fundamentals of Maxwell's Kinetic Theory of a Simple Monatomic Gas: Treated as a Branch of Rational Mechanics*
Vol. 84 Louis Halle Rowen, *Polynomial Identities in Ring Theory*
Vol. 85 Joseph J. Rotman, *An Introduction to Homological Algebra*
Vol. 86 Barry Simon, *Functional Integration and Quantum Physics*
Vol. 87 Dragos M. Cvetkovic, Michael Doob, and Horst Sachs, *Spectra of Graphs*
Vol. 88 David Kinderlehrer and Guido Stampacchia, *An Introduction to Variational Inequalities and Their Applications*
Vol. 89 Herbert Seifert, W. Threlfall, *A Textbook of Topology*
Vol. 90 Grezegorz Rozenberg and Arto Salomaa, *The Mathematical Theory of L Systems*
Vol. 91 Donald W. Kahn, *Introduction to Global Analysis*
Vol. 92 Eduard Prugovečki, *Quantum Mechanics in Hilbert Space, Second Edition*
Vol. 93 Robert M. Young, *An Introduction to Nonharmonic Fourier Series*
Vol. 94 M. C. Irwin, *Smooth Dynamical Systems*
Vol. 96 John B. Garnett, *Bounded Analytic Functions*
Vol. 97 Jean Dieudonné, A Panorama of Pure Mathematics: As Seen by N. Bourbaki
Vol. 98 Joseph G. Rosenstein, *Linear Orderings*
Vol. 99 M. Scott Osborne and Garth Warner, *The Theory of Eisenstein Systems*
Vol. 100 Richard V. Kadison and John R. Ringrose, *Fundamentals of the Theory of Operator Algebras:*
Volume 1, Elementary Theory; Volume 2, Advanced Theory
Vol. 101 Howard Osborn, *Vector Bundles: Volume 1, Foundations and Stiefel-Whitney Classes*

Vol. 102 Avraham Feintuch and Richard Saeks, *System Theory: A Hilbert Space Approach*

Vol. 103 Barrett O'Neill, *Semi-Riemannian Geometry: With Applications to Relativity*

Vol. 104 K. A. Zhevlakov, A. M. Slin'ko, I. P. Shestakov, and A. I. Shirshov, *Rings That Are Nearly Associative*

Vol. 105 Ulf Grenander, *Mathematical Experiments on the Computer*

Vol. 106 Edward B. Manoukian, *Renormalization*

Vol. 107 E. J. McShane, *Unified Integration*

Vol. 108 A. P. Morse, *A Theory of Sets, Revised and Enlarged Edition*

Vol. 109 K. P. S. Bhaskara-Rao and M. Bhaskara-Rao, *Theory of Charges: A Study of Finitely Additive Measures*

Vol. 110 Larry C. Grove, *Algebra*

Vol. 111 Steven Roman, *The Umbral Calculus*

Vol. 112 John W. Morgan and Hyman Bass, editors, *The Smith Conjecture*

Vol. 113 Sigurdur Helgason, *Groups and Geometric Analysis: Integral Geometry, Invariant Differential Operators, and Spherical Functions*

Vol. 114 E. R. Kolchin, *Differential Algebraic Groups*

Vol. 115 Isaac Chavel, *Eigenvalues in Riemannian Geometry*

Vol. 116 W. D. Curtis and F. R. Miller, *Differential Manifolds and Theoretical Physics*

Vol. 117 Jean Berstel and Dominique Perrin, *Theory of Codes*

Vol. 118 A. E. Hurd and P. A. Loeb, *An Introduction to Nonstandard Real Analysis*

Vol. 119 Charalambos D. Aliprantis and Owen Burkinshaw, *Positive Operators*

Vol. 120 William M. Boothby, *An Introduction to Differentiable Manifolds and Riemannian Geometry, Second Edition*

Vol. 121 Douglas C. Ravenel, *Complex Cobordism and Stable Homotopy Groups of Spheres*

Vol. 122 Sergio Albeverio, Jens Erik Fenstad, Raphael Høegh-Krohn, and Tom Lindstrøm, *Nonstandard Methods in Stochastic Analysis and Mathematical Physics*

Vol. 123 Alberto Torchinsky, *Real-Variable Methods in Harmonic Analysis*

Vol. 124 Robert J. Daverman, *Decomposition of Manifolds*

Vol. 125 J. M. G. Fell and R. S. Doran, *Representations of *-Algebras, Locally Compact Groups, and Banach *-Algebraic Bundles: Volume 1, Basic Representation Theory of Groups and Algebras*

Vol. 126 J. M. G. Fell and R. S. Doran, *Representations of *-Algebras, Locally Compact Groups, and Banach *-Algebraic Bundles: Volume 2, Induced Representations, the Imprimitivity Theorem, and the Generalized Mackey Analysis*

Vol. 127 Louis H. Rowen, *Ring Theory, Volume I*

Vol. 128 Louis H. Rowen, *Ring Theory, Volume II*

Vol. 129 Colin Bennett and Robert Sharpley, *Interpolation of Operators*

Vol. 130 Jürgen Pöschel and Eugene Trubowitz, *Inverse Spectral Theory*

Vol. 131 Jens Carsten Jantzen, *Representations of Algebraic Groups*

Vol. 132 Nolan R. Wallach, *Real Reductive Groups I*

Vol. 133 Michael Sharpe, *General Theory of Markov Processes*

Vol. 134 Igor Frenkel, James Lepowsky, and Arne Meurman, *Vertex Operators and the Monster*

Vol. 135 Donald Passman, *Infinite Crossed Products*

Vol. 136 Heinz-Otto Kreiss and Jens Lorenz, *Initial-Boundary Value Problems and the Navier-Stokes Equations*

Vol. 137 Jean-Dominique Deuschel and Daniel W. Stroock, *Large Deviations*